Molecular and Physiological Basis of Nematode Survival

MIX
Paper from
responsible sources
FSC® C013604

To Clare and Ann, whose support and patience have been essential for our careers in nematology.

MOLECULAR AND PHYSIOLOGICAL BASIS OF NEMATODE SURVIVAL

Edited by

Roland N. Perry

*Plant Pathology and Microbiology Department,
Rothamsted Research,
Harpenden, Hertfordshire,
UK and Biology Department,
Ghent University, Ghent, Belgium*

and

David A. Wharton

*Department of Zoology, University of Otago,
Dunedin, New Zealand*

www.cabi.org

CABI is a trading name of CAB International

CABI Head Office
Nosworthy Way
Wallingford
Oxfordshire OX10 8DE
UK

Tel: +44 (0)1491 832111
Fax: +44 (0)1491 833508
E-mail: cabi@cabi.org
Website: www.cabi.org

CABI North American Office
875 Massachusetts Avenue
7th Floor
Cambridge, MA 02139
USA

Tel: +1 617 395 4056
Fax: +1 617 354 6875
E-mail: cabi-nao@cabi.org

A catalogue record for this book is available from the British Library, London, UK.

Library of Congress Cataloging-in-Publication Data

Molecular and physiological basis of nematode survival/edited by Roland N. Perry and David A. Wharton.

p. cm.

Includes bibliographical references and index.

ISBN 978-1-84593-687-7 (alk. paper)

1. Nematodes--Physiology. 2. Nematodes--Adaptation. I. Perry, R. N. (Roland N.) II. Wharton, David A.

QL391.N4M55 2011
571.1′257--dc22

2010033015

ISBN-13: 978 1 84593 687 7

Commissioning editor: Nigel Farrar
Production editor: Fiona Chippendale

Typeset by SPi, Pondicherry, India.
Printed and bound in the UK by CPI Antony Rowe, Chippenham.

Contents

About the Editors

Roland N. Perry

Roland Perry's interests in nematode survival date from his PhD studies on physiological aspects of desiccation survival of *Ditylenchus* spp. His PhD was from Newcastle University, where he had previously graduated with a BSc (Hons) in Zoology. After a year's postdoctoral research at Newcastle, he moved to Keele University, UK, for 3 years, where he taught parasitology. He then moved to Rothamsted Research, where he is currently based. His research interests have centred primarily on nematode survival physiology, hatching, sensory perception and behaviour. Several of his past PhD and postdoctoral students are currently involved in nematology research.

He co-edited *The Physiology and Biochemistry of Free-living and Plant-parasitic Nematodes* (1997), the textbook *Plant Nematology* (2006), and *Root-knot Nematodes* (2009). He is author or co-author of over 40 book chapters and refereed reviews and over 100 refereed research papers. He is co-editor-in-chief of *Nematology* and chief editor of the *Russian Journal of Nematology*. He co-edits the book series *Nematology Monographs and Perspectives*. In 2001, he was elected Fellow of the Society of Nematologists (USA) in recognition of his research achievements, and in 2008 he was elected Fellow of the European Society of nematologists for outstanding contributions to the science of nematology. He is a visiting professor at Ghent University, Belgium, where he lectures on nematode biology.

David A. Wharton

David Wharton's PhD topic at the University of Bristol, after gaining a BSc (Hons) at the same university, was 'The Structure and Function of Nematode Eggshells'. This developed into an interest in nematode survival mechanisms, particularly how they survive freezing and extreme desiccation (anhydrobiosis). After postdoctoral positions at University College Cardiff and the University College of Wales, Aberystwyth, David was appointed to a lectureship in zoology at the University of Otago, New Zealand, in 1985, where he is now an associate professor. David was awarded a DSc by the University of Bristol in 1997 for his work on the environmental physiology of nematodes. His move to New Zealand gave him the opportunity to work in Antarctica, where he isolated and cultured an Antarctic nematode that is the only organism currently known to survive extensive intracellular freezing.

David is the author of two books: *A Functional Biology of Nematodes* (1986) and *Life at the Limits: Organisms in Extreme Environments* (2002). He has also published 92 refereed research papers and seven book chapters.

Contributors

Byron J. Adams, *Microbiology and Molecular Biology Department, and Evolutionary Ecology Laboratory, Brigham Young University, Provo, UT 84602-5253, USA. E-mail: bjadams@byu.edu*

Bishwo N. Adhikari *Microbiology and Molecular Biology Department, Brigham Young University, Provo, UT 84602-5253, USA. E-mail: adhikaribn@hotmail.com*

Xiaodong Bai *Department of Entomology, OARDC Research Internships Program, The Ohio State University, 1680 Madison Avenue, Wooster, OH 44691, USA. E-mail: bai.29@osu.edu*

John Barrett *Institute of Biological, Environmental and Rural Sciences, Edward Llwyd Building, Penglais Campus, Aberystwyth University, Aberystwyth, ST23 3DA, UK. E-mail: jzb@aber.ac.uk*

Ann M. Burnell *Department of Biology, National University of Ireland Maynooth, Maynooth, Co. Kildare, Ireland. E-mail: Ann.Burnell@nuim.ie*

Eileen Devaney *Parasitology Group, Veterinary Infection and Immunity, Institute of Comparative Medicine, Faculty of Veterinary Medicine, University of Glasgow, Bearsden Road, Glasgow, G61 1QH, UK. E-mail: E.Devaney@vet.gla.ac.uk*

Warwick Grant *Genetics Department, La Trobe University, Bundoora, Victoria 3086, Australia. E-mail: W.Grant@latrobe.edu.au*

Richard Grencis *Faculty of Life Sciences, University of Manchester, Manchester, M13 9PT, UK. E-mail: richard.grencis@manchester.ac.uk*

Parwinder S. Grewal *Department of Entomology, OARDC Research Internships Program, The Ohio State University, 1680 Madison Avenue, Wooster, OH 44691, USA. E-mail: grewal.4@osu.edu*

William Harnett *Strathclyde Institute of Pharmacy and Biomedical Sciences, Glasgow, G4 0NR, UK. E-mail: w.harnett@strath.ac.uk*

Matthias Herrmann *Max Planck Institute for Developmental Biology, Department for Evolutionary Biology, Spemannstrasse 37, 72076 Tübingen, Germany. E-mail: matthias.herrmann@tuebingen.mpg.de*

Ganpati B. Jagdale *Department of Plant Pathology, University of Georgia, Athens, GA 30605, USA. E-mail: gbjagdal@uga.edu*

Jose Lozano *Laboratory of Nematology, Wageningen University, PO Box 8123 6700 ES, Wageningen, The Netherlands. E-mail: jose.lozano@wur.nl*

Maurice Moens *Institute for Agriculture and Fisheries Research, Burg. Van Gansberghelaan 96, 9820 Merelbeke, Belgium. E-mail: maurice.moens@ilvo.vlaanderen.be*

Roland N. Perry *Plant Pathology and Microbiology Department, Rothamsted Research, Harpenden, Hertfordshire, AL5 2JQ, UK. E-mail: roland.perry@bbsrc.ac.uk*

Geert Smant *Laboratory of Nematology, Wageningen University, PO Box 8123 6700 ES, Wageningen, The Netherlands. E-mail: Geert.Smant@wur.nl*

Ralf J. Sommer *Max Planck Institute for Developmental Biology, Department for Evolutionary Biology, Spemannstrasse 37, 72076 Tübingen, Germany. E-mail: ralf.sommer@tuebingen.mpg.de*

Alan Tunnacliffe *Institute of Biotechnology, Department of Chemical Engineering and Biotechnology, University of Cambridge, Tennis Court Road, Cambridge, CB2 1QT, UK. E-mail: at10004@cam.ac.uk*

Mark Viney *School of Biological Sciences, University of Bristol, Woodland Road, Bristol, BS8 1UG, UK. E-mail: Mark.Viney@bristol.ac.uk*

David A. Wharton *Department of Zoology, University of Otago, PO Box 56, Dunedin 9054, New Zealand. E-mail: david.wharton@stonebow.otago.ac.nz*

Preface

Nematodes are a remarkable group of invertebrates; there are over 25,000 described species, including free-living, animal-parasitic and plant-parasitic species and, of all groups of animals on the planet, they are the most successful. Not only do species of nematodes live in a wide variety of habitats, from hot water springs and Antarctic tundra to habitats in plants and animals as parasites, but many species also show an astonishing ability to survive severe adverse environmental conditions. The early descriptions of nematodes date back over 3000 years and relate to nematode parasites of man. The damaging economic and social impacts of animal-parasitic species on man and other animals have long been recognized. The impact of plant-parasitic nematodes has been realized only relatively recently, but now the nematode pests of agricultural crops are known to cause considerable economic loss and, especially in developing countries, adverse social impact. One of the reasons for the success of nematodes as a group is their ability to survive adverse conditions by entering a resistant, dormant metabolic state. This survival ability has fascinated scientists for many years.

Parasitic species have to withstand periods outside the host, when they have to survive without food and in a situation where locating a host may be problematic. Free-living nematodes have to survive environmental fluctuations and also need to withstand adverse conditions during their dispersal phase. Different species of nematode have evolved similar methods to ensure survival, and the examples of convergent evolution to enhance survival are fascinating. Unfortunately, in the past, research on survival has been fragmented. In part this is because nematology as a scientific discipline has been separated into separate groups, the members of which rarely integrate with other groups, publish in separate journals and attend conferences dedicated solely to their group. Thus, there are the plant nematologists, animal nematologists (usually part of the wider animal parasitology community) and the group who work on free-living nematodes (often subdivided into marine, freshwater and soil ecology groups). A more recent addition is the group of scientists working on entomopathogenic species, a group of nematodes that are being commercialized as successful bioinsecticides. By far the largest group is the *Caenorhabditis elegans* community, whose contribution to our knowledge of nematodes is extensive. The vast amount of information on *C. elegans* and the increasing number of nematode genome sequences available is, to some extent, breaking down the scientific barriers that seem to have been a concomitant aspect of the research groupings. The burgeoning interest in comparative genomics is now a vital component in understanding survival attributes of nematodes and may lead to identifying novel targets for control options of parasitic species.

The above background led to the genesis of this book, but the defining impetus came from the 5th International Congress of Nematology held in Brisbane, Australia, in July 2008. The organizers invited us to arrange and coordinate a session entitled 'Survival, adaptation and tolerance of nematodes in extreme environments'. This gave us the opportunity of inviting speakers from different areas of nematology, and the discussion during the session, and subsequently, convinced us that there was a need for a book on nematode survival that combined information on nematodes from all groups. The duration of the session necessarily limited the number of speakers, so in this book we have taken the opportunity of expanding the number of authors from those who originally contributed to the session, to ensure that our coverage of this aspect of nematology is comprehensive. Research has basically progressed from investigating the physiological and biochemical methods utilized by some species of nematodes to ensure survival of adverse conditions to incorporate the more recent molecular advances. It is the intention of this book not only to reflect some of the older research that is still relevant and important but also to link it with the more recent advances facilitated by molecular biology. We have tried to avoid getting bogged down in terminology and definitions. One of the consequences of the historic organization of research along group lines is that there is a plethora of terms, many of which mean the same. Essentially we are examining the ways by which a nematode can suspend development during unfavourable conditions and ensure survival.

We are grateful to the chapter authors for their considerable time and effort in compiling their contributions; their expertise is the essential bedrock of this subject area. We hope that readers of this book will find the subject as intriguing and challenging as we do. It is certain that this subject will develop considerably with the information from comparative genomics and it is desirable for research on nematodes to become more integrated in the future.

Roland N. Perry and
David A. Wharton
April 2010

1 Survival of Parasitic Nematodes outside the Host

Roland N. Perry[1] and Maurice Moens[2]

[1]*Rothamsted Research, Harpenden, Hertfordshire, UK and Biology Department, Ghent University, Ghent, Belgium;* [2]*Institute for Agricultural and Fisheries Research, Merelbeke, Belgium and Laboratory for Agrozoology, Ghent University, Ghent, Belgium*

1.1 Introduction

The life cycle of parasitic nematodes essentially consists of two phases, the pre-parasitic and parasitic. The pre-parasitic phase, which may equate to the infective stage, occurs either as a free-living stage or inside, or transported by, an intermediate host. On locating and invading the definitive host, the parasitic phase commences. For obligate parasitic species there are situations where persistence of a population requires survival of the free-living stages. This may occur when the host is not available or environmental conditions exist that are not commensurate with continuing development. The requirements, first, to survive long enough to infect a host and, second, to ensure the survival of progeny when the host is no longer supportive, are the essential non-parasitic tasks of the life cycle. Survival of adverse environmental conditions may involve enduring temperature extremes (see Wharton, Chapter 8, and Devaney, Chapter 10, this volume), osmotic stress (see Wharton and Perry, Chapter 11, this volume) and dehydration, in addition to withstanding the absence of food.

The ability of some species of nematode to survive desiccation for periods considerably in excess of the duration of the normal life cycle has

 Molecular and Physiological Basis of Nematode Survival (eds R.N. Perry and D.A. Wharton)

been studied in detail, in part because in species with a direct life cycle this attribute is linked to effective dispersion of nematodes. In the past, research has focused primarily on the remarkable structural, physiological and behavioural adaptations that facilitate desiccation survival (Perry, 1999). However, more recently the molecular aspects have received considerable attention, and these are reviewed by Burnell and Tunnacliffe, Chapter 6, and Adhikari and Adams, Chapter 9, this volume. In this chapter, we examine the morphological, physiological and behavioural adaptations, focusing principally on desiccation survival and the link to nematode dispersion. This link and the need to understand the temporal factors involved in survival are clearly vital for effective management and control options for parasitic nematodes.

The pre-adult stages of nematodes are called juveniles by plant nematologists, and the term infective juvenile (IJ) is favoured by researchers working with entomopathogenic nematodes. However, the term larva(e) is the term of choice for animal nematologists and the *Caenorhabditis elegans* community. To ensure consistency throughout this chapter, larva(e) will be used.

1.2 Survival of Life Cycle Stages

There is no 'model' nematode that can be used as an example of the adaptations inherent in survival strategies, because various species show different combinations of adaptations, and cessation of development associated with survival of adverse conditions is not associated with any specific life cycle stage in the Phylum Nematoda, although the ability to survive desiccation is often commensurate with a dispersal phase of the life cycle. In the following sections, examples will be given of the survival and dispersion of parasitic forms at various phases of the life cycle.

1.2.1 The egg

There is little variation in the average size of eggs of nematodes, irrespective of the size of the adult. Wharton (1986) speculated that nematodes may increase the chances of survival and, thus, of infecting a host by providing a resistant eggshell rather than partitioning resources into increasing the size of the embryo. In most species, the eggshell typically consists of three layers: an outer vitelline layer, a middle chitinous layer and an inner lipid layer. The eggshell is more complex in structure, sometimes with up to five layers, in species such as *Ascaris suum*, where the egg is the stage responsible for direct transmission to the host. Eggs of several species of ascarids, including *Ascaris lumbricoides*, *Heterakis gallinarum* and *Ascaridia galli*, possess uterine layers, and the outer two layers of oxyurid eggshells are of uterine origin. Rogers and Sommerville (1968) pointed out that investigations of the *in vitro* hatch of *Ascaris* spp. have to be interpreted with care as some workers 'deshelled' eggs (i.e. removed the outer layers) in sodium hypochlorite before commencing hatching tests.

The lipid layer is the main permeability barrier of the eggshell and makes the egg very resistant to chemicals; as a consequence this stage is not sensitive to toxins such as common nematicides. In some species, such as *Nematodirus battus*, the eggshell protects against inoculative freezing (see Wharton, Chapter 8, this volume). The eggshell and perivitelline fluid also combine to protect the enclosed infective stage from water loss and to maintain the larva in a dormant state (see Section 1.3).

Trichostrongyle nematodes that parasitize sheep and cattle have direct life cycles, where eggs are voided in host faeces and have to withstand environmental extremes before ingestion by another host. In early studies on nematodes of this group, Waller and Donald (1970) demonstrated that eggs of *Haemonchus contortus* and *Trichostrongylus colubriformis* will survive dehydration provided that development can proceed to the infective larval stage during drying, and before the embryo loses a critical amount of water. The eggshell of *H. contortus* is more permeable to water loss than that of *T. colubriformis*, which may be correlated with the observations by Waller (1971) that the inner layer of the eggshell of *H. contortus* contains non-polar lipids of the hydrocarbon type, whereas the equivalent layer of *T. colubriformis* eggs contains either more polar unsaturated lipids or proteins.

Physiological adaptations that enhance survival, such as quiescence and diapause, are frequently associated with the unhatched larva (Perry, 1989). Quiescence and diapause are two forms of dormancy, both being induced by adverse environmental conditions, but whereas quiescence is readily reversible when favourable conditions return, diapause persists for a set period, even if favourable conditions return. If adverse conditions persist after diapause has ended, the larvae enter a quiescent state. In practice it is often difficult to differentiate between the different states and there have been several attempts to define the various types of dormancy and to integrate the definitions to include the concept of arrested development, a term preferred by animal nematologists. The induction and termination of diapause in relation to hatching of plant-parasitic cyst and root-knot nematodes have been discussed previously (Evans and Perry, 1976; Jones *et al.*, 1998; Perry, 2002) and are mentioned in Section 1.3.

When exposed to desiccation, the eggs of several species of nematode lose water very slowly, and the eggshell has been implicated in enabling the unhatched larvae to survive desiccation, the lipid layer providing the main permeability barrier to water loss (Wharton, 1980). In addition, the perivitelline fluid surrounding the unhatched larva may prevent it from losing all its body water. Thus, the eggshell and perivitelline fluid components of the egg combine to afford protection to the unhatched infective stage. However, it is important to realize that extrapolating data from *in vitro* desiccation experiments to the field ignores the interaction of factors prevalent in the natural environment. The infective larva of *Ascaris* is protected by the eggshell until ingestion by the host. However, the rate of water loss of unhatched larvae increased as an exponential function of increasing temperature (Wharton, 1979) and, although *Ascaris* eggs lose water very slowly relative to their surface–volume ratio (Wharton, 1979), they do not survive long-term desiccation. Roepstorff

(1997) considered that mortality of unhatched larvae due to dehydration was responsible for the complete lack of transmission of *A. suum* under intensive indoor production systems. On grass plots, high temperature in combination with severe dehydration in faecal samples may have contributed to the large mortality of *A. suum* (Larsen and Roepstorff, 1999).

1.2.2 Egg packaging

In some species of plant-parasitic nematodes, the nematodes themselves provide the packaging for groups of eggs to form an 'ecological unit' that enhances survival and distribution.

Root-knot nematodes (*Meloidogyne* spp.) are obligate plant endoparasites. Females lay eggs into a gelatinous matrix secreted through the anus by six large rectal glands, which comprises an irregular meshwork of glycoprotein material (Sharon and Spiegel, 1993). The gelatinous matrix surrounds the eggs and retains them in a package termed an egg mass. A female may lay 30–40 eggs per day into the matrix, and in a favourable host several hundred eggs are produced by each female; a mean of 770 ± 190 eggs per egg mass of *Meloidogyne incognita* on cotton has been recorded (Starr, 1993). Within each egg, the embryo develops to the first-stage larva (L1), which moults to the infective second-stage larva (L2) and, under suitable environmental conditions, the L2 hatches and emerges from the egg mass. Hatched L2 are vulnerable to environmental stresses and they are viable in the soil for periods much shorter than if they had remained unhatched.

The gelatinous matrix forms the first line of defence against predators and parasites; for example, Orion *et al.* (2001) demonstrated that the gelatinous matrix of *Meloidogyne javanica* protects the enclosed eggs from invasion of some microorganisms. The gelatinous matrix also protects against adverse soil conditions, especially the desiccating effects of low soil moisture. If the matrix is exposed on the root surface, low soil moisture causes it to shrink and harden as the outer layers dry, resulting in mechanical pressure on the eggs, which inhibits hatch of L2, thus ensuring that hatch occurs mainly when conditions are favourable for movement of L2 through the soil (Wallace, 1968; Bird and Soeffky, 1972).

In addition to the gelatinous matrix, the eggshell affords protection to the enclosed L2. The eggshell protects the embryo and L1 from water loss, and these stages survive drying conditions more effectively than L2 that are about to hatch (Wallace, 1968), because, immediately prior to hatch, enzyme activity erodes layers of the eggshell, resulting in a change in permeability and a loss of desiccation protection.

Cyst nematodes have a different type of ecological unit for egg packaging. Mature females of these obligate plant-parasitic nematodes are spherical (e.g. *Globodera* spp.) or lemon-shaped (*Heterodera* spp.) and, after death of the fertilized female, polyphenol oxidase tanning of the cuticle results in a hard, brown cyst, often containing several hundred eggs. Over 60 years ago, Ellenby (1946) demonstrated that, during exposure to drying conditions, the cyst wall of *Globodera rostochiensis* dries faster than the rate at which water can be replaced from within the cyst, and this permeability change results in an effective barrier to further water loss. The eggshell also becomes differentially permeable as it dries, resulting in a reduced rate of water loss of unhatched L2 compared with free L2 (Ellenby, 1968a). Ultimately, the unhatched L2 becomes as dry as the hatched L2, yet the former survives but the latter perishes; clearly, as discussed in Section 1.5.1, the rate of water loss is a decisive survival factor. Unhatched L2 within the cysts of *Globodera* spp. will survive and remain infective for many years, although unhatched L2 of *Heterodera* are less resistant to desiccation extremes.

Similar types of egg packaging units are not found in animal-parasitic or free-living nematodes. However, although the cyst wall or gelatinous matrix and the eggshell enhance the survival of unhatched larvae of cyst and root-knot nematodes, different species do not survive equally well. Long-term survival seems to be associated primarily with species that have a very restricted host range, such as *G. rostochiensis* (Perry, 2002). It is also evident that species of cyst nematodes with sophisticated host-stimulated hatching mechanisms have very restricted host ranges, and the hatching response ensures that the nematode is able to survive unhatched in the absence of a host but will hatch when suitable hosts are available (see Section 1.3).

1.2.3 Larval stages

It is clear from the preceding sections that larvae survive effectively when protected by the eggshell and, in a limited number of species, by the egg packaging. In many species, it is the L1 that hatches, but in most plant-parasitic nematodes the larva moults within the egg and the resulting L2 hatches. In some animal-parasitic species, there is a further moult in the egg and it is the third-stage larva (L3) that hatches. Hatched larvae are very vulnerable to environmental stresses but some species have remarkable abilities to survive, using a variety of behavioural, physiological and morphological adaptations.

Anguina spp. inhabit the aerial parts of cereals and forage grasses and invade ovules, where they induce galls, mate and lay eggs, and the L2 accumulate in the galls, where they can survive dry for many years. By contrast, the survival stage of *Ditylenchus dipsaci* is the L4, and in adverse conditions, especially at the end of the growing season, when food is limiting, development stops at the L4 and large numbers of this stage aggregate. The L4 have several behavioural, morphological and physiological attributes that combine to provide an astonishing ability to survive extreme desiccation (Perry 1977a,b,c; see Section 1.5). The rice stem nematode, *Ditylenchus angustus*, is adapted to more humid habitats and there is no specific survival stage. L3, L4 and adults have only limited survival attributes, although the presence of viable, dry *D. angustus* on harvested rice seeds may be important for the dissemination of this species (Ibrahim and Perry, 1993).

Some species of nematode retain the moulted cuticles as sheaths to aid survival. Exsheathed L3 of *T. colubriformis* will survive transfer to 0% relative humidity if they are first dried slowly at high humidity (Allan and Wharton, 1990). Under suitable environmental conditions, the L1 of *H. contortus* hatches from the egg and develops to the L2 and then to the infective L3, which retains the cuticle of the L2 as a sheath. Development is arrested until the L3 is ingested and exsheathment occurs in the rumen of the host. *In vitro* experiments by Ellenby (1968b) demonstrated that the ensheathed L3 survives desiccation better than the exsheathed form; when exposed to desiccation, the sheath dried first and became increasingly impermeable, thus slowing down the rate of water loss of the enclosed L3 and enabling it to survive. Similarly, the sheath surrounding the infective larva of the entomopathogenic nematode *Heterorhabditis megidis* slows down the rate of drying of the enclosed larva (Menti *et al.*, 1997).

O'Leary and Burnell (1997) isolated mutant lines of *H. megidis* with an increased tolerance to desiccation at low humidities. The surface of the sheaths of mutant lines is more negatively charged than that of the wild-type and removal of the outer layer, possibly the epicuticle, resulted in loss of the mutant phenotype (O'Leary *et al.*, 1998). Murrell *et al.* (1983) found a strongly negative charge on the epicuticle of larvae of *Strongyloides ratti* and related it to desiccation tolerance. O'Leary *et al.* (1998) suggested that the presence of a strongly ionized or polar coat on the surface of nematodes could facilitate the maintenance of a film of water over the cuticle.

The retention of moulted cuticles is found in other species of soil-dwelling nematodes but their presence does not necessarily indicate a role in desiccation survival; a sheath or sheaths also may afford protection against antagonistic organisms such as pathogenic fungi (Timper *et al.*, 1991). Species of *Steinernema*, another genus of entomopathogenic nematodes, have soil-dwelling, ensheathed infective larvae but there is no evidence that the sheath aids desiccation survival (Campbell and Gaugler, 1991; Patel *et al.*, 1997). The sheath of *Steinernema* spp. fits very loosely and is readily lost during movement through the soil, whereas the sheath of *Heterorhabditis* spp. is closely associated with the nematode's body and may have a role in enhancing desiccation survival (Menti *et al.*, 1997); the sheath of *Steinernema* may have no role in protection of the infective stage. Survival of entomopathogenic nematodes, viewed in terms of longevity under different conditions and advances in molecular information, is reviewed by Grewal *et al.*, Chapter 7, this volume, and has important relevance to commercial formulations of these bioinsecticides.

1.2.4 Adults

Although survival is primarily associated with larval stages, there are examples of species where it is the adult that survives unfavourable conditions. As rice grains infected with *Aphelenchoides besseyi* ripen, reproduction of the nematode stops, and adults aggregate and coil in clumps beneath the hull of grains. The nematodes can remain viable for 2–3 years in dry grains. L2 of the sedentary plant semi-endoparasite *Rotylenchulus reniformis* hatch in the soil and moult to the adult without feeding, resulting in a decrease in body volume from L2 to adult (Gaur and Perry, 1991a). The young adults are enclosed in all three moulted cuticles, retained as sheaths, from the previous stages. They remain inactive in dry soil until favourable wet conditions return, when soil moisture facilitates movement and frictional forces against the soil result in exsheathment; the adult then locates a host root and starts to feed. Gaur and Perry (1991b) showed that the exsheathed adults survived poorly compared with ensheathed adults and, as with *H. contortus*, the sheaths aided desiccation survival by slowing the rate of drying of the enclosed individual.

However, the reduced rate of water loss only assisted individuals of *R. reniformis* to survive for periods over which water loss was controlled; they showed no ability for prolonged survival once their water content had been reduced to less than 10% (Gaur and Perry, 1991b). The larvae of *H. megidis*, discussed in the preceding section, are also unable to survive for extended periods. Thus, whilst control of water loss enables some species to enter anhydrobiosis and survive for years, *R. reniformis* and *H. megidis* are examples of nematodes that show little intrinsic ability for anhydrobiotic survival; control of water loss merely prolongs the time taken for the nematode's water content to reach lethal low levels.

1.2.5 Dauer forms

The term dauer comes from the German for enduring and describes an alternative developmental stage enabling nematodes to survive adverse environmental conditions. The dauer stage may be an obligate part of the life cycle or may occur in response to adverse conditions. There has been extensive research on the dauer larva in *C. elegans*, which represents a developmental arrest (Riddle and Albert, 1997) similar to that found in some animal-parasitic nematodes, such as *S. ratti*, that can switch between free-living and parasitic life cycles in response to environmental cues (Viney, 1996; see Grant and Viney, Chapter 5, this volume). Dauer larvae are specialized L3 enclosed by a dauer-specific cuticle and exhibit several characteristics including reduced metabolism, elevated levels of several heat shock proteins and an enhanced resistance to desiccation (Kenyon, 1997). The factors initiating dauer formation act on the L1 and early L2 and include food availability, temperature and levels of a *C. elegans*-specific pheromone (Riddle and Albert, 1997). Grant and Viney (Chapter 5, this volume) discuss the dauer phenomenon in the context of nematode life history strategies and evolution, with particular emphasis on animal-parasitic nematodes.

The formation of the infective larvae of entomopathogenic nematodes encompasses developmental adaptations similar to dauer formation (Womersley, 1993), and Bird and Bird (1991) suggested that the survival forms of some plant-parasitic nematodes, such as L2 of species of *Anguina*, may be regarded as dauers. In *D. dipsaci*, the cessation of development beyond L4 and the accumulation of this stage in response to adverse conditions is accompanied by other changes: the L4 produced in response to adverse conditions are larger, have more lipid reserves and show a propensity to aggregate compared with L4 in a population feeding and developing under ideal conditions (Perry, unpublished results). High levels of lipid reserves are maintained in many anhydrobiotic species and provide energy reserves before they can resume feeding. The indications in *D. dipsaci* and some other species of parasitic nematodes of an alternative developmental stage similar to a dauer larva are persuasive.

Although more work is needed to validate the dauer condition in plant- and animal-parasitic species, it is evident that dauers are present in species of nematodes other than *Caenorhabditis* spp. In some species of the genus *Bursaphelenchus* a dauer form is present as a specialized survival and dispersal stage of the life cycle. *Bursaphelenchus xylophilus* is a migratory endoparasitic nematode that feeds on live trees as well as fungi and may represent an intermediate form in the evolution of plant parasitism. Plant-feeding nematodes may have evolved from those feeding on fungi, and Jones *et al.* (2005) suggested that, within the *Bursaphelenchus* group, genes acquired on two separate occasions by horizontal gene transfer from bacteria and fungi have helped to facilitate the two feeding strategies: fungal and plant tissue feeding. *Bursaphelenchus xylophilus* has a complex life cycle involving beetles of the genus *Monochamus* as the vector (Mota and Vieira, 2008). *Bursaphelenchus xylophilus* has a dauer stage, which uses the insect for transport and is associated with the dispersal mode of the life cycle; the second mode of the life cycle is the propagative mode. Dauer larvae are transported to susceptible hosts, where they enter the shoots through the feeding wounds caused by the vector. Although the nematodes can move within the tree, they cannot move from tree to tree without their vectors.

A dead tree is ideal for the insect hosts to breed. During oviposition by the insect vector into dead or dying trees or recently cut logs, dauers migrate out of the beetle tracheal system. During this propagative mode, the nematodes feed, reproduce and greatly increase their population densities on fungi. Predauer L3 of *B. xylophilus* develop at the same time as pupae of its vector and aggregate in large numbers around the pupal chambers, where they overwinter with the beetle. A large proportion of the predauer L3 moult to dauer larvae and are attracted to the vectors on which they are transported. During this phase the nematodes have to survive without food and are exposed to drying conditions. Late embryogenesis abundant (LEA) proteins have been associated with tolerance to desiccation in seeds, pollen, desiccation-resistant plants and some nematodes, including *C. elegans* (Gal *et al.*, 2004; see Burnell and Tunnacliffe, Chapter 6, and Barrett, Chapter 12, this volume). Homologues of LEA genes were identified in *B. xylophilus* (Kikuchi *et al.*, 2007). LEA proteins may protect cellular components against the effects of desiccation (Goyal *et al.*, 2005).

There are few molecular genetic studies on dauer formation in migratory endoparasitic nematodes. Kikuchi *et al.* (2007) analysed more than 13,000 expressed sequence tags from *B. xylophilus*, looking for homologues of 37 genes involved in dauer entry and maintenance in *C. elegans*. They identified

31 homologues of 18 *C. elegans* genes, including nine homologues for *daf* (dauer formation) genes. Genes known to be associated with longevity, such as glutathione peroxidase, superoxide dismutase and catalase, were also identified. The sedentary endoparasitic nematode *Meloidogyne hapla* has 14 orthologues of *C. elegans daf* genes as well as three further matches that are weak (Abad *et al.*, 2008; Abad and Opperman, 2009). However, it does not carry the *daf-28* orthologue, which is key in the signal transduction pathway, and Abad and Opperman (2009) concluded that basic development mechanisms are conserved, although signalling is not. Comparison of expression profiles of dauer genes in *C. elegans* and in survival stages of parasitic nematodes (Elling *et al.*, 2007) reveals marked differences in expression patterns between *C. elegans* and other nematodes. Thus, there may be differences between free-living and parasitic nematodes in developmental response to adverse changes in the environment.

Such studies provide initial evidence that the dauer phenomenon may be more widespread than currently recognized. Certainly, the indications in some species of plant-parasitic nematodes of an alternative developmental stage similar to a dauer larva are convincing. However, there are difficulties in relating information on dauer formation in *C. elegans* to parasitic nematodes. Survival biology of parasitic nematodes is complex and there is insufficient information to be able to link individual *daf* genes to specific survival traits.

1.3 Hatching and Dormancy

In species such as *G. rostochiensis* and *A. suum*, the unhatched larva can remain viable for many years. The synchronization of host and parasite life cycles is often predicated on the stimulus for hatching being provided by the host itself. This synchronization favours persistence of the nematode in the absence of hosts and ensures that when hosts are present infective larvae hatch inside or close to the host. Many animal-parasitic nematodes, including *A. lumbricoides, A. suum, H. gallinarum* and *Trichuris suis*, hatch inside the host in response to conditions in the host's alimentary tract. In some species of plant-parasitic nematodes, including *G. rostochiensis* and *Globodera pallida*, the stimulus for hatching emanates from host roots, as hatching factors in root diffusates (Perry, 2002; Wright and Perry, 2006). With some other species of plant-parasitic nematodes that hatch freely in water, root diffusates enhance the rate of hatching (Perry, 2002). The requirement for diffusates to stimulate hatch is most common among species of cyst nematodes, but other species, such as *M. hapla* and *R. reniformis*, also hatch in response to host root diffusates. Hatching of intestinal animal-parasitic nematodes may be close to 100% during a single exposure to the hatching stimulus, probably because no advantage accrues by delaying eclosion once the eggs have been ingested. By contrast, 60–80% of larvae of *G. rostochiensis* hatch in the presence of diffusates from host crops, but some larvae are in diapause and this 'carry over' population enables the species to persist and remain viable in the field for a number of years.

In a crop growing season, later generations of some species of polycyclic cyst nematodes show an increased dependence on root diffusates for hatch, reflecting a change of priority during the host plant growing season from rapid re-infection and population build-up to survival after host senescence (Perry and Gaur, 1996). As well as retaining eggs inside the cyst, some species of *Heterodera* extrude a proportion of their eggs into an 'egg sac', which remains attached to the cyst. Working with *Heterodera glycines*, Ishibashi *et al.* (1973) were the first to note differences in hatching response. Under favourable conditions most eggs were laid into the egg sacs, and larvae from these eggs hatched in water without the need for host stimulus, providing a secondary inoculum for rapid re-infestation of the host plant and permitting a rapid population increase in one season. Under less favourable conditions, at the end of the growing season for example, more eggs were retained within the protection of the cysts and a large proportion of these encysted eggs required host root diffusates (Ishibashi *et al.*, 1973) or artificial hatching factors (Thompson and Tylka, 1997) to stimulate hatch. Similar phenomena have been reported for *Heterodera carotae* (Greco, 1981), *Heterodera goettingiana* (Greco *et al.*, 1986), *Heterodera sacchari* (Ibrahim *et al.*, 1992) and *Meloidogyne triticoryzae* (Gaur *et al.*, 2000). Three main types of hatching response were identified in encysted eggs of *Heterodera schachtii* (Zheng and Ferris, 1991) and *Heterodera sorghi* (Gaur *et al.*, 1995): some larvae hatched very readily and infected any host plants present; in others hatching was delayed, which reduces intraspecific competition in roots, and the remaining larvae did not hatch for a considerable period, thus increasing their chances of survival in the absence of a host.

In a more detailed study with *Heterodera cajani*, which produces multiple generations during a crop season, Gaur *et al.* (1992) found that larvae in cysts of the first four generations hatched well in water with no enhancement of hatch by root diffusates, but in the fifth and sixth generations produced on senescing plants, 18–22% of the unhatched larvae required root diffusate to stimulate hatch; in the final generation, the encysted larvae contained more lipid reserves. Compared with eggs that hatch in water, the eggs of species that are dependent on root diffusates for hatch contain larvae that may be in a modified physiological state, perhaps involving the induction of obligate quiescence (Evans and Perry, 1976). These species of plant endoparasitic nematodes feed from a nematode-induced feeding site within the host root, and Perry and Gaur (1996) speculated that possible changes in the feeding site with age of the host plant may be crucial to the feeding female, which, in turn, may trigger biochemical changes in the larvae to enhance survival.

Although there is considerable variation between species of nematodes in behavioural responses and the sequence of events during the hatching process, in general the hatching process can be divided into: (i) changes in the eggshell permeability; (ii) activation of the larva; and (iii) eclosion. In many species, activation of the larva appears to precede, and may even cause, changes in the eggshell; in others, alteration of eggshell permeability characteristics appears to be a necessary prerequisite for the ending of quiescence and resumption of normal metabolism and locomotory activity.

The eggshell of *A. suum* is permeable to water but impermeable to water-soluble molecules such as trehalose, which is present in the perivitelline fluid surrounding the unhatched larva, until the hatching sequence is initiated; then the eggshell permeability alters to allow the escape of trehalose into the external medium (Clarke and Perry, 1980). In *G. rostochiensis* the permeability change and subsequent leakage of trehalose out of the egg is caused by host root diffusates. Clarke and Perry (1988) considered that the permeability change in eggshells of *A. suum* is a Na^+-mediated process, with the Na^+ forming a complex with the lipid layer. By contrast, the change in permeability of the lipid layer of *G. rostochiensis* involves Ca^{2+} (Clarke and Hennessy, 1983; Clarke and Perry, 1985). Tefft and Bone (1985) considered that Zn^{2+} mediated hatch of *H. glycines*.

The majority of studies on the trehalose content of the perivitelline fluid have focused on its role in nematode hatching and survival. In species of nematodes where the trehalose concentration of the perivitelline fluid has been estimated, it varies from 0.1M to 0.5M (reviewed by Perry, 2002). In the egg of species dependent on host stimulus for hatch, trehalose provides an osmotic stress on the unhatched larva, which causes a reduction in larval water content to levels where locomotion is inhibited and quiescence is induced, with a concomitant reduction in utilization of energy reserves (Ellenby and Perry, 1976; Clarke and Perry, 1980; Perry *et al.*, 1983; Ash and Atkinson, 1984). Trehalose is involved in the desiccation protection of the unhatched larvae of *N. battus* (Ash and Atkinson, 1983) and *G. rostochiensis* (Perry, 1983). The role of trehalose in protecting nematodes from adverse environmental conditions is discussed by Burnell and Tunnacliffe, Chapter 6, this volume.

1.4 Behavioural Adaptations

Study of the behavioural adaptations commensurate with persistence and dispersal has focused on desiccation survival. As with the morphological features discussed in Section 1.2, the behavioural adaptations associated with desiccation survival serve primarily to reduce the rate of drying. In species such as *R. reniformis* and *H. megidis*, the morphological adaptation of cuticle retention prolongs the time taken for the nematode's water content to reach levels that are lethal to these species. There is an additional role of morphological adaptations such as sheaths and egg packaging, which is to ensure that a high humidity is retained around the nematode to prevent complete drying out. By contrast, species able to survive anhydrobiotically for a considerable period without any detectable internal water have several behavioural characteristics to slow the rate of drying, in order to provide sufficient time for the necessary structural and biochemical changes to take place.

Nematode anhydrobiotes can be grouped into those that rely on environmental factors to control water loss and those that have intrinsic abilities to control water loss; we consider that these two groups should be termed **external dehydration strategists** and **innate dehydration strategists**, respectively.

Womersley (1987) called these groups slow- and fast-dehydration strategists, respectively, but these terms are somewhat misleading as both groups require controlled drying in order to survive, the first group to prolong the time to lethal low water content and the second group to enable biochemical changes to take place to facilitate long-term survival. Control of the rate of drying is the first phase; successful entry into long-term anhydrobiosis depends on subsequent biochemical and molecular adaptations (see Burnell and Tunnacliffe, Chapter 6, and Adhikari and Adams, Chapter 9, this volume).

The majority of free-living stages of animal- and plant-parasitic nematodes belong to the external dehydration strategists and show little intrinsic ability to control water loss and survive desiccation, being dependent on high relative humidity within soil pores or plant material to slow or inhibit water loss. Some species of plant ecto- and endoparasitic nematodes that attack roots avoid drying out of the upper layers by moving downwards in the soil. By contrast, species in the innate dehydration strategists have intrinsic adaptations to control their rate of water loss; this control is associated with biochemical adaptations enabling long-term survival in extreme environments (see Barrett, Chapter 12, this volume). Plant endoparasitic species such as *D. dipsaci* and *Anguina tritici*, inhabiting aerial parts of plants, demonstrate the most spectacular intrinsic abilities to withstand severe desiccation for many years (Moens and Perry, 2009); these species epitomize the attributes of innate dehydration strategists.

Coiling is a behavioural response to dehydration that reduces the surface area of the nematode that is exposed to drying conditions. When exposed to desiccation, the infective larvae of *T. colubriformis* form tight coils (Wharton, 1981). Coiling has been shown to reduce the rate of water loss of *Ditylenchus myceliophagus* (Womersley, 1978). However, unlike the related species *D. dipsaci*, *D. myceliophagus* has only very limited ability to survive desiccation (Perry, 1977a) and the coiling response cannot be used to distinguish between the external and innate dehydration strategists.

Aggregation, or clumping, occurs in very few species. L4 of *D. dipsaci* can survive extreme desiccation for many years and thousands of L4 can be found as aggregations, called eelworm wool, on the basal plate of infected narcissus bulbs or inside bean pods at the end of the growing season. L4 on the periphery of the aggregations die and apparently provide a protective coat that enables survival of the L4 in the centre of the aggregation by slowing their rate of drying (Ellenby, 1969). In addition to coiling and clumping, individual L4 of *D. dipsaci* have a marked ability to withstand environmental extremes (see Section 1.5.1). *Anguina amsinckia* and *A. tritici* aggregations occur within modified seeds of the host inflorescence, called galls. Inside the galls induced by *A. amsinckia* are hundreds of desiccated adults and larvae of all stages, many of which are coiled. By contrast, the galls induced by *A. tritici* contain tightly packed aggregates of L2 only, each of which remains uncoiled when dry. Similarly, anhydrobiotic juvenile stage 2 (J2) of *Anguina pacificae* remain uncoiled (McClure *et al.*, 2008). Thus, coiling is not a prerequisite of survival. Like L4 of *D. dipsaci*, L2 of *A. tritici* can survive severe desiccation as individuals (Ellenby, 1969). The combination of this intrinsic ability and the behavioural adaptation of clumping, plus the protection of the gall tissue, enables L2 to survive anhydrobiotically for many years.

1.5 Water Dynamics

It is clear that control of water loss is central to desiccation survival of nematodes. Nematodes protected by physical structures, such as cysts, eggshells or sheaths, may lose all their body water but the rate of water loss is much slower than that of unprotected individuals. However, it is worth repeating that only a few species are able to survive beyond the period during which water loss is controlled. With these species, additional adaptations are required for long-term survival. It is also important to realize that, although survival can be for many years in anhydrobiotic nematodes, the induction into anhydrobiosis occurs over a relative short period of time and the critical control of water loss can occur over minutes only (Perry, 1999; Fig. 1.1a). The physiological and morphological correlates of water dynamics during desiccation and rehydration have been investigated in detail, particularly in *D. dipsaci*.

1.5.1 Dehydration

Nematodes able to survive desiccation for long periods also have the ability to tolerate rapid dehydration regimes and repeated cycles of dehydration and rehydration. For example, L4 of *D. dipsaci* have remained viable in dry plant material for 23 years, yet the total duration of the life cycle ranges from only 19 to 23 days at 15°C (Evans and Perry, 1976). In a series of *in vitro* experiments to examine the mechanisms of survival, Perry (1977a,b) subjected non-coiled, non-clumped individuals of the hatched stages of *D. dipsaci* to various humidity extremes, including exposure to 0% relative humidity (RH). In general, the survival of individual L4, L3 and L2 can be expressed in weeks, days and minutes, respectively; in all cases survival increased with an increase in humidity, especially in adults, where survival was for hours at humidities under 50% but for days at higher humidities. During drying, L4 lost water more slowly than L3, and both stages lost water more slowly than L2 and adults. Thus, the slower dryers are the best survivors. Although L4 can survive low RH, where the drying regime is rapid, their survival ability is linked to an intrinsic property of the cuticle to reduce the rate of water loss. The cuticle of the L4 dries more rapidly than deeper layers of the nematode and slows down the rate of water loss of internal, and perhaps more vital, structures (Ellenby, 1969; Perry, 1977b). The remarkable ability of L4 of *D. dipsaci* to withstand extreme environments is further demonstrated by the revival of 30% of individuals after exposure to vacuum desiccation of 800 Pa for 1.5 h (Perry, 1977b).

Evidence indicates that the permeability properties of the cuticle that control water loss are linked to lipid components. In desiccated L2 of *A. tritici*, the outermost osmiophilic layer of the external cortical layer of the cuticle was doubled in thickness compared with that of hydrated individuals, indicating

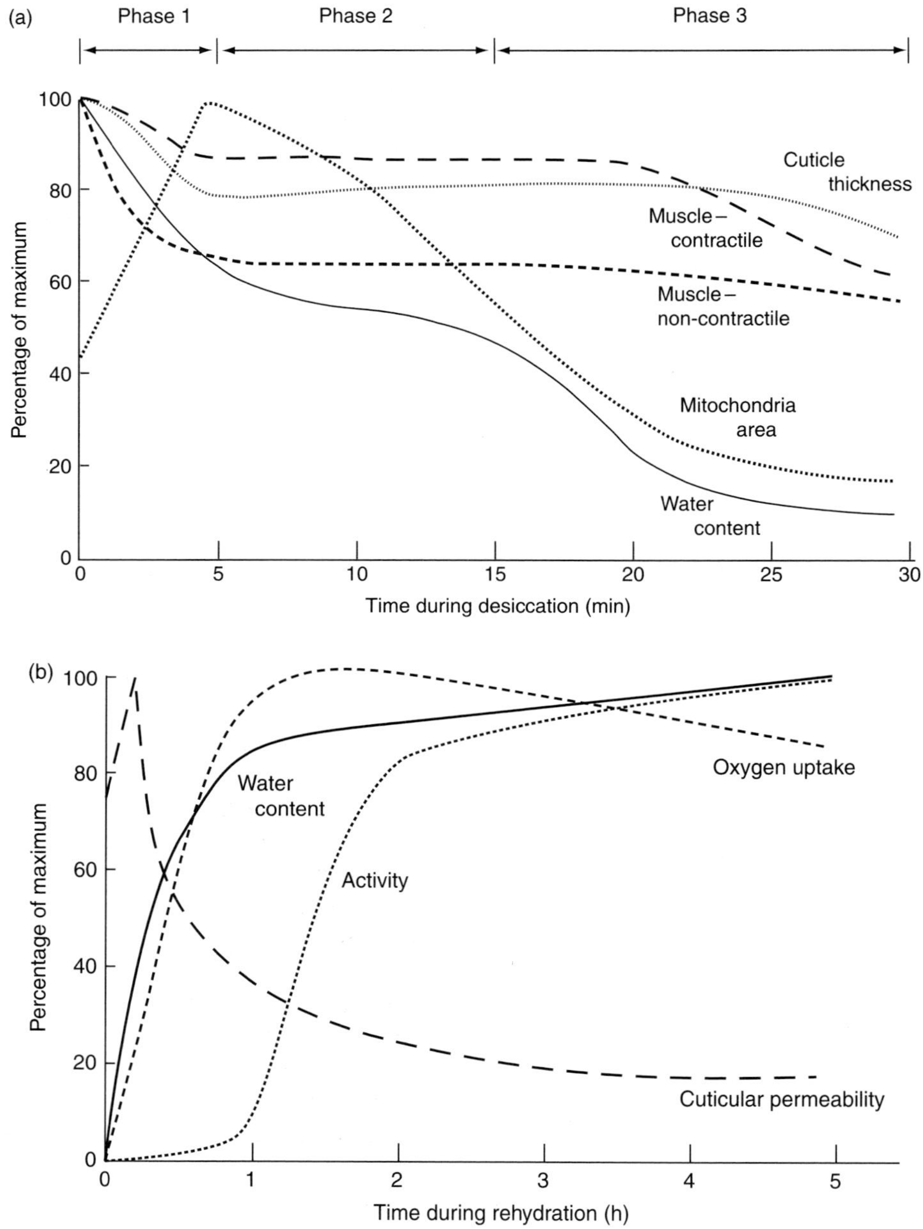

Fig. 1.1. (a) Changes accompanying desiccation of fourth-stage larvae (L4) of *Ditylenchus dipsaci* following placement of hydrated individuals in 50% relative humidity at time zero. Nematode water content data were calculated from Perry, 1977b. Data for cuticle thickness, muscle region thickness and mitochondrial profile area were calculated from Wharton and Lemmon, 1998. The three phases reflect differences in the rate of water loss (Perry, 1977b). (From Perry, 1999.). (b) Changes accompanying rehydration of L4 following placement of desiccated individuals in water at time zero. Nematode water content data were calculated from Perry, 1977b. Cuticular permeability data were calculated from Wharton *et al.*, 1988. Oxygen uptake and activity data were calculated from Barrett, 1982. Activity is defined as the percentage of L4 showing movement. The time difference between water uptake and activity is the 'lag phase' (Barrett, 1982). (Redrawn from Barrett, 1991.)

an increase in lipids (Bird and Buttrose, 1974). An outer layer was found to be present in desiccated L2 of *A. amsinckia* (Womersley *et al.*, 1998). The permeability barrier of the cuticle of *Anguina agrostis* was also considered to be associated with lipoprotein in the epicuticle (Preston and Bird, 1987; Bird and Zuckerman, 1989). The cuticular permeability barrier of L4 of *D. dipsaci* is heat labile and is destroyed by brief extraction with diethyl ether, indicating that an outer lipid layer, possibly the epicuticle, is involved (Wharton *et al.*, 1988). Infective L3 of *Nippostrongylus brasiliensis* became coated with a thin monolayer of lipid, probably from the skin and hairs of the prospective host, which may reduce the rate of water loss from the nematode (Lee, 1972).

It is important to note that control of the rate of drying does not, by itself, guarantee survival. Repeated cycles of desiccation and rehydration of L4 of *D. dipsaci* resulted in a decrease in the percentage surviving each cycle; however, after the initial cycle the rate of water loss of previously desiccated and revived individuals remained constant, irrespective of the number of cycles (Perry, 1977c). Thus, death caused by repeated cycles of desiccation and rehydration is not associated with a decreased ability to control water loss.

Water loss of individual L4 of *D. dipsaci* exposed to 0% and 50% RH occurred in three distinct phases: an initial rapid loss of water, followed by a plateau phase, or 'permeability slump' (Wharton, 1996), of very slow water loss, and a final phase of rapid water loss to leave individuals with no detectable water content (Perry, 1977b; Fig. 1.1a). The permeability of the cuticle alters after the first phase to reduce the rate of water loss during the second phase (Perry, 1977b; Wharton, 1996). Wharton *et al.* (2008) considered that this change in permeability was associated with an extracuticular layer of surface lipid and showed that the material was a triglyceride. An extracuticular layer containing lipid was reported in some of the larval stages of *Trichinella spiralis* and *Trichinella pseudospiralis* (Lee, 2002) and several species of *Heterodera* produce exudates through the cuticle that contain lipids (Endo and Wyss, 1992). Material on the surface of the cuticle has also been noted in desiccated larvae of *A. amsinckia* (Womersley *et al.*, 1998) and *D. myceliophagus* (Perry, 1999). The production of the surface lipid by desiccated L4 of *D. dipsaci* resulted in 'cuticle prints' of lipid material adhering to the cover slip (Wharton *et al.*, 2008; Fig. 1.2), a phenomenon previously noted by Bird (1988) in L2 of *A. agrostis*.

The control of water loss during the plateau phase appears to allow orderly packing and stabilization of structures to maintain functional integrity during desiccation. The mitochondria swell and then shrink during desiccation, which may indicate disruption of the permeability of the mitochondrial membrane. During the first phase, the cuticle, the lateral hypodermal cords and the muscle cells shrink rapidly, followed by a slower rate of shrinkage during the second phase (Wharton and Lemmon, 1998). The contractile region of the muscle cells resists shrinkage until the third phase of water loss (Fig. 1.1a). The large lipid reserves found in some nematodes, such as *D. dipsaci* and *A. tritici*, may prevent structural damage. Wharton and Lemmon (1998) found that intestinal cells of *D. dipsaci* changed little during desiccation, possibly because the lipid droplets they contain resist shrinkage. The contraction of the cuticle and muscles results in a decrease in diameter of L4 that is more marked

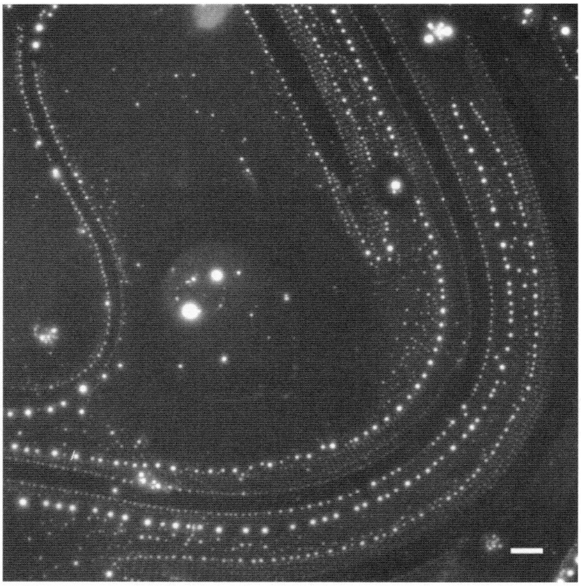

Fig. 1.2. Cuticle print from fourth-stage larvae of *Ditylenchus dipsaci* desiccated at 50% relative humidity and 20°C for 5 min. The material adhering to the cover slip was then observed by epifluorescence microscopy. Cuticle prints stained with Nile Red leave clear impressions of cuticular annulations. Scale bar = 10 μm. (From Wharton *et al.*, 2008.)

than the change in length. By contrast, reduction in the rate of water loss of *Rotylenchus robustus* is achieved by controlled contraction of cuticular annuli, resulting in decreased length, but not diameter, of the nematode (Rössner, 1973; Rössner and Perry, 1975). This 'concertina' response effectively reduces the surface area of the nematode exposed to drying.

The control of water loss enables biochemical changes to take place that ensure, in true anhydrobiotes, long-term survival. However, there is much speculation but only limited experimental evidence about the necessary biochemical adaptations. Barrett (1982) found that desiccation of these nematodes did not result in any appreciable denaturation of metabolic enzymes. At water contents below about 20%, there is no free water in the cells. This 20%, usually referred to as 'bound water', is involved in the structural integrity of macromolecules and macromolecular structures, such as membranes (see Barrett, Chapter 12, this volume). In desiccated, anhydrobiotic nematodes it is probable that the bound water has been lost, although there is no experimental evidence that nematodes can survive the complete loss of structural water. Research on biochemical attributes of organisms that may be associated with anhydrobiosis has centred on molecules that might replace bound water and preserve structural integrity.

The accumulation of the disaccharide trehalose, the only naturally occurring non-reducing disaccharide of glucose, during water loss of anhydrobiotic organisms has been reported frequently. In nematodes, L4 of *D. dipsaci* and L2 of *A. tritici* sequester trehalose, and other carbohydrates, such as myo-inositol and ribitol, may be involved (Womersley, 1987). The reported accumulation of glycerol in *Aphelenchus avenae* during desiccation (Madin and Crowe, 1975) is considered by Higa and Womersley (1993) to be an artefact due to the anaerobic conditions produced in large aggregates and not an adaptation to anhydrobiosis. There is no evidence that glycerol is preferentially synthesized during desiccation of *A. tritici* or *D. dipsaci* (Womersley and Smith, 1981; Womersley *et al.*, 1982). Womersley *et al.* (1998) consider that this is consistent with the fact that glycerol is highly fusogenic in dry membrane systems and thus would have an adverse effect on membrane stability during desiccation. Some nematodes, such as *Ditylenchus phyllobius*, accumulate no extra polyols during water loss yet can survive very rapid drying (Robinson *et al.*, 1984). Barrett (1991) considered it possible that such species may normally have large amounts of tissue polyols when active.

There are several possible roles for the involvement of trehalose in desiccation protection. Trehalose may replace bound water by attaching to polar side groups on proteins and phospholipids, thus maintaining the balance between hydrophilic and hydrophobic forces acting on the molecules and preventing their collapse. Preventing cross-linkage of molecules and fusion of membranes as bulk water is removed also preserves membrane stability. Stabilizing the membranes allows them to remain in a liquid crystalline phase and prevents a phase change to a gel state, which would cause loss of the contents of cells and membrane vesicles during rehydration. Stabilization of molecules in the dry state also requires vitrification, which keeps membranes in a glass-like state to prevent a variety of deterioration processes (Levine and Slade, 1992; Crowe *et al.*, 1998). Trehalose also may prevent protein denaturation. Glucose reacts with the amino acid side chains of proteins to form brown pigments called melanoidins. By contrast, trehalose does not react with proteins in this way and also appears to suppress this adverse reaction of other sugars with proteins (Loomis *et al.*, 1979). Trehalose also can act as a free-radical scavenging agent to reduce random chemical damage (Barrett, 1991).

Synthesizing trehalose during dehydration may indicate preliminary preparation for a period in the dry state, but it does not necessarily mean that survival during subsequent severe desiccation is assured. Research by Higa and Womersley (1993) contradicts the view that, once trehalose synthesis is complete, nematodes can survive further desiccation irrespective of the subsequent rate of water loss. It appears that, following trehalose synthesis, other, at present unknown, adaptations are required at the cellular and subcellular levels for nematode survival, and the rate of drying still has to be controlled (Higa and Womersley, 1993). By contrast to *D. dipsaci*, all stages of the mycophagous *D. myceliophagus* survive desiccation poorly, even at high humidities, and show no intrinsic ability to control water loss (Perry, 1977a,b). When raised on different food sources and exposed to various desiccation regimes, aggregates of *D. myceliophagus* contained different amounts of trehalose (~3–16% dry weight), depending on treatment, yet the nematodes are unable to survive direct exposure to low RH (Womersley and Higa, 1998). The survival of *D. myceliophagus* was unrelated to their trehalose content, and elevated levels of trehalose did not enhance anhydrobiotic survival of this species. The survival of *Steinernema carpocapsae* was also not enhanced by elevated trehalose content (Womersley, 1990).

For bioinsecticides, the importance of enhancing survival of entomopathogenic nematodes in commercial formulations during storage and after foliar application is vital. Future work may focus on genetic transformation of entomopathogenic nematodes to improve their environmental tolerance (reviewed by Burnell and Dowds, 1996; see Grewal *et al.*, Chapter 7, this volume). A transgenic approach was used by Gaugler *et al.* (1997) to introduce a heat shock protein gene, *hsp70A*, from *C. elegans* into *Heterorhabditis bacteriophora* to enhance thermotolerance. Jagdale *et al.* (2005) examined the relationship between heat (35°C) or cold shock (1°C and 10°C) and trehalose metabolism in *H. bacteriophora*. Their results showed that the trehalose concentrations were increased by both heat and cold shocks and are regulated by the action of two trehalose-metabolizing enzymes, trehalose 6-phosphate synthase (T6PS) and trehalase. The trehalose may provide protection against desiccation that may result from freezing or evaporation during cold and warm conditions, respectively. If trehalose is central for survival of species and/or strains of entomopathogenic nematodes, then the use of genes for enzymes involved in the synthesis of trehalose, such as *tps1* coding for T6PS, may cause trehalose overproduction and enhanced survival (Vellai *et al.*, 1999).

1.5.2 Rehydration

Changes during the revival process need to be ordered and controlled to complete successful survival of adverse conditions, such as desiccation. In anhydrobiotic survival, the transformation to normal activity reverses those changes that occurred during drying; however, they occur at different rates (Fig. 1.1b). L2 of *G. rostochiensis* and L4 of *D. dipsaci* took up water at the same rate (Ellenby, 1968a) and there were no differences between stages of *D. dipsaci* in the rate of water uptake, irrespective of the period of desiccation (Perry, 1977b). In all cases, the initial rate of rehydration was rapid, with 50% water content being achieved in only a few minutes. The water content of L4 of *D. dipsaci* increased logarithmically for up to 2.4 h of rehydration (Wharton *et al.*, 1985), whereas, during rehydration of L2 of *A. agrostis*, cuticle permeability initially increased slightly, followed by a sharp decrease in permeability between 1 h and 8 h, after which there were two successive slower declines in permeability up to 24 h (Preston and Bird, 1987).

Although L4 of *D. dipsaci* rehydrate very rapidly, there is a delay of several hours before the onset of locomotory activity (Fig. 1.1b). Barrett (1982) termed this delay the 'lag phase' and considered that it may be necessary to restore membrane function. The permeability barrier of the cuticle of *D. dipsaci* and *A. agrostis* is restored during the lag phase (Wharton *et al.*, 1985; Preston and Bird, 1987). The length of the lag phase in *D. dipsaci* (Wharton and Aalders, 1999) and *T. colubriformis* (Allan and Wharton, 1990) increased with increase in the severity of the desiccation stress during dehydration. The restoration of the cuticular permeability barrier during rehydration can be prevented by inhibitors that block enzyme activity and post-transcriptional protein synthesis (Wharton *et al.*, 1988), indicating an active repair mechanism. Leakage of inorganic ions during rehydration has been demonstrated in *A. tritici* (Womersley, 1981). The leakage ceases during the lag phase, indicating the repair of damaged membranes or the restoration of the permeability barrier due to a physical change associated with rehydration.

During the lag phase, morphological changes occur gradually. In *D. dipsaci* and *A. agrostis*, small lipid droplets coalesce within the intestine to form large droplets, and muscle cells of L4 of *D. dipsaci* resume normal hydrated thickness (Wharton and Barrett, 1985; Wharton *et al.*, 1985; Preston and Bird, 1987). There is a decrease in body length of *D. dipsaci* (Wharton *et al.*, 1985) and *T. colubriformis* (Allan and Wharton, 1990) during the lag phase, which may indicate a temporary contraction of the muscle cells as they recover. In *T. colubriformis* there is evidence of a change in the arrangements of muscle filaments in the contractile region of the muscle cells (Allan and Wharton, 1990).

Analyses of metabolic changes during rehydration have been confined almost entirely to L4 of *D. dipsaci*. Metabolism of L4, as measured by heat output, oxygen uptake or $^{14}CO_2$ production from labelled substrates, begins immediately after hydration (Barrett, 1982). The metabolite profiles recover quickly during hydration, with noticeable changes after 10 min and completion by 1 h. However, the ATP content does not recover as rapidly as those of the other metabolites; after 10 min there is little change and even after 1 h it is still low (Barrett, 1982). The slow trehalose depletion (up to 48 h to return to pre-desiccation levels) may be associated with the slow recovery of ATP levels. Immediately after hydration, the mitochondria are essentially uncoupled and there is no oxidative phosphorylation (Barrett, 1982); the mitochondria gradually swell during rehydration before adopting a normal morphology (Wharton and Barrett, 1985). Barrett (1982) suggested that, during the dehydration–rehydration cycle, membrane function is disrupted and the lag phase reflects the time required to restore metabolic and ionic gradients.

Protein synthesis during the first 2 h of rehydration is negligible and L4 of *D. dipsaci* revive successfully in the presence of inhibitors of protein and RNA synthesis (Barrett, 1982). However, there is an increase in the activity of certain enzymes involved in prevention of cellular ageing through free-radical scavenging reactions and negation of lipid peroxidation. For example, during rehydration of *A. avenae* an increase in superoxide dismutase activity occurs (Womersley, 1987) and catalase activity essentially triples during the first 4 h (Gresham and Womersley, 1991).

1.6 Implications for Control Options

As dry individuals or in eggs, nematodes can be dispersed by wind and other agents and can withstand other environmental stresses, such as extremes of temperature and nematicides. Many species of animal-parasitic nematodes in a quiescent (arrested) state show greater tolerance of anthelmintics (Prichard, 1988; Sargison *et al.*, 2007) and plant-parasitic nematodes are less sensitive to nematicides (Thomason and McKenry, 1974). Schroeder and MacGuidwin (2010) demonstrated that quiescent L2 of *H. glycines* survived higher concentrations of ethanol and the plant-derived compound allyl isothiocyanate compared with active L2. There was also a reduced penetration of fluorescein isothiocyanate (FITC) in quiescent L2 compared with active L2. Schroeder and MacGuidwin (2010) concluded that behavioural quiescence is correlated with exclusion of the compound from the nematode's body. These authors demonstrated that the route of entry of FITC in active L2 of *H. glycines* was via the cephalic region, and they suggested that future research should use microautoradiography techniques to examine toxin penetration in quiescent nematodes compared with active ones.

The ability of plant-parasitic nematodes to survive between successive crops has necessitated the development of various control and management strategies to reduce the number of nematodes and their adverse impact on crops. Several control methods are species-specific and, thus, correct identification is necessary to ensure the relevant control option is used. Traditionally, nematode identification has relied on morphological characters. However, morphological identification is not always straightforward; the majority of immature specimens (survival stages such as eggs and larvae) cannot be identified by morphological traits. Even when identifiable characteristics are available, identification to species level may still be difficult for some survival stages (e.g. cysts). In these situations, DNA-based tools and other molecular techniques (e.g. antibodies and isozymes) are valuable aids for species identification (Perry *et al.*, 2007).

Preventing the introduction and dissemination of plant-parasitic nematodes within a country and between countries is an essential control tactic used by both farmers and authorities. Plant-parasitic nematodes can be disseminated via plant material (endoparasitic stages) and adhering soil (soil stages). They are spread via seed (*Anguina*, *Aphelenchoides* and *Ditylenchus*), leaves (*Aphelenchoides*), tubers (*Meloidogyne*) and bulbs (*Ditylenchus*). In all of these circumstances the nematodes are in a state of anhydrobiosis, thereby surviving desiccation. Soil is another possible reservoir of nematode survival stages that resist dry conditions (e.g. cysts of *Globodera* and *Heterodera*). It is clear that all of these situations require appropriate nematode extraction techniques and the subsequent identification of the nematode in its survival stage. Desiccation-resistant stages of plant-parasitic nematodes can also be dispersed by wind (Orr and Newton, 1971; Gaur, 1988) and water (Faulkner and Bolander, 1966, 1970).

Some cultural methods aim to reduce the nematode density by starvation. Delayed planting of a host crop can enable the plants to avoid mass invasion of nematodes as many nematodes that hatch in response to rising soil temperatures will no longer have the protection of the eggshell and will die in the absence of a host. This strategy, however, will not be effective against plant-parasitic nematodes in diapause or those that depend on a stimulus from the host for hatching (e.g. *Globodera*; Perry, 2002). Rotation on the same area of land with different crops is one of the key methods for nematode management. The basic premise of crop rotation is to distance in time hosts with susceptibility to the same nematode species using poor, inhibitory or non-hosts, in order that population densities do not increase to damaging levels but decline below damaging thresholds before the next fully susceptible crop is grown (Viaene *et al.*, 2006). This strategy is very useful with nematodes that have a restricted host range. Although the potato cyst nematodes, *G. rostochiensis* and *G. pallida*, have an extremely narrow host range of commercially grown hosts (potato, tomato and aubergine), their control by rotation with non-host crops is not always effective because in their survival stage (eggs in cysts) these nematodes can endure extremely long periods without a host (Turner, 1996). Even in the presence of a host, not all larvae hatch out at the same time; a proportion is retained in the cyst body and can survive until the next crop (Turner and Rowe, 2006). Similarly, fallow will not reduce populations of plant-parasitic nematodes if they are in a survival stage. Flooding drastically reduces soil oxygen concentration and increases carbon dioxide, and is associated with an increase in toxic substances and a reduction in pH. Few nematodes survive short periods of flooding. Extended periods, therefore, provide almost nematode-free conditions (Viaene *et al.*, 2006). The rice root nematodes, *Hirschmanniella* spp., are notable exceptions (Fortuner and Merny, 1979). *Hirschmanniella oryzae* survives poorly in dried fields but may overwinter in dead roots as eggs if kept moist.

Some control methods use high temperatures to kill plant-parasitic nematodes. In high-yielding crops steaming was frequently used to reduce the impact of soil-borne pathogens. However, steaming often fails to give satisfactory results for root-knot nematode control, especially when survivors in the deeper layers of soil can build up infestations (Karssen and Moens, 2006).

1.7 Conclusions and Future Directions

The urgent need for environmentally acceptable methods to control pests has provided the impetus for studies on aspects of the survival of parasitic nematodes outside their hosts and research on the use of entomopathogenic nematodes as bioinsecticides. In turn, this has generated further research on the morphological and biochemical adaptations associated with anhydrobiosis. For example, the role of trehalose in survival and life cycle physiology of animal- and plant-parasitic nematodes has received much attention, and Behm (1997) has suggested that, if trehalose is important for the survival of animal-parasitic nematodes, enzymes of trehalose metabolism may offer molecular control targets, as trehalose metabolism appears not to be important in mammals.

Only a limited number of nematode species have been used as the basis for detailed research and there is still much to be understood about the genetic control associated with induction of the dormant state. The increasing number of available genome sequences of entomopathogenic and animal- and plant-parasitic nematodes has important implications for studies on nematode biology, and comparative genomics using other non-parasitic nematode genomes, such as *C. elegans* and *Caenorhabditis briggsae*, will provide information on the genetic basis of survival attributes. In particular, molecular studies on responses to adverse environmental cues, in conjunction with further studies on physiological and behavioural aspects, will provide extensive information about survival biology and, especially, how extensive is the dauer larvae phenomenon in the Phylum Nematoda.

1.8 References

Abad, P. and Opperman, C.H. (2009) The complete sequence of the genomes of *Meloidogyne incognita* and *Meloidogyne hapla*. In: Perry, R.N., Moens, M. and Starr, J.L. (eds) *Root-knot Nematodes*. CAB International, Wallingford, UK, pp. 363–379.

Abad, P., Gouzy, J., Aury, J.-M. *et al.* (2008) Genome sequence of the metazoan plant-parasitic nematode *Meloidogyne incognita. Nature Biotechnology* 8, 909–915.

Allan, G.S. and Wharton, D.A. (1990) Anhydrobiosis in the infective juveniles of *Trichostrongylus colubriformis* (Nematoda: Trichostrongylidae). *International Journal for Parasitology* 20, 183–192.

Ash, C.P.J. and Atkinson, H.J. (1983) Evidence for a temperature-dependent conversion of lipid reserves to carbohydrates in quiescent eggs of the nematode *Nematodirus battus. Comparative Biochemistry and Physiology B* 76, 603–610.

Ash, C.P.J. and Atkinson, H.J. (1984) *Nematodirus battus*: permeability changes, calcium binding, and phosphorylation of the eggshell during hatching. *Experimental Parasitology* 58, 27–40.

Barrett, J. (1982) Metabolic responses to anabiosis in the fourth stage juveniles of *Ditylenchus dipsaci* (Nematoda). *Proceedings of the Royal Society of London B* 216, 157–177.

Barrett, J. (1991) Anhydrobiotic nematodes. *Agricultural Zoology Reviews* 4, 161–176.

Behm. C.A. (1997) The role of trehalose in the physiology of nematodes. *International Journal for Parasitology* 27, 215–229.

Bird, A.F. (1988) Cuticle printing of nematodes. *International Journal of Parasitology* 18, 869–871.

Bird, A.F. and Bird, J. (1991) *The Structure of Nematodes*, 2nd edn. Academic Press, San Diego, California.

Bird, A.F. and Buttrose, M.S. (1974) Ultrastructural changes in the nematode *Anguina tritici* associated with anhydrobiosis. *Journal of Ultrastructural Research* 48, 177–189.

Bird, A.F. and Soeffky, A. (1972) Changes in the ultrastructure of the gelatinous matrix of *Meloidogyne javanica* during dehydration. *Journal of Nematology* 4, 166–169.

Bird, A.F. and Zuckerman, B.M. (1989) Studies on the surface coat (glycocalyx) of the dauer larva of *Anguina agrostis. International Journal for Parasitology* 19, 235–247.

Burnell, A. and Dowds, B.C.A. (1996) The genetic improvement of entomopathogenic nematodes and their symbiotic bacteria: phenotypic targets, genetic limitations and an assessment of possible hazards. *Biocontrol Science and Technology* 6, 435–447.

Campbell, J.F. and Gaugler, R. (1991) Role of the sheath in desiccation tolerance of two entomopathogenic nematodes. *Nematologica* 37, 324–332.

Clarke, A.J. and Hennessy, J. (1983) The role of calcium in the hatching of *Globodera rostochiensis. Revue de Nématologie* 6, 247–255.

Clarke, A.J. and Perry, R.N. (1980) Egg-shell permeability and hatching of *Ascaris suum. Parasitology* 80, 447–456.

Clarke, A.J. and Perry, R.N. (1985) Egg-shell calcium and the hatching of *Globodera rostochiensis*. *International Journal of Parasitology* 15, 511–516.

Clarke, A.J. and Perry, R.N. (1988) The induction of permeability in egg-shells of *Ascaris suum* prior to hatching. *International Journal for Parasitology* 18, 987–990.

Crowe, J.H., Carpenter, J.F. and Crowe, L.M. (1998) The role of vitrification in anhydrobiosis. *Annual Review of Physiology* 60, 73–103.

Ellenby, C. (1946) Nature of the cyst wall of the potato-root eelworm *Heterodera rostochiensis* Wollenweber, and its permeability to water. *Nature* 157, 302.

Ellenby, C. (1968a) Desiccation survival in the plant parasitic nematodes, *Heterodera rostochiensis* Wollenweber and *Ditylenchus dipsaci* (Kuhn) Filipjev. *Proceedings of the Royal Society of London B* 169, 203–213.

Ellenby, C. (1968b) Desiccation survival of the infective larva of *Haemonchus contortus*. *Journal of Experimental Biology* 49, 460–475.

Ellenby, C. (1969) Dormancy and survival in nematodes. *Symposium of the Society for Experimental Biology* 23, 83–97.

Ellenby, C. and Perry, R.N. (1976) The influence of the hatching factor on the water uptake of the second stage larva of the potato cyst nematode *Heterodera rostochiensis*. *Journal of Experimental Biology* 64, 141–147.

Elling, A.E., Mitreva, M., Recknor, J. *et al.* (2007) Divergent evolution of arrested development in the dauer stage of *Caenorhabditis elegans* and the infective stage of *Heterodera glycines*. *Genome Biology* 8, R211.

Endo, B.Y. and Wyss, U. (1992) Ultrastructure of cuticular exudations in parasitic juvenile *Heterodera schachtii* as related to cuticle structure. *Protoplasma* 166, 67–77.

Evans, A.A.F. and Perry, R.N. (1976) Survival strategies in nematodes. In: Croll, N.A. (ed.) *The Organisation of Nematodes*. Academic Press, London, pp. 383–424.

Faulkner, L.R. and Bolander, W.J. (1966) Occurrence of large nematode populations in irrigation canals in South Central Washington. *Nematologica* 12, 591–600.

Faulkner, L.R. and Bolander, W.J. (1970) Acquisition and distribution of nematodes in irrigation waterways of the Columbia Basin in eastern Washington. *Journal of Nematology* 2, 362–367.

Fortuner, R.C. and Merny, C. (1979) Root-parasitic nematodes of rice. *Revue de Nématologie* 2, 79–102.

Gal, T.Z., Glazer, I. and Koltai, H. (2004) An LEA group 3 family member is involved in survival of *C. elegans* during exposure to stress. *FEBS Letters* 577, 21–26.

Gaugler, R., Wilson, M. and Shearer, P. (1997) Field release and environmental fate of a transgenic entomopathogenic nematode. *Biological Control* 9, 75–80.

Gaur, H.S. (1988). Dissemination and mode of survival of nematodes in dust storms. *Indian Journal of Nematology* 18, 94–98.

Gaur, H.S. and Perry, R.N. (1991a) The biology and control of the plant parasitic nematode *Rotylenchulus reniformis*. *Agricultural Zoology Reviews* 4, 177–212.

Gaur, H.S. and Perry, R.N. (1991b) The role of the moulted cuticles in the desiccation survival of adults of *Rotylenchulus reniformis*. *Revue de Nématologie* 14, 491–496.

Gaur, H.S., Perry, R.N. and Beane, J. (1992) Hatching behaviour of six successive generations of the pigeon-pea cyst nematode, *Heterodera cajani*, in relation to growth and senescence of cowpea, *Vigna unguiculata*. *Nematologica* 38, 190–202.

Gaur, H.S., Beane, J. and Perry, R.N. (1995) Hatching of four successive generations of *Heterodera sorghi* in relation to the age of sorghum, *Sorghum vulgare*. *Fundamental and Applied Nematology* 18, 599–601.

Gaur, H.S., Beane, J. and Perry, R.N. (2000) The influence of root diffusate, host age and water regimes on hatching of the root-knot nematode, *Meloidogyne triticoryzae*. *Nematology* 2, 191–199.

Goyal, K., Walton, L.J. and Tunnacliffe, A. (2005) LEA proteins prevent protein aggregation due to water stress. *Biochemical Journal* 388, 151–157.

Greco, N. (1981) Hatching of *Heterodera carotae* and *H. avenae*. *Nematologica* 27, 366–371.

Greco, N., Vito, M.D. and Lamberti, F. (1986) Studies on the biology of *Heterodera goettingiana* in southern Italy. *Nematologia Mediterranea* 14, 23–29.

Gresham, A. and Womersley, C.Z. (1991) Modulation of catalase activity during the enforced induction of and revival from anhydrobiosis in nematodes. *FASEB Journal* 5A, 682.

Higa, L.M. and Womersley, C.Z. (1993) New insights into the anhydrobiotic phenomenon: the effects of trehalose content and differential rates of evaporative water loss on the survival of *Aphelenchus avenae*. *Journal of Experimental Zoology* 267, 120–129.

Ibrahim, S.K. and Perry, R.N. (1993) Desiccation survival of the rice stem nematode *Ditylenchus angustus*. *Fundamental and Applied Nematology* 16, 31–38.

Ibrahim, S.K., Perry, R.N., Plowright, R.A. and Rowe, J. (1992) Hatching behaviour of the rice cyst nematode *Heterodera sacchari* and *H. oryzicola* in relation to age of host plant. *Fundamental and Applied Nematology* 16, 23–29.

Ishibashi, N., Kondo, E., Muraoka, M. and Yokoo, T. (1973) Ecological significance of dormancy in plant parasitic nematodes. I. Ecological difference between eggs in gelatinous matrix and cysts of *Heterodera glycines* Ichinohe (Tylenchida: Heteroderidae). *Applied Entomology and Zoology* 8, 53–63.

Jagdale, G.B., Grewal, P.S. and Salminen, S.O. (2005) Both heat-shock and cold-shock influence trehalose metabolism in an entomopathogenic nematode. *Journal of Parasitology* 91, 988–994.

Jones, J.T., Furlanetto, C. and Kikuchi, T. (2005) Horizontal gene transfer from bacteria and fungi as a driving force in the evolution of plant parasitism in nematodes. *Nematology* 7, 641–646.

Jones, P., Tylka, G. and Perry, R.N. (1998) Hatching. In: Perry, R.N. and Wright, D.J. (eds) *The Physiology and Biochemistry of Free-living and Plant-parasitic Nematodes*. CAB International, Wallingford, UK, pp. 181–212.

Karssen, G. and Moens, M. (2006) Root-knot nematodes. In: Perry, R.N. and Moens, M. (eds) *Plant Nematology*. CAB International, Wallingford, UK, pp. 59–90.

Kenyon, C. (1997) Environmental factors and gene activities that influence life span. In: Riddle, D.L., Blumenthal, T., Meyer, B.J. and Priess, J.R. (eds) C. elegans *II*. Cold Spring Harbor Laboratory Press, New York, pp. 791–813.

Kikuchi, T., Aikawa, T., Kosaka, H., Pritchard, L., Ogura, N. and Jones, J.T. (2007) Expressed sequence tag (EST) analysis of the pine wood nematodes *Bursaphenelchus xylophilus* and *B. mucronatus*. *Molecular and Biochemical Parasitology* 155, 9–17.

Larsen, M.N. and Roepstorff, A. (1999) Seasonal variation in development and survival of *Ascaris suum* and *Trichuris suis* eggs on pastures. *Parasitology* 119, 209–220.

Lee, D.L. (1972) Penetration of mammalian skin by the infective larva of *Nippostrongylus brasiliensis*. *Parasitology* 65, 499–505.

Lee, D.L. (2002) Cuticle, moulting and exsheathment. In: Lee, D.L. (ed.) *The Biology of Nematodes*. Taylor & Francis, London, pp. 171–209.

Levine, H. and Slade, L. (1992) Another view of trehalose for drying and stabilizing biological material. *BioPharm* 5, 36–40.

Loomis, S.H., O'Dell, S.J. and Crowe, J.H. (1979) Anhydrobiosis in nematodes: inhibition of the browning reaction of reducing sugars with dry protein. *Journal of Experimental Zoology* 208, 355–360.

Madin, K.A.C. and Crowe, J.H. (1975) Anhydrobiosis in nematodes: carbohydrate and lipid metabolism during rehydration. *Journal of Experimental Zoology* 193, 335–342.

McClure, M.A., Schmitt, M.E. and McCullough, M.D. (2008) Distribution, biology and pathology of *Anguina pacificae*. *Journal of Nematology* 40, 226–239.

Menti, H., Wright, D.J. and Perry, R.N. (1997) Desiccation survival of populations of the entomopathogenic nematodes

Steinernema feltiae and *Heterorhabditis megidis* from Greece and the UK. *Journal of Helminthology* 71, 41–46.

Moens, M. and Perry, R.N. (2009) Migratory plant endoparasitic nematodes: a group rich in contrasts and diversity. *Annual Review of Phytopathology* 47, 313–332.

Mota, M. and Vieira, P.R. (eds) (2008) *Pine Wilt Disease: a Worldwide Threat to Forest Ecosystems*. Springer, Berlin.

Murrell, K.D., Graham, C.E. and McGreevy, M. (1983) *Strongyloides ratti* and *Trichinella spiralis*: net charge of epicuticle. *Experimental Parasitology* 55, 331–339.

O'Leary, S.A. and Burnell, A.M. (1997) The isolation of mutants of *Heterorhabditis megidis* (strain UK211) with increased desiccation tolerance. *Fundamental and Applied Nematology* 20, 197–205.

O'Leary, S.A., Burnell, A.M. and Kusel, J.R. (1998) Biophysical properties of the surface of desiccation-tolerant mutants and parental strain of the entomopathogenic nematode *Heterorhabditis megidis* (strain UK211). *Parasitology* 117, 337–345.

Orion, D., Kritzman, G., Meyer, S., Erbe, E. and Chitwood, D. (2001) A role of the gelatinous matrix in the resistance of root-knot nematode (*Meloidogyne* spp.) eggs to microorganisms. *Journal of Nematology* 33, 203–207.

Orr, C.C. and Newton, O.H. (1971) Distribution of nematodes by wind. *Plant Disease Reporter* 55, 61–63.

Patel, M.N., Perry, R.N. and Wright, D.J. (1997) Desiccation survival and water contents of entomopathogenic nematodes, *Steinernema* spp. (Rhabditida: Steinernematidae). *International Journal for Parasitology* 27, 61–70.

Perry, R.N. (1977a) Desiccation survival of larval and adult stages of the plant parasitic nematodes, *Ditylenchus dipsaci* and *D. myceliophagus*. *Parasitology* 74, 139–148.

Perry, R.N. (1977b) The water dynamics of stages of *Ditylenchus dipsaci* and *D. myceliophagus* during desiccation and rehydration. *Parasitology* 75, 45–70.

Perry, R.N. (1977c) The effect of previous desiccation on the ability of the fourth stage larva of *Ditylenchus dipsaci* to control its rate of water loss and survive drying. *Parasitology* 75, 215–231.

Perry, R.N. (1983) The effect of potato root diffusate on the desiccation survival of unhatched juveniles of *Globodera rostochiensis*. *Revue de Nématologie* 6, 99–102.

Perry, R.N. (1989) Dormancy and hatching of nematode eggs. *Parasitology Today* 5, 377–383.

Perry, R.N. (1999) Desiccation survival of parasitic nematodes. *Parasitology* 119, S19–S30 (published January, 2001).

Perry, R.N. (2002) Hatching. In: Lee, D.L. (ed.) *The Biology of Nematodes*. Taylor & Francis, London, pp. 147–169.

Perry, R.N. and Gaur, H.S. (1996) Host plant influences on the hatching of cyst nematodes. *Fundamental and Applied Nematology*, 19, 505–510.

Perry, R.N., Clarke, A.J., Hennessy, J. and Beane, J. (1983) The role of trehalose in the hatching mechanism of *Heterodera goettingiana*. *Nematologica* 29, 324–335.

Perry, R.N., Subbotin, S.A. and Moens, M. (2007) Molecular diagnostics of plant-parasitic nematodes. In: Punja, Z.K., De Boer, S.H. and Sanfaçoni, H. (eds) *Biotechnology and Plant Disease Management*. CAB International, Wallingford, UK, pp. 195–226.

Preston, C.M. and Bird, A.F. (1987) Physiological and morphological changes associated with recovery from anabiosis in the dauer larva of the nematode *Anguina agrostis*. *Parasitology* 44, 125–133.

Prichard, R.K. (1988) Anthelmintics and control. *Veterinary Parasitology* 27, 97–109.

Riddle, D.L. and Albert, P.S. (1997) Genetic and environmental regulation of dauer larva development. In: Riddle, D.L., Blumenthal, T., Meyer, B.J. and Priess, J.R. (eds) C. elegans *II*. Cold Spring Harbor Press, New York, pp. 739–768.

Robinson, A.F., Orr, C.C. and Heintz, C.E. (1984) Some factors affecting survival of desiccation by infective juveniles of *Orrina phyllobia*. *Journal of Nematology* 16, 86–91.

Roepstorff, A. (1997) Helminth surveillance as a prerequisite for anthelmintic treatment in intensive sow herds. *Veterinary Parasitology* 73, 139–151.

Rogers, W.P. and Sommerville, R.I. (1968) The infection process, and its relation to the development of early parasitic stages of nematodes. *Advances in Parasitology* 6, 327–348.

Rössner, J. (1973) Anpassung wandernder Wurzelnematoden an Austrocknung im Boden. *Nematologica* 19, 366–378.

Rössner, J. and Perry, R.N. (1975) Water loss and associated surface changes after desiccation in *Rotylenchus robustus*. *Nematologica* 21, 438–442.

Sargison, N.D., Wilson, D.J., Bartley, D.J. and Penny, C.D. (2007) Haemonchosis and teladorsagiosis in a Scottish sheep flock putatively associated with the overwintering of hypobiotic fourth stage larvae. *Veterinary Parasitology* 147, 326–331.

Schroeder, N.E. and MacGuidwin, A.E. (2010) Behavioral quiescence reduces the penetration and toxicity of exogenous compounds in J2 *Heterodera glycines*. *Nematology* 12, 277–287.

Sharon, E. and Spiegel, Y. (1993) Glycoprotein characterization of the gelatinous matrix in the root-knot nematode *Meloidogyne javanica*. *Journal of Nematology* 25, 585–589.

Starr, J.L. (1993) Recovery and longevity of egg masses of *Meloidogyne incognita* during simulated winter survival. *Journal of Nematology* 25, 244–248.

Tefft, P.M. and Bone, L.W. (1985) Plant-induced hatching of the soybean cyst nematode *Heterodera glycines*. *Journal of Nematology* 17, 275–279.

Thomason, I.J. and McKenry, M.V. (1974) 1,3-dichloropropene and 1,2-dibromoethane compounds: 1. Movement and fate as affected by various conditions in several soils. 2. Organism dosage-response studies in laboratory with several nematode species. *Hilgardia* 42, 393–438.

Thompson, J.M. and Tylka, G.L. (1997) Differences in hatching of *Heterodera glycines* egg-mass and encysted eggs *in vitro*. *Journal of Nematology* 29, 315–321.

Timper, P., Kaya, H.K. and Jaffee, B.A. (1991) Survival of entomogenous nematodes in soil infested with the nematode-parasitic fungus *Hirsutella rhossiliensis* (Deuteromycotina: Hyphomycetes). *Biological Control* 1, 42–50.

Turner, S.J. (1996) Population decline of potato cyst nematodes (*Globodera rostochiensis*, *G. pallida*) in field soils in Northern Ireland. *Annals of Applied Biology* 129, 315–322.

Turner, S.J. and Rowe, J.A. (2006) Cyst nematodes. In: Perry, R.N. and Moens, M. (eds) *Plant Nematology*. CAB International, Wallingford, UK, pp. 91–122.

Vellai, T., Molnár, A., Lakatos, L., Bánfalvi, Z., Fodor, A. and Sáringer, G. (1999) Transgenic nematodes carrying a cloned stress resistance gene from yeast. In: Glazer, P., Richardson, P., Boemare, N. and Coudert, F. (eds) *Survival of Entomopathogenic Nematodes*. Luxembourg, Office for Official Publications of the European Communities, pp. 105–119.

Viaene, N., Coyne, D.N. and Kerry, B.R. (2006) Biological and cultural management. In: Perry, R.N. and Moens, M. (eds) *Plant Nematology*. CAB International, Wallingford, UK, pp. 346–369.

Viney, M. (1996) A genetic analysis of reproduction in *Strongylus ratti*. *Parasitology* 109, 511–515.

Wallace, H.R. (1968) The influence of soil moisture on the survival and hatch of *Meloidogyne javanica*. *Nematologica* 14, 231–242.

Waller, P.J. (1971) Structural differences in the egg envelope of *Haemonchus contortus* and *Trichostrongylus colubriformis* (Nematoda: Trichostrongylidae). *Parasitology* 62, 157–160.

Waller, P.J. and Donald, A.D. (1970) The response to desiccation of eggs of *Trichostrongylus colubriformis* and *Haemonchus contortus* (Nematoda: Trichostrongylidae). *Parasitology* 61, 195–204.

Wharton, D.A. (1979) *Ascaris lumbricoides*: water loss during desiccation of embryonating eggs. *Experimental Parasitology* 48, 398–406.

Wharton, D.A. (1980) Studies on the function of the oxyurid egg-shell. *Parasitology* 81, 103–113.

Wharton, D.A. (1981) The initiation of coiling behaviour prior to desiccation in the infective larvae of *Trichostrongylus colubriformis. International Journal for Parasitology* 11, 353–357.

Wharton, D.A. (1986) *A Functional Biology of Nematodes*. Croom Helm, London and Sydney.

Wharton, D.A. (1996) Water loss and morphological changes during desiccation of the anhydrobiotic nematode *Ditylenchus dipsaci. Journal of Experimental Biology* 199, 1085–1093.

Wharton, D.A. and Aalders, O. (1999) Desiccation stress and recovery in the anhydrobiotic nematode *Ditylenchus dipsaci* (Nematoda: Anguinidae). *European Journal of Entomology* 96, 199–203.

Wharton, D.A. and Barrett, J. (1985) Ultrastructural changes during recovery from anabiosis in the plant parasitic nematode, *Ditylenchus dipsaci. Tissue and Cell* 17, 79–96.

Wharton, D.A. and Lemmon, J. (1998) Ultrastructural changes during desiccation of the anhydrobiotic nematode *Ditylenchus dipsaci. Tissue and Cell* 30, 312–323.

Wharton, D.A., Barrett, J. and Perry, R.N. (1985) Water uptake and morphological changes during recovery from anabiosis in the plant parasitic nematode, *Ditylenchus dipsaci. Journal of Zoology* 206, 391–402.

Wharton, D.A., Preston, C.M., Barrett, J. and Perry, R.N. (1988) Changes in cuticular permeability associated with recovery from anhydrobiosis in the plant parasitic nematode, *Ditylenchus dipsaci. Parasitology* 97, 317–330.

Wharton, D.A., Petrone, L., Duncan, A. and McQuillan, A.J. (2008) A surface lipid may control the permeability slump associated with entry into anhydrobiosis in the plant parasitic nematode *Ditylenchus dipsaci. Journal of Experimental Biology* 211, 2901–2908.

Womersley, C. (1978) A comparison of the rate of drying of four nematode species using a liquid paraffin technique. *Annals of Applied Biology* 90, 401–405.

Womersley, C. (1981) The effect of dehydration and rehydration on salt loss in the second-stage larvae of *Anguina tritici. Parasitology* 82, 411–419.

Womersley, C. (1987) A reevaluation of strategies employed by nematode anhydrobiotes in relation to their natural environment. In: Veech, J.A. and Dickson, D.W. (eds) *Vistas on Nematology*. Society of Nematologists Inc., Hyattsville, Maryland, pp. 165–173.

Womersley, C.Z. (1990) Dehydration survival and anhydrobiotic survival. In: Gaugler, R. and Kaya, H.K. (eds) *Entomopathogenic Nematodes in Biological Control*. CRC Press, Boca Raton, Florida, pp. 117–137.

Womersley, C.Z. (1993) Factors affecting physiological fitness and modes of survival employed by dauer larvae and their relationship to pathogenicity. In: Bedding, R.A., Akhurst, R. and Kaya, H.K. (eds) *Nematodes and the Biological Control of Insect Pests*. CRC Press, Boca Raton, Florida, pp. 79–88.

Womersley, C.Z. and Higa, L.M. (1998) Trehalose: its role in the anhydrobiotic survival of *Ditylenchus myceliophagus. Nematologica* 44, 269–291.

Womersley, C. and Smith, L. (1981) Anhydrobiosis in nematodes. 1. The role of glycerol, myoinositol and trehalose during desiccation. *Comparative Biochemistry and Physiology* 70B, 579–586.

Womersley, C., Thompson, S.N. and Smith, L. (1982) Anhydrobiosis in nematodes. 2. Carbohydrate and lipid analysis in undesiccated and desiccated nematodes. *Journal of Nematology* 14, 145–153.

Womersley, C.Z., Wharton, D.A. and Higa, L.M. (1998) Survival biology. In: Perry, R.N. and Wright, D.J. (eds) *The Physiology and Biochemistry of Free-living and Plant-parasitic Nematodes*. CAB International, Wallingford, UK, pp. 271–302.

Wright, D.J. and Perry, R.N. (2006) Reproduction, physiology and biochemistry. In: Perry, R.N. and Moens, M. (eds) *Plant Nematology*. CAB International, Wallingford, UK, pp.187–209.

Zheng, L. and Ferris, H. (1991) Four types of dormancy exhibited by eggs of *Heterodera schachtii. Revue de Nématologie* 14, 419–426.

2 Survival of Plant-parasitic Nematodes inside the Host

Jose Lozano and Geert Smant

Laboratory of Nematology, Wageningen University, Wageningen, The Netherlands

2.1 Introduction

Plant parasitism arose at least three times as an independent evolutionary innovation within the Phylum Nematoda during a period of ~350 million years (Blaxter *et al.*, 1998). The most advanced lineage of plant parasites includes the sedentary root-knot nematodes and cyst nematodes, i.e. members of the family Heteroderidae (Endo, 1975). After a migratory phase of host finding and invasion, root-knot and cyst nematodes transform host cells into complex metabolically active structures, from which they acquire their nutrients. Feeding site formation is associated with loss of mobility in root-knot and cyst nematodes. Thus, these parasites have evolved towards an absolute dependency on the food provided by a single feeding structure. Although perhaps not so advanced, several parasites from the other two lineages also transform host cells to some degree prior to feeding.

Feeding on plants has led to convergent morphological adaptations in all three lineages of plant-parasitic nematodes. Herbivory in nematodes is, for example, always associated with specialized morphological adaptations in the outer surface of the nematode and, most notably, in the feeding apparatus. In this chapter we review the recent literature on various aspects of nematode survival in plants. We briefly summarize morphological, molecular and physiological adaptations to plant parasitism in nematodes (see also Blaxter *et al.*, 1998; Baldwin *et al.*, 2004; Dieterich and Sommer, 2009). We also describe the life support structures induced by nematodes in host plant tissues and, because plants appear to 'tolerate' feeding nematodes, we examine evidence that nematodes must have evolved ways to modulate the innate immunity of the host. We summarize current concepts of innate immunity in plants, which are mostly founded on work done with plant-pathogenic fungi and bacteria. Finally, we address the molecular tools used by plant-parasitic nematodes to modulate host defence responses.

2.2 Morphological Adaptations to Plant Parasitism

2.2.1 Cuticle, surface coat and cuticular camouflage

The external cuticle of nematodes is an unusual and complex multilayered structure that acts as an exoskeleton (Bird, 1968; Lee, 1972; Wright, 1987). The cuticle is made of collagens and a variety of other proteins, lipids and carbohydrates (Spiegel and McClure, 1995). The transitions through successive developmental stages in all members of the Phylum Nematoda are marked by a moult. During each moult the cuticle is replaced with a new one, which is assembled from components produced in the underlying hypodermis. In plant-parasitic nematodes the rigid cuticular lining of the pharynx, including the feeding apparatus, is also renewed during moulting.

Besides maintaining the shape of the nematode, the cuticle also constitutes a strong protective barrier. The cuticle provides an impervious interface between nematode and host cells, which may provide protection against host defence responses. The outer surface of the nematode cuticle is covered with a negatively charged carbohydrate-rich surface coat (Spiegel and McClure, 1995). Glandular cells in the head and tail region of the nematode are believed to secrete the components that make up the surface coat. However, the exact origin and composition of the surface coat is not clear. Binding of lectins, human red blood cells and gold-labelled glycoproteins to the surface coat of root-knot nematodes indicates that the surface coat includes carbohydrate-binding proteins (Spiegel, 1995; Spiegel and McClure, 1995; Sharon and Spiegel, 1996; Spiegel *et al.*, 1997).

The surface coat of animal-parasitic nematodes is implicated in two distinct immune-evading strategies. The first strategy involves the shedding of the surface coat to ward off immune cells attacking the worm. For example, when immune cells attack *Toxocara canis*, the nematode sheds its mucin-based surface coat to escape from attached killer cells (Badley *et al.*, 1987; Theodoropoulos *et al.*, 2001). A question that needs further research is whether plant-parasitic nematodes also deploy surface coat shedding to evade host innate immunity. Plants do not have mobile immune cells that can be directed towards an invader. Instead, many sedentary plant parasites lie embedded within host tissue and have to deal with the defence responses from surrounding host cells. So it is difficult to envisage how surface coat shedding could provide plant-parasitic nematodes protection from host immunity. However, labelling experiments suggest that root-knot nematodes none the less shed their surface coat during host invasion. By shedding the surface coat during the transition from migratory to sedentary invader, the nematode may change the composition of the surface coat to adapt it to a new and hostile microenvironment (Sharon *et al.*, 2002). Mucin-like proteins that have been identified in the pharyngeal glands of plant-parasitic nematodes could be involved in changes of the surface coat (Wang *et al.* in GenBank accession AAC62109).

The second strategy deployed by animal-parasitic helminths to evade the host's immune system is cuticular camouflage. To achieve cuticular camouflage, animal parasites cover themselves with host molecules so that they are no longer recognized as non-self in the host. In the current concepts of cuticular camouflage a key role is assigned to C-type lectins (Loukas and Maizels, 2000). C-type lectins are capable of binding carbohydrate moieties of glycosylated host proteins, such as major histocompatibility complex class I antigens, C3 complement proteins and IgG immunoglobulins (Loukas and Maizels, 2000). Because host C-type lectins control diverse immunity-related processes, parasite C-type lectins could also compete with the natural glycosylated ligands of the host C-type lectins. It has, for example, been proposed that parasite C-type lectins sequester alarm-signalling ligands of host C-type lectin receptors. Secretory C-type lectins have also been found in the pharyngeal glands of the soybean cyst nematode, *Heterodera glycines*, suggesting that these nematodes have at least the potential to use a type of cuticular camouflage in plants (De Boer *et al.*, 2002b). Some support for this hypothesis stems from a knock-down by RNA interference (RNAi) of a C-type lectin in *H. glycines*, which significantly reduced nematode survival inside host plants (Urwin *et al.*, 2002).

Another interesting case of cuticular camouflage occurs in the interaction between the entomopathogenic nematode *Steinernema feltiae* and its lepidopteran host *Galleria mellonella* (Brivio *et al.*, 2004, 2006; Mastore and Brivio, 2008). *Steinernema feltiae* penetrates the haemocoel of lepidopteran insects, wherein it releases endosymbiotic bacteria that kill the host. The innate immune system of the lepidoptera consists of at least three components, including antibacterial peptides, the prophenoloxidase activation system (humoral responses) and parasite encapsulation (cellular response). In an immunocompetent lepidopteran host, innate immunity is fully equipped to nip a bacterial outbreak in the haemocoel in the bud. However, *S. feltiae* manages to give its unleashed endosymbionts in the haemocoel a head start with the deployment of immune evasion and depression tactics centred on cuticular lipids. These cuticular lipids specifically bind various host haemolymph proteins involved in the synthesis of antimicrobial peptides and the proteolytic activation of prophenoloxidase. Further aspecific coating with host factors is believed to make the *S. feltiae* virtually undetectable as non-self, thus also preventing proper humoral and cellular responses that normally lead to parasite encapsulation.

2.2.2 The oral stylet – a multi-tool for nematodes

All plant cells are insulated by a rigid cell wall, which constitutes a protective container for the protoplast (Carpita and Gibeaut, 1993). In order to gain access to the host cell cytoplasm, nematodes penetrate plant cell walls with a protrusible cuticular stylet, which is also used to inject secretions into host cells and to take up plant nutrients (Baldwin *et al.*, 2004). Fierce outward movements of the stylet provide the necessary physical impact to perforate cell walls during host invasion. A more subtle behaviour is associated with feeding from plant cells, presumably to avoid the collapse of the host cell protoplasm (Wyss, 1992). The feeding routine of the nematode includes repeated cycles of stylet insertion into the host cell, release of secretions and uptake of plant solutes (Wyss and Zunke, 1986). Whether the stylet actually penetrates the cell membrane during feeding or the nematode uses an alternative mode of bidirectional transport over the cell membrane without disturbing the integrity of the membrane is still a matter of debate. Passive diffusion over the cell membrane into the host cytoplasm can be excluded as a mode of transport, because the molecular mass of many of the nematode secretory proteins is too high.

Work on the plant-pathogenic oomycete *Phytophthora infestans* has revealed a possible mechanism for the delivery of pathogen proteins into the host cells (Birch *et al.*, 2006; Morgan and Kamoun, 2007). *Phytophthora infestans* forms haustoria in between the cell wall and the cell membrane of host cells. The haustoria are enveloped by, but do not penetrate, the cell membrane of host cells. *Phytophthora infestans* releases secretions, so-called effectors, via its haustoria into the extracellular matrix of recipient host cells. One large class of *P. infestans* effectors, named RxLR-DEER effectors after a conserved sequence motif, are translocated over the cell membrane into the cytoplasm via a specific carrier/receptor protein. Although the translocation mechanism of RxLR-DEER effectors is not completely clear, there is a striking similarity with the translocation of effectors by the malaria parasite *Plasmodium* into human cells. Because biotrophic nematodes do not evidently penetrate the cell membrane, a similar translocation system could deliver the nematode secretions into host cytoplasm. One approach to test this hypothesis is to scan nematode secretory proteins systematically for RxRL-DEER-like or other motifs that could function as tags for a translocation pathway.

2.2.3 Pharyngeal glands – the source of all evil

A quick glance through the lens of a microscope by students often raises questions about the huge nuclei in the pharyngeal region of plant-parasitic nematodes (Raski *et al.*, 1969; Endo, 1987; Hussey and Mims, 1990). These nuclei mark the position of large single-celled pharyngeal (or oesophageal) glands alongside the anterior section of the digestive tract. The sheer size of the nuclei and gland cells is a remarkable adaptation to plant parasitism in nematodes. It was for this reason alone that secretions from the pharyngeal glands were associated with parasitism long before the nature of these secretions became known (Bird, 1983; Perry *et al.*, 1989). The pharyngeal glands in the advanced cyst and root-knot nematodes have been studied in particular detail. Therefore, most of our current knowledge of the pharyngeal gland cells and their secretions derives from these sedentary endoparasites (see the sections below for more details).

The pharyngeal glands are flask-shaped cells with a large, highly active nucleus, which lies embedded in an elaborate endoplasmic reticulum with numerous Golgi bodies (Endo, 1987; Hussey and Mims, 1990). The nuclei are positioned in the posterior wider section of the gland cells, which narrows down into long extensions towards the head of the nematode. The pharyngeal gland cells are filled with secretory granules floating forwards from the Golgi bodies to the ampulla. In the collecting reservoir of the ampullae, the secretory granules release their contents by means of exocytosis into the lumen of the pharynx. Inside the pharyngeal lumen the secretions move to the stylet base, and finally emanate from the orifice in the stylet tip.

Cyst and root-knot nematodes have three pharyngeal gland cells, of which one is positioned in the dorsal sector and two are located in the subventral sector of the pharyngeal region. Earlier studies have revealed that the subventral pharyngeal gland cells are mostly, but not exclusively, active in migratory stages (Hussey, 1987; Davis *et al.*, 1994). The dorsal pharyngeal gland is mainly active when the nematode is feeding. Remarkably, potato cyst nematodes can be fooled to believe that they are inside plants with a brief exposure to collected potato root exudates (Perry *et al.*, 1989). These exudates activate gene transcription and the synthesis of secretions in both types of pharyngeal glands *in vitro* (Jones *et al.*, 1997).

2.3 Molecular and Physiological Adaptations to Plant Parasitism

2.3.1 Host invasion

The first real direct encounter of the nematode and host plant, and more specifically the host's defences, is during invasion of the host. Vertebrate animals have mobile defender cells to respond to invaders, and a nearly infinite diversity in binding potential in immunoglobulins to tag invaders as being foreign so that the defender cells can exterminate them (Abbas and Lichtman, 2005). The immune system in vertebrates also builds on a memory of previous encounters with invaders, which allows them to respond quicker the next time the same type of invader makes an attempt to attack the host animal. Plants do not have such an adaptive immune system that can be directed towards invading parasites and, as far as is known, they do not have the capabilities to build a memory of previous encounters with microbes. Instead plants have evolved other unique features to protect themselves against the ingress of pathogens (Jones and Dangl, 2006).

The first line of defence in plants is the cell wall, which is a nearly impenetrable physical barrier around each individual cell. The plant cell wall is an extremely complex structure including a variety of highly diverse carbohydrate polymers, mixed with extended hydroxy-proline-rich glycoproteins and aromatic compounds such as lignins (Carpita and Gibeaut, 1993). The main structural components of the cell wall are the cellulose microfibrils that, together with the xyloglycans and glucoarabinoxylans, provide the cell wall's rigidity. The structural network of cellulose and cross-linking glycans is embedded in a gel matrix made of pectins. The backbone of pectic polysaccharides consists of blocks of α-1,4-linked polygalacturonic acid residues interspersed with regions of alternating galacturonic acid and rhamnose residues, of which the rhamnose residues might be decorated with short galactans and arabinans (Willats *et al.*, 2001).

The finding of various endogenous cell-wall-degrading enzymes in nematodes (Table 2.1) over the last 10 years implies that the physical impact of the stylet alone is not sufficient to perforate the cell wall. All nematodes feeding on plants studied to date appear to use cellulases to breakdown cellulose. Cellulases hydrolyse the cellulose polymers into oligosaccharides, making the micofibrils significantly weaker. The cellulose microfibrils are tethered together by xyloglycans or glucoarabinoxylans, depending on the

Table 2.1. Cell-wall-modifying proteins in plant-parasitic nematodes and their substrates.[a]

Substrate	Enzyme class	Genus
Cellulose	Cellulases	*Globodera*, *Heterodera*, *Meloidogyne*, *Bursaphelenchus*, *Ditylenchus*, *Radopholus*, *Pratylenchus*
	Cellulose-binding proteins	*Globodera*, *Heterodera*, *Meloidogyne*
	Expansin (or expansin-like)	*Globodera*, *Heterodera*, *Bursaphelenchus*, *Meloidogyne*, *Ditylenchus*
Xyloglycans and glucoarabinoxylans	Endoxylanases	*Meloidogyne*, *Radopholus*
Pectins	Pectatelyases	*Globodera*, *Heterodera*, *Meloidogyne*, *Bursaphelenchus*
	Polygalacturonase	*Meloidogyne*

[a]De Meutter *et al.*, 1998; Ding *et al.*, 1998; Smant *et al.*, 1998; Rosso *et al.*, 1999; Wang *et al.*, 1999; Goellner *et al.*, 2000; Popeijus *et al.*, 2000; Dautova *et al.*, 2001; De Boer *et al.*, 2002a; Doyle and Lambert, 2002; Jaubert *et al.*, 2002; Kikuchi *et al.*, 2004, 2009; Qin *et al.*, 2004; Ledger *et al.*, 2006; Mitreva-Dautova *et al.*, 2006; Kudla *et al.*, 2007; Abad *et al.*, 2008; Haegeman *et al.*, 2008, 2009a,b; Hewezi *et al.*, 2008; Kyndt *et al.*, 2008; Rehman *et al.*, 2009a; Vanholme *et al.*, 2009a.

plant species. Cross-linking hemicelluloses in dicotyledons have a glycan backbone, whereas non-commelinoid monocots use xylan-based polymers. Dicot xyloglycans are a substrate of cellulases, but xylan hydrolysis in monocots requires specific endoxylanases. The discovery of endoxyalanases in root-knot nematodes and burrowing nematodes (*Radopholus* spp.) feeding on monocots is therefore a possible adaptation to the cell wall composition of monocots.

A novel class of cell-wall-modifying proteins targeting the hemicellulose/cellulose network are the expansins (Choi *et al.*, 2006). Although the biochemistry of expansin activity is not completely clear, they are believed to weaken the non-covalent interactions between cellulose microfibrils and associated xyloglycan and glucoarabinoxylans. Expansins induce a measurable relaxation of the polymer interactions in the plant cell wall, allowing it to expand as a result of hydrostatic forces (Cosgrove *et al.*, 2002). For some time expansins were considered to be a plant-specific evolutionary innovation (Kende *et al.*, 2004), but the recent discovery of functional expansins in pharyngeal gland secretions of plant-parasitic nematodes challenges this view (Qin *et al.*, 2004; Kudla *et al.*, 2005; Kikuchi *et al.*, 2009). So far, all nematodes secreting cellulases also appear to release expansins, which suggests that the concerted action of cellulases and expansins may both be required to weaken the structural rigidity of the cell wall. The idea is that expansins open up the hemicellulose/cellulose network to make it more accessible for cellulases.

The cellulose-binding proteins in nematode secretions are also associated with cellulose degradation. Cellulose-binding proteins have a type II cellulose-binding domain attached to a short stretch of the amino acids with no match in the current sequence databases. The function of the ancillary domain in cellulose-binding proteins is not clear, other than that it likely acts on plant cell walls. Recent work on a cellulose-binding protein from *H. glycines* suggests that it may help to recruit plant cell-wall-degrading enzymes, not so much for host invasion but for cell wall degradation during feeding site development (Hewezi *et al.*, 2008).

The structural hemicellulose/cellulose scaffold is embedded in pectic polysaccharides. Pectins are important for water retention, for determining the size exclusion limit of cell walls and for the defence against a variety of plant pathogens. Pectins have extremely diverse decorations, and the degradation of pectins often involves enzymes to remove these decorations (i.e. esterases) and so-called backbone cutters (i.e. lyases and hydrolases). Two types of pectin-degrading enzymes have been found in pharyngeal gland secretions of plant-parasitic nematodes at present. Both types, pectate lyases and polygalacturonases, are backbone cutters. Most of the plant parasites studied to date secrete pectate lyases, while polygalacturonases have only been found in root-knot nematodes. Moreover, the overall picture arising from the current sequencing projects is that root-knot nematodes deploy a more diverse repertoire of cell-wall-modifying proteins than cyst nematodes. This would make sense given that, in contrast to cyst nematodes, the majority of root-knot nematodes have extremely wide host ranges and are stealthy invaders of plants (Hansen *et al.*, 1996).

2.3.2 Feeding behaviour and structures

In plant pathology a classical division is made between necrotrophs feeding on dead cells of the host and biotrophs requiring living host tissues to feed on. The most advanced biotrophs are the obligate sedentary plant parasites such as the cyst and root-knot nematodes. They transform host cells into specialized feeding structures to feed on for several weeks. One of the great mysteries in plant nematology is how these parasites can take up vast quantities of food from these feeding structures without actually killing the host cells. The answer probably is that the nematodes take full control over gene expression in these host cells, which changes them into life support systems and essentially governs the host's innate immunity.

For more detailed descriptions of nematode-induced feeding structures the reader is referred to excellent reviews by Gheysen and Fenoll (2002), Gheysen and Mitchum (2009) and Sobczak and Golinowski (2009). Here we will only briefly summarize current insights in cyst and root-knot nematode feeding structure development to emphasize the distinctive nature of these sites as a unique cellular and molecular phenomenon contributing to the survival and development of the parasite.

Shortly after host invasion, cyst and root-knot nematodes start probing host cells for their competence to be adequate feeding structures. Following positional or development cues in the plant, the nematodes carefully perforate the cell wall of a selected host cell and inject secretions into it. This behaviour sets off a series of cellular and molecular responses in the recipient host cells, resulting in the formation of either a syncytium in the case of cyst nematodes or giant cells in the case of root-knot nematodes. The ontogeny of the two types of feeding structures is fundamentally different, but both involve early manipulation of the mitotic cell cycle (Jones, 1981; Gheysen *et al.*, 1997; Engler *et al.*, 1999; Goverse *et al.*, 2000b). The observed expression of cell cycle genes suggests that host cells prepare for a mitotic cell division while developing into a feeding structure. The chromosomes and the whole cellular machinery are duplicated to provide a viable legacy for the two daughter cells. However, the preparations for mitotis do not end in the completion of cell division. Instead, a shortcut in the cell cycle forces the cell into another round of preparations for cell division, which again is not completed. This process is repeated a couple of times, resulting in large cells with high DNA contents. A key difference between a syncytium and a giant cell is the stage at which the mitotic cell cycle is aborted. In a syncytium, abortion takes place just before nuclear division, whilst in giant cells the cell cycle progresses past nuclear division to be aborted just prior to cellular division.

The typical cellular phenotype of a syncytium then arises through progressive local cell wall degradation and subsequent fusion of the protoplasts (Jones and Northcote, 1972). After a few weeks the syncytium consists of a large fusion complex of hundreds of hypertrophic cells. By contrast, the giant cells of root-knot nematodes remain as discrete cellular units while expanding to gigantic proportions over weeks (Bird, 1974). Typically, a single root-knot nematode transforms 5–12 host cells into giant cells. Both the syncytium and the giant cells acquire the social status of a metabolic sink, which implies that the plant redirects much of its resources to these structures.

The cellular changes in the nematode-induced feeding structures are the outcome of extensively reprogrammed gene transcription in host cells (Gheysen and Fenoll, 2002). Our understanding of host gene regulation by nematodes has leapt forwards over the last couple years because of amazing technological advances (Khan *et al.*, 2004; Ramsay *et al.*, 2004; Klink *et al.*, 2005, 2007a,b; Ithal *et al.*, 2007a,b; Jolivet *et al.*, 2007; Puthoff *et al.*, 2007; Caillaud *et al.*, 2008; Klink and Matthews, 2008; Fosu-Nyarko *et al.*, 2009; Gheysen and Mitchum, 2009; Portillo *et al.*, 2009; Swiecicka *et al.*, 2009; Szakasits *et al.*, 2009). The *summum* of gene expression analysis in nematode-induced feeding structures at the moment is the application of laser-capture technologies to isolate individual host cells from microscopic cross sections of nematode-infected root material. From these and other studies it has become clear that nematodes regulate the expression patterns of hundreds of host genes. It is a challenge to put all the pieces of this complex puzzle together and to separate host genes that are under the direct control of the feeding nematode from those that are merely responding indirectly to molecular and cellular changes. None the less, the nematode-regulated host genes roughly fit into six functional categories, which help to draw an overall picture of the molecular phenomenon underlying feeding structures formation (Gheysen and Fenoll, 2002).

2.3.3 Plant innate immunity

It is remarkable that a parasite is capable of redirecting fundamental developmental programmes in host cells towards its nutritional and developmental needs. It is also remarkable that a host 'permits' a parasite to survive, to develop and to reproduce while feeding from feeding structures. One would expect that the nematode and its feeding site are readily recognized as foreign bodies inside the plant and that, by default, the plant responds to this with a series of deadly defence reactions. The fact that plant-parasitic nematodes remain inside the host for weeks suggests that they have evolved the means to modulate the innate immune system of the plant. Before detailing the tools used by nematodes to evade or suppress host defences, we will first address current concepts of plant innate immunity to pathogenic microbes in general.

2.3.4 PAMP-triggered immunity

Despite the continuous presence of pathogenic microbes in their environment, disease in plants is still an exception. To protect themselves against the threats of pathogens and parasites, plants deploy a multilayered innate immune system (Jones and Dangl, 2006). Basal defence responses – the first line of active defence – in plants are activated following the detection of non-self epitopes by extracellular pattern recognition receptors (Zipfel, 2008; Boller and Felix, 2009; Boller and He, 2009). These immune receptors perceive highly conserved pathogen-associated molecular patterns (PAMPs), which are parts of molecules accidently released by an invading pathogen or present as a structural component at the pathogen's surface (Fig. 2.1). Typically, PAMPs are required for the survival of the pathogen, which implies that they are not easily modified by mutations. PAMPs are therefore highly conserved across different taxonomic classes, which explains why basal defences in a plant provide protection against a wide range of pathogens. Examples of PAMPs are chitin and lipopolysaccharides in the cell walls of fungi and bacteria (Dangl and Jones, 2001; Boller and Felix, 2009; Boller and He, 2009).

A classical example illustrating that there is also a significant degree of convergent evolution in innate immune systems is the recognition of bacterial flagellin (Gomez-Gomez and Boller, 2000, 2002). Flagellin is a principal motility component of flagellate bacteria, which is recognized by the extracellular leucine-rich repeat domain of both the receptor flagellin-sensing 2 in plants and a Toll-like receptor in vertebrates (Nurnberger *et al.*, 2004). Flagellin detection in both plants and vertebrates activates signalling pathways, leading to a type of immunity referred to as PAMP-triggered immunity (PTI). PAMP-triggered immunity involves, among others, altered ion fluxes, an increase in intracellular Ca^{2+} concentration, an oxidative burst, mitogen-activated protein kinase (MAPK) activation, protein phosphorylation, receptor endocytosis, defence gene induction, changes in protein–protein interactions and callose deposition on cell walls (Schwessinger and Zipfel, 2008; Zipfel, 2009). These defence reactions will be discussed in more detail later on in this chapter. PTI activating signals can also originate from host tissues as products of the lytic activity of microbial enzymes. These host-borne elicitors are known as damage-associated molecular patterns (DAMPs). Thus, basal defences can be activated by pathogen molecules and by the perturbations pathogens induce in host molecules.

To date there is no nematode epitope identified as a PAMP in plant innate immunity, which is because scientists have not so far systematically addressed the role of PTI in nematode–plant interactions. None the less, it seems indisputable that nematodes present many conserved epitopes on their cuticles that could act as PAMPs in plant innate immunity. Perhaps PAMP-triggered immunity is not so effective against plant-parasitic nematodes because the response is simply not rapid enough to prevent host invasion and migratory nematodes are able outrun this line of defence. In their migratory phase, even biotrophic nematodes exhibit necrotrophic behaviour, which makes them less vulnerable to PTI, but in the sedentary biotrophic stage they need to have evolved other means to evade or suppress PTI. Perhaps, like some of the animal parasites, plant-parasitic nematodes use cuticular camouflage using host molecules to avoid being detected by extracellular pattern recognition receptors. Alternatively, the biotrophic plant parasites may have evolved the means to modulate innate immunity downstream of PAMP recognition in host plants.

2.3.5 Effector-triggered immunity

Some strains of bacterial and fungal pathogens avoid PAMP-triggered immunity by deliberately releasing effectors into host cells (Fig. 2.1). These effectors

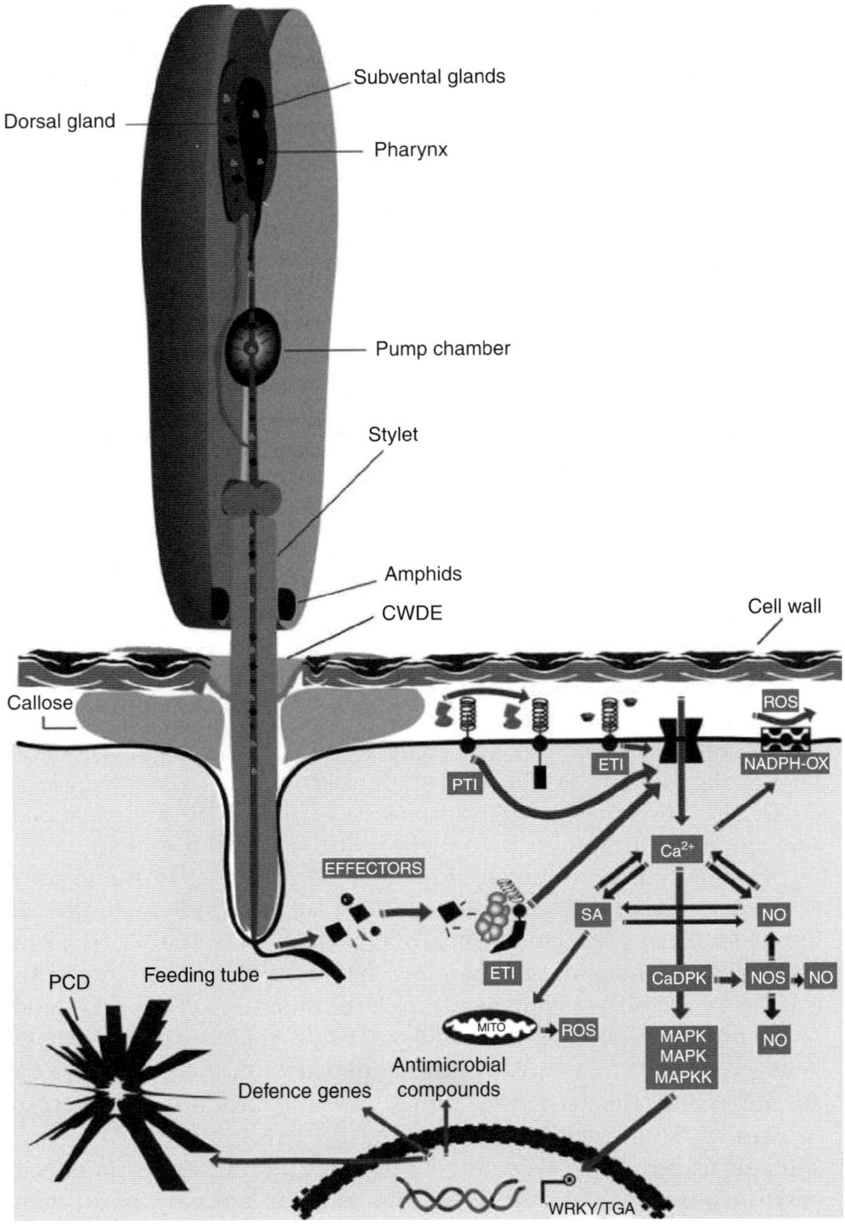

Fig. 2.1. The anterior section of a plant-parasitic nematode injecting effectors produced in the pharyngeal glands into a host cell. The nematode uses cell-wall-degrading enzymes (CWDE) to penetrate the host cell wall. Some of the nematode molecules may be recognized by the receptors of PAMP-triggered immunity (PTI) or effector-triggered immunity (ETI) in the host cell, which leads to defence signalling involving ion fluxes (Ca^{2+}), salicylic acid (SA), reactive oxygen species (ROS) and nitric oxide (NO). Defence signalling in innate immunity in plants often operates via mitogen-activated protein kinase (MAPK) cascades into the nucleus, wherein specific transcription factors (WRKY and TGA) regulate the expression of defence genes, pattern-recognition proteins and antimicrobial compounds, eventually culminating in a hypersensitive response and programmed cell death (PCD).

can intercept PAMP detection by receptor proteins or suppress downstream PAMP-triggered signalling, which, either way, restores full virulence of the pathogen. Because PTI suppression poses an immediate threat to plants, they have evolved a second layer in their innate immune system, referred to as effector-triggered immunity (ETI) (Chisholm *et al.*, 2006; Jones and Dangl, 2006). While PTI can be effective against a wide range of pathogens, ETI is highly specific to certain isolates of a single pathogen species. ETI receptors (formerly known as R proteins) are located as membrane-bound proteins on the cell surface and as intracellular receptors in the cytoplasm. The plant genes coding for the receptors in ETI are so-called major resistance genes. Thus, ETI occurs only in cases where the plant harbours an R protein that matches a specific effector in a pathogen.

Interestingly, some bacterial and fungal pathogens deliver 'second generation' effectors into host cells, which suppress the effector recognition or the signal transduction in ETI signalling pathways. These pathogens acquire full virulence again with their ETI suppressors. However, in their turn, some plant genotypes have evolved new recognition specificities in their repertoire of R proteins, such that they are able to detect these ETI-suppressing effectors, which restores disease resistance in these plants again. This co-evolutionary battle between pathogens and their host plants follows a zig-zag pattern of reciprocal adaptations either in the pathogen to acquire virulence or in the host plants to restore disease resistance (Jones and Dangl, 2006).

R proteins are multi-domain receptor molecules that can be divided into several distinct structural classes (Bent and Mackey, 2008; Van Ooijen *et al.*, 2008). The most abundant class consists of proteins with a nucleotide-binding domain and a leucine-rich repeat domain (NB-LRR). Some R proteins carry an additional amino-terminal domain with similarity to the *Drosophila* Toll or human interleukin-1 receptor (i.e. TIR-NB-LRR), while others include an amino-terminal coiled-coil structure (i.e. CC-NB-LRR). All current members of the NB-LRR class are located in the cytoplasm. A second class of R proteins consists of an extracellular leucine-rich repeat domain linked to a transmembrane domain. Based on their domain structures the extracellular LRR proteins are divided further as receptor-like proteins or, in cases where they carry an additional carboxy-terminal kinase domain, they are named receptor-like kinases.

How exactly R proteins detect pathogen effectors and then activate ETI is still not fully understood. Initially the idea was that effector–receptor interactions followed classical direct ligand-receptor binding (Van Der Biezen and Jones, 1998). However, this model is supported by experimental data for only a handful of effectors and matching R proteins. The prevailing opinion at the moment is that effector recognition in most of the cases does not involve direct binding between the effector and the receptor. Recognition by indirect interaction between effector and receptor is described in the guard model, which assumes that R proteins detect the perturbations brought about by pathogen effectors to other host proteins (Dangl and Jones, 2001). The finding of various R proteins recognizing effector-induced cleavage or other types of modifications on host proteins supports the guard model. Remarkably, several R proteins may guard the same host protein for different types of modifications brought about by diverse bacterial effectors (e.g. RIN4 in *Arabidopsis thaliana*) (Marathe and Dinesh-Kumar, 2003).

Mutant analysis and protein–protein interaction studies in *Arabidopsis* suggest that ETI constitutes an accelerated and amplified form of PAMP-triggered immunity. ETI also leads to changes in ion fluxes, elevated intracellular Ca^{2+} concentrations, the production of reactive oxygen species and defence genes expression (Jones and Dangl, 2006). However, in most cases ETI ultimately directs the host cell into a hypersensitive response and programmed cell death (see below for more details). The actual molecular components in defence signalling downstream of R proteins and PAMP receptors have not been mapped out in great detail at the moment, but a critical component in early defence signalling of both PAMP-triggered immunity and ETI seems to be the plant hormone salicylic acid (SA) (Loake and Grant, 2007). Exogenous application of SA to plants regulates defence gene expression and induces disease resistance to biotrophic pathogens. Salicylic acid is a derivative of chorismate in the shikimate biosynthesis pathway. It is stored and can be released as different types of conjugates in various subcellular compartments. There is also substantial evidence for the involvement of MAPK cascades in the early defence signalling (Pedley and Martin, 2005). It has been shown that salicylic acid signalling in plant defences depends on various components of MAPK cascades. Moreover, some of the MAPK pathways have been shown to feed into WRKY transcription factors controlling defence genes expression (Eulgem and Somssich, 2007).

Breeding for resistance to plant-parasitic nematodes thus essentially aims to find genes encoding receptors for ETI in wild plants and to introduce these receptors into important food crops by genetic selection (Williamson and Kumar, 2006). At present six nematode resistance genes conditioning ETI have been cloned, and some of these were introduced into commercial cultivars. Most of the nematode resistance genes result in a hypersensitive response, with increased defence gene expression and local cell death as the end result (see section below).

2.4 Molecular and Cellular Phenomena in Plant Innate Immunity to Nematodes

2.4.1 Defence genes: phytoalexins, pathogenesis-related proteins and protease inhibitors

Plants harbour a rich collection of chemical compounds capable of killing invading microbes (Bednarek and Osbourn, 2009). Here we briefly discuss three categories of defence genes, including phytoalexins, pathogenesis-related (PR) proteins and protease inhibitors. Phytoalexins are low-molecular-weight antimicrobial compounds, which are produced in the secondary metabolic pathways. The presence of these antimicrobial compounds in plants sometimes correlates with resistance to a pathogen. The virulence of pathogens in its turn can be determined by the ability to breakdown (pre)formed antimicrobial compounds. A textbook case of this type of chemical warfare in plants is the interaction of the fungal pathogen *Gaeumannomyces graminis* var. *avenae* and oat (Crombie *et al.*, 1986). Oat coleoptiles accumulate the broad-spectrum antimicrobial avenacin, which permeabilizes cellular membranes. Virulence of the fungus *G. graminis* var. *avenae* on oat depends on the production of an avenacin-hydrolysing enzyme.

Our current knowledge on the role of antimicrobial compounds in nematode–plant interactions is very limited. A few studies have addressed the importance of the isoflavonoid glyceollin in soybean during infection with *H. glycines*. Glyceollin was found to accumulate close to the nematode's head in a resistant cultivar but not in susceptible plants (Huang and Barker, 1991). Elliger *et al.* (1988) have studied the accumulation of α-tomatine in susceptible and resistant tomato cultivars and found no correlation with nematode resistance. However, α-tomatine is constitutively produced in tomato and more recent studies suggest that for virulence on tomato several fungal pathogens require the enzyme tomatinase (Pareja-Jaime *et al.*, 2008).

Thus, experimental data pointing at a direct role of antimicrobial compounds in nematode–plant interactions is limited, mainly because these compounds have not received much attention from the scientific community. Indeed, recent comprehensive transcriptome analyses show that many key enzymes in the biosynthetic pathways of antimicrobial compounds are regulated following nematode infections (Gheysen and Fenoll, 2002; Jammes *et al.*, 2005; Ithal *et al.*, 2007a,b; Fosu-Nyarko *et al.*, 2009; Szakasits *et al.*, 2009). Phenylalanine ammonia lyase, chalcone synthase, myrosinases, and hydroxy-methyl-glutaryl-CoA reductase are consistently upregulated in nematode-infected plants. Phenylalanine ammonia lyase is the key regulatory enzyme into the phenylpropanoid biosynthetic pathway, leading to the production of precursors of four major classes of phenylpropanoid derivates (i.e. salicylates, coumarins, monolignols and flavonoids). Chalcone synthase operates downstream of phenylalanine ammonia lyase in the production of flavonoids. The flavonoids and the coumarins include many phytoalexins, whilst the central defence signalling molecule salicylic acid belongs to the group of salicylates. Part of the phenylpropanoid pathway is dedicated to the synthesis of lignin, lignan and suberine, which are used in plants to fortify the cell walls. Myrosinases are involved in the hydrolysis of glucosinolates into extremely toxic isothiocyanates. Lastly, hydroxy-methyl-glutaryl-CoA reductase is the rate-limiting enzyme in a separate biosynthetic route into isoprenoids, which include the antimicrobial sesquiterpene phytoalexins. Thus, plants at least have the capability to deploy a broad arsenal of chemicals against parasitic nematodes. To what extent the phytoalexins contribute to immunity to nematodes is not clear and needs to be investigated in more detail.

2.4.2 Pathogenesis-related proteins

PR proteins are loosely defined as microbe-induced proteins associated with plant defences (van Loon, 1983; van Loon *et al.*, 2006). The concept of PR proteins was inspired by the finding of proteins specifically expressed during a hypersensitive response (see below) in tobacco plants resistant to tobacco mosaic virus. The key identifier for a PR protein is indeed its expression pattern, and not an evolutionary relatedness or a similarity in biochemical activity. As a consequence of this relaxed criterion, the list of PR proteins now includes 18 different families but is likely to grow further. The PR proteins are believed to restrict the pathogenicity of microbes on plants, but experimental evidence to substantiate their contribution in immunity is often weak, if present at all. None the less, *in vitro* several PR proteins do have antimicrobial activities. For instance, some PR protein families break down fungal and bacterial cell walls, inhibit proteases, degrade ribonucleases or display antimicrobial toxicity. Given this wide spectrum of potential antimicrobial activity, and even though there is no robust experimental data to support this for many PR proteins, it seems likely that they provide at least partial protection against invaders.

In Table 2.2, we have summarized the results of several transcriptome analyses of nematode-infected plant tissues, including a category of defence-related genes. This category, in fact, includes several PR proteins that are locally and systemically expressed in plants infected with nematodes. However, the expression patterns in the direct vicinity of the site of infection may be different from a systemic response in a plant (Bowles *et al.*, 1991). For example, *Arabidopsis* plants infected with the beet cyst nematode, *Heterodera schachtii*, show elevated levels of PR-2 and PR-5 in infected roots, but not of PR-1 (Wubben *et al.*, 2008). However, in the shoots of these nematode-infected plants PR-1 expression was strongly induced. PR-1, PR-2 and PR-5 are all salicylic acid-induced defence genes and are often simultaneously induced by microbes. The absence of PR-1 induction in roots, and the upregulation of

Table 2.2. Functional categories of nematode-regulated genes in feeding structures and examples of associated cellular and molecular processes. (Adapted from Gheysen and Fenoll, 2002.)

Category	Examples of molecular processes and activities
Defence-related	PR proteins, oxidative burst, wound-inducible proteins, lignin biosynthesis
Cell cycle and organization	Cyclin-dependent kinases, cyclins, tubulins
Plant hormones	Auxin-responsive genes, auxin transport, ethylene synthesis, jasmonic acid biosynthesis
Cell wall	Cell-wall-degrading enzymes, expansins, cellulose synthesis, cell wall proteins
Metabolism and water status	Metabolic enzymes, sugar transport
Gene expression	Transcription factors, protein turnover

PR-1 in shoots, led to the hypothesis that perhaps cyst nematodes suppress local accumulation of PR-1 (Wubben *et al.*, 2008). Again, a systematic analysis of local PR-protein expression in the infection site, and in shoots at some distance from the infection site, could shed some light on the role of PR proteins in nematode resistance.

2.4.3 Protease inhibitors

The third group of defence genes involved in plant defence to nematodes encodes protease inhibitors. Microbe-induced protease inhibitors fit into the definition of a PR protein but are often dealt with separately. The expression of protease inhibitors is regulated through signalling pathways that are activated by tissue injury (Koiwa *et al.*, 1997). Plant protease inhibitors are also expressed in response to herbivorous animals, including plant-parasitic nematodes, possibly as a generic reaction to wounding. Animals feeding on plant tissue ingest host protease inhibitors, which are then believed to interact with proteases in the digestive tract. Inhibition of intestinal proteases probably disrupts normal uptake of protein fragments in the gut. Enzyme activity assays and proteomics on collected stylet secretions further suggest that plant-parasitic nematodes also secrete proteases that could be targeted by protease inhibitors (Robertson *et al.*, 1999; Bellafiore *et al.*, 2008). To understand the role of these secreted proteases in nematode–plant interaction, further investigations are needed. Because secreted proteases are among the few enzymes present in secretions from both animal- and plant-parasitic nematodes, novel insights may be acquired in a cross-disciplinary approach.

Efforts to engineer nematode-resistant plants by constitutive overexpression of specific protease inhibitors demonstrates that proteases contribute to success in parasitism (Atkinson *et al.*, 2003). Overexpression of a cysteine protease inhibitor, cystatin, from rice, targeting intestinal proteases in cyst nematodes, reduced the fecundity of female worms. Similarly, studies targeting the intestinal cysteine proteases of root-knot nematodes with RNAi led to a significant reduction in parasite development and reproduction (Shingles *et al.*, 2007). It may not be appropriate to translate the results obtained with transgenic protease inhibitors and RNAi to the natural situation in plants, but they do show that protease inhibitors could constitute one more layer in plant defence to nematodes. However, a recent expression analysis of protease inhibitors in plants did not show a correlation with natural resistance, which challenges the view that protease inhibitors have a key role in nematode resistance (Turrà *et al.*, 2009).

2.4.4 Cell wall fortifications with callose deposits and lignin

Microbial plant pathogens, nematodes included, have evolved a repertoire of cell-wall-degrading enzymes to breakdown plant cell wall polymers.

Plants sometimes respond to these invasions with the local deposition of callose (a β-1,3-glucan polymer) and lignin in between the cell wall and the cell membrane (Hardham *et al.*, 2007; Bhuiyan *et al.*, 2009). For example, cell wall fortifications with callose/lignin papillae in monocots can provide basal resistance against a variety of fungal pathogens (Maor and Shirasu, 2005; Hardham *et al.*, 2007). Cell wall depositions make the cell wall more penetration-resistant for two reasons. The deposits provide extra strength to resist the physical impact of the stylet, and lignin makes the cell wall less susceptible to cell-wall-degrading enzymes.

To what extent cell wall fortifications contribute to the penetration resistance to nematodes has not been studied in great detail. Callose deposits have been observed on plant cell walls penetrated by the stylet of biotrophic plant parasites (Hussey *et al.*, 1992; Grundler *et al.*, 1997). Callose accumulates on the cell wall between the site where the stylet is inserted and the invaginated cell membrane around the stylet tip (Fig. 2.1). However, callose deposits seem to occur in both susceptible and resistant plants, and there is no correlation between resistance to nematodes and callose deposition.

Lignin is one of the products of the phenylpropanoid biosynthesis pathways (see section above). Transcriptome analysis of susceptible soybean roots suggests that most of the rate-limiting enzymes in the phenylpropanoid pathways are upregulated in nematode-infected root tissue (Ithal *et al.*, 2007a,b). Lignification of cell walls is observed in defence responses to biotrophic cyst nematodes in *A. thaliana*, but it is not clear to what extent lignification contributes to nematode resistance (Grundler *et al.*, 1997). Resistance to the necrotrophic nematode *Radopholus similis* in banana is correlated with high lignin content in cell walls (Wuyts *et al.*, 2007). Besides having a higher constitutive level of lignin, resistant banana plants also respond to nematode infection with further lignification of the cell walls. Increasing penetration resistance with cell wall fortifications may affect migratory necrotrophic plant parasites more significantly. Biotrophs can be hindered by lignified cell walls during their short migratory phase, but a slowdown may not yield a major effect on their development and reproduction.

2.4.5 Hypersensitive response and programmed cell death

The ultimate defence layer in plant innate immunity is the generation of a hypersensitive response leading to local programmed cell death (HR-PCD) at the site of infection (Heath, 2000; Hofius *et al.*, 2007; Mur *et al.*, 2008). The hypersensitive response in plant cells prevents further ingress of fungal and bacterial pathogens and, in the case of nematodes, the development of a proper feeding structure (Fig. 2.2). It is still debated whether the programmed cell death or a barrage of cytotoxic compounds halts the pathogen. Cell death could also be induced by neighbouring cells, to limit the damage caused by cytotoxic compounds to those cells that are in direct contact with the pathogen. So, local cell death could therefore be initiated in a cell by its neighbours to prevent a runaway autoactive response.

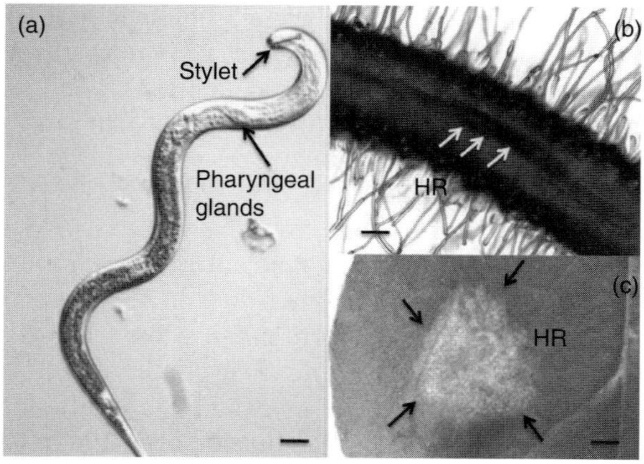

Fig. 2.2. (a) An infective second-stage juvenile of the potato cyst nematode *Globodera rostochiensis*. The translucent anterior section of the nematode includes the pharynx, pharyngeal glands and the stylet. Scale bar = 20 μm. (b) A potato root harbouring the H1 resistance gene showing a hypersensitive response (HR) in host cells close to the nematode. Scale bar = 300 μm. (c) Transient co-expression of a nematode effector and a matching resistance gene in a *Nicotiana benthamiana* leaf results in a hypersensitive response and programmed cell death of plant cells (arrows). Scale bar = 0.25 cm.

The signalling cascades underlying the hypersensitive response are strictly controlled by highly specific immune receptors (Van Ooijen *et al.*, 2008). These immune receptors, encoded by R genes, only activate downstream signalling when they detect the presence of matching pathogen-derived effectors (see Section 2.3.5). The pathways connecting activated immune receptors and the hypersensitive response are currently the subject of intense research. Genetic studies with signalling mutants have revealed several critical nodes in signal-transduction routes downstream of immune receptors (Feys and Parker, 2000; McDowell and Dangl, 2000; Thomma *et al.*, 2001; Martin *et al.*, 2003; Pieterse and Van Loon, 2004; Wiermer *et al.*, 2005; Shirasu, 2009). For example, RAR1 and Hsp90 are thought to act as chaperones to stabilize various immune receptors and to maintain them in an active configuration. A third component required for resistance, referred to as SGT1, is associated with linking pathogen recognition complexes to ubiquitination pathways. Similarly, EDS1 and PAD4 seem to operate downstream of TIR-NB-LRR receptors, whereas the NDR1 is required for and acts downstream of the CC-NB-LRR class of receptors. So, a growing number of molecular components in disease-signalling pathways have been discovered, but there still is a significant gap in our knowledge spanning these components and the early molecular and cytological phenomena associated with a hypersensitive response.

The earliest phenomena signalling the activation of a hypersensitive response are rapid ion fluxes across the cell membrane, the production of reactive oxygen species outside the cell and nitric oxide inside the cytoplasm, and the accumulation of salicylic acid (Garcia-Brugger *et al.*, 2006; Ma and Berkowitz, 2007). Tests with specific inhibitors of ion channels and radio-labelled ions demonstrate that rapid changes in ion fluxes are required for the activation of a hypersensitive response. Of particular importance are influxes of Ca^{2+} ions that act as secondary messengers in a wide variety of signalling pathways in plants (Lecourieux *et al.*, 2006). ETI signalling apparently feeds into ion channels in the cell membrane, which increases the Ca^{2+} permeability of the cell membrane. The concentration of Ca^{2+} ions in the apoplast is in the order of 10,000-fold higher than in the cytoplasm. Rapid influxes of Ca^{2+} ions convey messages from outside the cell into the cytoplasm and the nucleus of the recipient cell. Spatiotemporal changes are believed to generate a Ca^{2+} signature in the cytoplasm that is decoded by specific Ca^{2+} 'sensors'. Next, these Ca^{2+} sensors translate oscillations in cytoplasmic Ca^{2+} concentrations into the activation of downstream signalling molecules involving calmodulins, calcium-dependent kinases and MAPK cascades.

Calcium influx is a key regulator of plant defences. For example, calcium-dependent phosphorylation activates membrane-associated NADPH oxidase, which generates extracellular superoxide radicals (Garcia-Brugger *et al.*, 2006; Ma and Berkowitz, 2007). Apoplastic superoxide dismutases convert superoxide into hydrogen peroxide, and indirectly into hydroxyl radicals, resulting in what is known as oxidative burst. Interestingly, apoplastic hydrogen peroxide also acts in a positive feedback loop, which further increases Ca^{2+} permeability and Ca^{2+} influx. The reactive oxygen species (ROS) are potentially toxic to invading microbes, but they also induce fortifications of the plant cell walls by lignification and cross-linking cell wall proteins.

Perhaps the strongest effect of ROS occurs through peroxidation of lipids in cellular membranes. Hydroxyl radicals abstract a proton from unsaturated phospholipids to generate a lipid hydroxyperoxide radical. It is not difficult to see how extensive lipid peroxidation of lipids could lead to a loss of membrane integrity. Moreover, lipid peroxidation could also result from an increase in lipoxygenase action, which is strongly upregulated in cells undergoing a hypersensitive response.

Influx of Ca^{2+} also induces the synthesis of nitric oxide through a calcium-dependent nitric oxide synthase (Wendehenne *et al.*, 2004; Besson-Bard *et al.*, 2008). Specific inhibitors of nitric oxidase and scavengers of nitric oxide point at an important role for nitric oxide in plant innate immunity. Exogenously applied nitric oxide to *Arabidopsis* cells regulates the expression of several pathogenesis-related proteins and other defence genes. To further add to the complexity evolving around signal transduction cascades in plant defences, evidence suggests that nitric oxide is capable of mobilizing Ca^{2+} ions from intracellular calcium pools, which again creates an amplification cycle in the accumulation of secondary messengers.

Salicylic acid (SA) is also a key compound in early signalling of a hypersensitive response (Loake and Grant, 2007). Exogenous application of SA to plant cells can elicit a cell death response. SA is associated with the generation of cytoplasmic ROS, possibly through derailing the oxidative phosphorylation pathway in mitochondria such that ATP is no longer produced in the cells. SA is also held accountable for the changes in the cytoplasmic redox state which activate other proteins downstream in the SA signalling cascade, such as NPR1. SA-activated NPR1 moves into the nucleus, wherein it interacts with TGA-type transcription factors that regulate the transcription of various defence genes. Some of the redox-based reactions induced by SA probably involve post-translational modifications via a mechanism called *S*-nitrosylation, which uses nitric oxide (Lindermayr and Durner, 2009). In conclusion, it is a complex interplay among at least four critical messengers (i.e. Ca^{2+}, ROS, nitric oxide and SA) that determines the onset of the hypersensitive response.

The cellular changes associated with the hypersensitive response and local programmed cell death seem to vary somewhat among individual pathosystems. Nevertheless, the HR-PCD roughly proceeds through the following sequence of events (Heath, 2000). The first observable change in cells undergoing a hypersensitive response is cessation of the cytoplasmic streaming, which coincides with the reorganization of the cytoskeleton. Next, the cytoplasm acquires a more granular appearance and shrinks in volume. The mitochondria inside the cytoplasm swell and then terminate normal oxidative phosphorylation and the production of metabolic energy. Chromatin DNA condensation is also observed in the nuclei of cells undergoing cell death. Increasing intracellular concentrations of hydrogen peroxide and lipoxygenase are the likely cause of lipid breakdown in cellular membranes. This irreversible damage to the membranes leads to a loss of semi-permeability, which is quickly followed by disintegration of the nucleus and collapse of the protoplast (Fig. 2.2).

The hypersensitive response is an extremely fast and powerful defence strategy to ward off biotrophic microbes. This seems to be especially true for sedentary endoparasitic nematodes (Fig. 2.2), because the transition from the migratory to the sedentary stage involves the breakdown of locomotory muscles, which leaves them completely dependent on the resources provided by their living feeding structure (Williamson and Kumar, 2006; Fuller *et al.*, 2008). Plants seem to have exploited this vulnerability as they use the nematode-induced feeding structure as their primary battlefield to deploy a hypersensitive response to these nematodes (Fig. 2.2). The hypersensitive responses mediated by nematode resistance genes can be roughly divided into two types (Tomczak *et al.*, 2009). First are the hypersensitive responses culminating in a classical fast and local cell death in and around the feeding structure. In addition, several nematode resistance genes induce a delayed-type hypersensitive response, mostly involving cell death in cells surrounding the feeding structure and in cells between the feeding structure and nearby vascular tissue of the plant. The delayed-type hypersensitive response develops over days to weeks and allows the infective nematodes to feed on their feeding structures for a significant amount of time.

Root-knot nematodes invading tomato harbouring the *Mi-1* gene induce a rapid and local hypersensitive response and cell death in the initial feeding site within 24 h (Milligan *et al.*, 1998; Williamson, 1998, 2000). As a consequence, the formation of giant cells is completely suppressed by the

Mi-mediated hypersensitive responses. The *H1* resistance gene to cyst nematodes in potato also induces a fast hypersensitive response and cell death in the cells surrounding the initial feeding structure, but not in the feeding structure itself (Rice *et al.*, 1985). The 'ring of death' induced by the *H1* gene isolates the feeding structure from nearby vascular tissue, preventing transfer of nutrients from the flow of assimilates to the feeding nematode. The *Gpa2* resistance gene to cyst nematodes in potato confers a delayed-type hypersensitive response that develops over weeks and allows some nematodes to mature but not to reproduce.

2.5 Immune Modulation by Nematodes in Plants

While embedded in host tissues, the sedentary endoparasitic nematodes are continuously exposed to the innate immune system of the plant, yet some of these parasites are able to live inside hundreds of different plant species for long periods. Even among biotrophic bacterial and fungal plant pathogens this kind of success in parasitism is exceptional, and it suggests that nematodes are very efficient in governing plant innate immunity. There are three main strategies, which are not mutually exclusive, for a parasite to deal with immunity in a host. The nematode can first avoid being recognized by the immune system. When that fails, it can actively suppress immune signalling triggered by activated recognition complexes or, as a last resort, it can neutralize antimicrobial compounds that are part of activated defence responses. Immune evasion by cuticular camouflage has been discussed in Section 2.2.1. In this section we will discuss the possible approaches for nematodes to modulate immunity in plants, with a focus on the role of recently discovered effectors.

2.5.1 Detoxification of reactive oxygen species (ROS) and modulation of ROS signalling

One of the earliest responses to pathogen infections in plants is the production of ROS. ROS have two roles in plant defences: (i) as antimicrobial compounds in a 'chemical warfare'; and (ii) as a critical messenger in defence signalling. In both susceptible and resistant tomato plants the root-knot nematode *Meloidogyne incognita* induces a fast oxidative burst in the migratory tracks and in the feeding cells. However, in plants resistant to nematodes a second oxidative burst associated with a hypersensitive response occurs, while in susceptible plants the ROS concentration declines in feeding sites after a few hours (Melillo *et al.*, 2006). A similar biphasic oxidative burst has been observed in *A. thaliana* infected with the soybean cyst nematode *H. glycines* (Waetzig *et al.*, 1999). The absence of a biphasic ROS response in susceptible plants suggests that plant-parasitic nematodes may modulate the signalling that leads to the second-phase oxidative burst. As this second wave of ROS is specific for resistant plants, it is likely to be part of the highly specific

ETI. By the same reasoning, the initial phase of ROS production could then perhaps be produced following non-specific PAMP-triggered immune signalling. So the absence of a biphasic oxidative burst could be either evasion by the nematode of immune responses by avoiding recognition by the ETI receptors or true modulation of the signalling pathways inducing the biphasic oxidative burst downstream of ETI receptors.

A significant outcome of these studies is that, even in susceptible plants, invading nematodes encounter an oxidative burst. Therefore, protection against ROS seems extremely important for parasite survival in plants, because ROS could probably diffuse through the cuticle of the nematodes and cause significant damage to DNA, proteins and cellular membranes in the underlying tissues. Plant-parasitic nematodes acquire some protection with surface antioxidants and ROS scavengers such as secreted glutathione peroxidases and thioredoxin peroxidases (Robertson *et al.*, 2000; Jones *et al.*, 2004). Thioredoxin peroxidase specifically metabolizes hydrogen peroxide. Although nematode glutathione peroxidases are also capable of converting hydrogen peroxide, they seem to have a higher affinity for the products of lipid peroxidation (see Section 2.4.5). The root-knot nematode, *M. incognita*, produces glutathione *S*-transferase as one of the components of its pharyngeal gland secretions, which are probably injected into the host cells during feeding (Dubreuil *et al.*, 2007). Knocking-down the glutathione *S*-transferase with RNAi reduced the egg production by females but not the number of females in a host. A plausible explanation for this effect on nematode fecundity could be that glutathione *S*-transferase supports sustained feeding on host cells rather than host invasion and the establishment of the feeding site. It is tempting to speculate that antioxidants at the nematode surface provide protection to apoplastic antimicrobial ROS, while antioxidant enzymes in pharyngeal secretions may target hydrogen peroxide in the host cell cytoplasm to intercept ROS-mediated signalling.

2.5.2 Modulation of plant hormone balance and secondary metabolism

Plant hormones are key players in the regulation of plant development and defence responses to biotic and abiotic stress (reviewed in Goverse *et al.*, 2000a). Ethylene-insensitive *Arabidopsis* mutants are less susceptible to *H. schachtii*, while ethylene-overproducing mutants are hypersusceptible. The nematode-induced feeding sites in ethylene-overproducing mutants are significantly larger than in wild-type plants. Similarly, auxin-insensitive mutants have shown resistance to *Globodera rostochiensis* and *H. schachtii.* Resistance in auxin-insensitive plants was characterized by disturbed feeding site formation. So the establishment and maintenance of proper feeding sites are strongly dependent on ethylene and auxin. The dominating theory at the moment assumes that nematodes induce a local increase in auxin levels in feeding site initials, which is followed by the auxin-responsive synthesis of ethylene. The gross phenotype of auxin- and ethylene-insensitive mutants is indeed an increase in nematode resistance, but this may point to a lack of susceptibility rather than to specific immune responses.

While there is some consensus on the importance of plant hormones in feeding site formation, the actual trigger from the nematode driving a rise in auxin levels is not yet known. Chorismate mutase enzymes secreted from the pharyngeal glands of plant-parasitic nematodes have been linked to auxin balances in plant cells (Lambert *et al.*, 1999; Doyle and Lambert, 2003; Jones *et al.*, 2003; Vanholme *et al.*, 2009b). Systemic overexpression of a nematode chorismate mutase in soybean hairy roots gives a disturbed root morphology that can be rescued with the exogenous application of auxin (Doyle and Lambert, 2003). It is not clear if these overexpression studies reflect the natural situation correctly, but this finding suggests that nematodes injecting chorismate mutase into host cells could reduce local auxin levels, which is not in agreement with earlier mutant analyses. Chorismate mutase is a key enzyme in the shikimate pathway and catalyses the conversion of chorismate into prephenate. Chorismate mutase directs the shikimate pathway away from tryptophane, to favour the production of tyrosine and phenylalanine. Auxin derives from tryptophane, and chorismate mutase could thus affect local auxin levels. However, auxin is mainly produced in the aerial parts of the plant and then transported to the roots, wherein the nematodes induce their feeding sites. Therefore, nematodes are more likely to raise local auxin levels by either enhancing the influx or reducing the efflux of auxin in feeding site initials. Hence, secreted chorismate mutases may not increase auxin biosynthesis but their deployment in a wide variety of plant-parasitic nematodes suggests that the modulation of secondary metabolism is none the less crucial in parasitism. The chorismate mutases could, for example, interfere with the generation of antimicrobial flavanoid derivates of aromatic amino acids.

2.5.3 Modulation of lipid-based defences

Lipids are important in plant–microbe interactions in the chemical defence against invading pathogens as lipid peroxides but also as second messengers in defence signalling. Hydrogen peroxides produced in the oxidative burst can react with unsaturated lipids into cellular membranes to produce toxic lipid hydroperoxides and other free radicals. Similarly, the plant defence responses often involve the upregulation of lipoxygenases that convert lipids into bioactive lipid hydroperoxides. Besides being highly toxic, lipid hydroperoxides are also rapidly converted into the precursors for lipid-based signalling molecules such as jasmonic acid. A recent paper reports that the lipoxygenase gene *ZmLOX3* is involved in regulating susceptibility to root-knot nematodes in maize (Gao *et al.*, 2008). A comparison of syncytial cells in susceptible and resistant plants also revealed a tenfold upregulation of lipoxygenase genes in plants resistant to cyst nematodes (Klink *et al.*, 2007b). Furthermore, in nematode-resistant pea, lipoxygenases were highly induced in cells surrounding the feeding sites and inside feeding sites undergoing a hypersensitive response (Veronico *et al.*, 2006). Several studies have shown that the exogenous application of methyl-jasmonate or synthetic jasmonic acid reduces host susceptibility to root-feeding nematodes (Cooper *et al.*, 2005). Thus, there is some experimental support for a significant role of lipoxygenase, bioactive lipids and lipid-based signalling in plant–nematode interactions.

Plant-parasitic nematodes secrete a specific class of fatty-acid and retinol-binding (FAR) proteins that may interfere with lipid-based defences in the host. Recombinantly produced FAR protein of the cyst nematode *Globodera pallida* binds to linolenic and linoleic acids (Prior *et al.*, 2001). Further enzyme activity assays showed that the recombinant FAR protein also inhibits the lipoxygenase-mediated conversion of unsaturated linolenic and linoleic acids. The FAR protein is located at the interface of the nematode and the host cells, where it may neutralize toxic lipid hydroxyperoxides or intercept lipid-based defence signalling.

A second group of lipid-binding proteins in nematodes with possible immunomodulatory properties are the so-called annexins. Annexins bind phospholipids, the main component of cell membranes, in a calcium-dependent manner. Annexins are expressed in the pharyngeal and the amphidial glands of plant-parasitic cyst nematodes (Fioretti *et al.*, 2001; Gao *et al.*, 2003). They have been implicated in vesicle transport during endo- and exocytosis, and in providing a membrane-anchored docking scaffold for other molecules. As such, the annexins could have a role in vesicle transport inside nematode gland cells; however, the finding of annexins in the secretory–excretory products of plant-parasitic nematodes suggests that they also may have a role inside host cells.

2.5.4 Modulation of calcium signalling

Ca^{2+} is an important secondary messenger, capable of conveying all sorts of external signals to responsive developmental and cellular processes inside plant cells. A rapid Ca^{2+} influx precedes both basal and specific disease resistance. The exogenous application of a specific Ca^{2+} channel inhibitor (i.e. La_3Cl) also points to a role for Ca^{2+} signalling in potato roots susceptible to nematodes (Sheridan *et al.*, 2004). Although the authors of this study could not exclude a direct effect of the inhibitor on nematodes, their results suggest that Ca^{2+} signalling is required for successful invasion and feeding site formation by cyst nematodes.

Plant-parasitic nematodes have evolved the means to interfere with Ca^{2+} signalling in host plants. It has been shown that *M. incognita* secretes calreticulin during host invasion and feeding (Jaubert *et al.*, 2005). The nematode-secreted calreticulin was localized close to the stylet tip *in planta* and along the cell walls of the giant cells. However, their role in plant–nematode interactions is not well understood at present. Calreticulins are conserved in all multicellular organisms and carry a sorting signal for secretion and a C-terminal endoplasmic reticulum (ER) retention signal sequence (Jia *et al.*, 2009). They act as molecular chaperones during protein folding in the ER. Although the ER is considered the main residence of calreticulins, several studies have reported their presence outside the ER. Cytoplasmic calreticulins have been implicated in modulating Ca^{2+} homeostasis and signalling, gene expression and cell adhesion (Jia *et al.*, 2009).

2.5.5 Modulation of host protein turnover rate

The addition of small ubiquitin monomers to proteins in eukaryotic cells can target these proteins to the 26S proteasome for degradation (Mukhopadhyay and Riezman, 2007). Ubiquitination of proteins proceeds via a multistep process that is controlled by an E1–E2–E3 enzyme cascade (Craig *et al.*, 2009). First in line are the E1 ubiquitin-activating enzymes involved in the recruitment of free ubiquitin in an ATP-dependent manner. The bound ubiquitin is then rapidly transferred to an E2 ubiquitin-conjugating enzyme. The E2 conjugating enzyme next delivers the activated ubiquitin monomer to the targeted substrate protein. However, the actual target of a ubiquitination complex is determined by the binding specificity of a third component in the cascade, the so-called E3 ubiquitin ligases. Eukaryotic genomes include hundreds of different E3 ubiquitin ligases, each having unique substrate specificity for a particular target protein. Ubiquitination has been implicated in the regulation of a wide variety of processes in plants, such as innate immunity, cell death, cell cycle regulation, hormone signalling, circadian rhythms and many more.

In the last couple of years it has become clear that many plant pathogens hijack the ubiquitination system of the host to take over control of various cellular processes. For example, the plant pathogenic bacterium *Pseudomonas syringae* delivers an effector protein, AvrPtoB, with novel E3 ubiquitin ligase activity into the host cell to target the ubiquitination machinery to the protein kinase Fen and suppress innate immunity (Rosebrock *et al.*, 2007). This demonstrates that bacterial plant pathogens are capable of redirecting the specificity of host ubiquitination complexes so that the plant's innate immunity is no longer effectively controlling pathogen ingress. The recent discovery of several secreted variants of ubiquitination complex components in pharyngeal glands of plant-parasitic nematodes indicates that plant-parasitic nematodes may also exploit the host's ubiquitination system (Davis *et al.*, 2004). Sequencing of pharyngeal-gland-specific cDNA libraries revealed homologues of SKP1 and RING-H2 zinc finger proteins. SKP1 is a subunit of an SKP1-Cullin-F-box (SCF) E3 ubiquitin ligase complex, whereas RING-H2 finger proteins form an E3 ligase complex together with Cullin and a variable substrate recognition subunit. Thus, it is likely that nematodes inject components of E3 ligases into host cells to redirect the ubiquitin–proteasome degradation pathway. However, SKP1 and RING-H2 finger proteins have not been implicated in determining the substrate specificity of the E3 ligase complexes. It is therefore not evident what the roles of parasite-secreted SKP1 and RING-H2 finger proteins are and how these components could redirect E3 ligase complexes. Perhaps other, not yet identified components with similarity to F-box proteins or other types of variable substrate recognition subunits are present in nematode secretions that could confer novel substrate recognition specificities of host E3 ligase complexes.

2.5.6 Modulation of host immune receptors

The recently discovered SPRYSEC proteins in the stylet secretions of cyst nematodes could have a role as variable substrate recognition subunits in E3 ligase complexes (Rehman *et al.*, 2009b). Secreted SPRYSECs have a similar architecture to the GUSTAVUS protein in *Drosophila*, consisting only of the SPRY/30.2 domain. GUSTAVUS includes a small C-terminal BC box and functions as an adaptor subunit in an E3 ubiquitin ligase complex. It is not clear whether the SPRYSECs also have a functional BC-box-like structure. The SPRY/B30.2 domain occurs in many proteins combined with a variety of other domains. It is associated with protein–protein interactions, often involving a receptor molecule.

SPRYSECs have only been found in cyst nematodes of solanaceous plants so far. In these species of cyst nematode, the SPRYSECs occur as large gene families with many highly diverse members. Evolutionary studies on SPRYSECs suggest that they are subjected to strong diversifying selection, which implies that they either tend to evolve new recognition specificities to host targets or that they change to avoid being recognized by host immune receptors (Rehman *et al.*, 2009b). Recent experimental data have not provided much clarity with regard to these two models. One of the SPRYSECs of *G. rostochiensis* was shown to interact physically with the LRR domain of a classical CC-NB-LRR receptor protein from the SW5 resistance gene cluster in tomato. However, the tomato cultivar harbouring this CC-NB-LRR is fully susceptible to *G. rostochiensis*. Furthermore, co-expression of SPRYSEC19 with the CC-NB-LRR protein in *Nicotiana benthamiana* did not result in a hypersensitive response. So, although there is binding and possible recognition of a nematode effector by an immune receptor in a host plant, this does not lead to activation of ETI and resistance. Because the interaction of SPRYSEC19 with a CC-NB-LRR protein does not activate plant defences, it could have the opposite effect on innate immunity through modulation of immune receptors.

However, more recent data on an SPRYSEC homologue in a closely related cyst nematode question this model (Sacco *et al.*, 2009). The SPRYSEC homologue RBP-1 in *G. pallida* was found to induce a *Gpa-2*-dependent hypersensitive response in *N. benthamiana* leaves. *Gpa-2* encodes a CC-NB-LRR protein that confers resistance to specific avirulent strains of *G. pallida* in potato. This implies that RBP-1 of *G. pallida* is recognized by Gpa-2 and activates effector-triggered immunity. Obviously, the nematode does not inject RBP-1 proteins into host cells to betray its presence in the host. RBP-1 will have another intrinsic function, which perhaps involves the interaction with other CC-NB-LRR proteins to modulate their activation. Alternatively, SPRYSEC19 may be an evolutionary intermediate that binds to CC-NB-LRRs, but this binding is not yet or no longer sufficient to elicit a defence response in the plant. Further investigations addressing the primary role of the SPRYSECs will clarify the importance of immune receptor modulation in nematode–plant interactions.

2.5.7 Cross-kingdom modulation

Animal-parasitic nematodes are renowned for their ability to modulate the innate and adaptive immunity of their host (see Grencis and Harnett, Chapter 3, this volume). A wide range of modulation mechanisms involving the secretory–excretory products of the parasite seem to operate at the animal-parasitic nematode–host interface. The immune systems in plants and animals are fundamentally different and it is difficult to predict whether plant- and animal-parasitic nematodes use similar entry points to target host immunity. In a review of possible similarities between plant- and animal-parasitic nematodes, Jasmer *et al.* (2003) noted a remarkable conservation in the effector repertoire that could point at overlapping principles of immune modulation. Highly conserved in all parasitic nematodes are the so-called secreted venom-allergen-like proteins (VAPs or VALs). The VAPs belong to the SCP/TAPS protein family, which is a subclass within the cysteine-rich secretory proteins superfamily (CRISP; Cantacessi *et al.*, 2009). The name SCP/TAPS is short for the acronym SCP/Tpx-1/Ag-5/PR-1/Sc7, referring to several of its members, such as sperm-coating proteins (SCP), testis-specific extracellular proteins (Tpx), glioma pathogenesis-related proteins, venom-allergen from wasps and ants (Ag), and plant PR-1 proteins.

A rather typical example of a member of the SCP/TAPS protein family is the pathogenesis-related protein PR-1. PR-1 is one of the most abundantly expressed secretory proteins following pathogen infection in plants. PR-1 accumulates locally in the apoplast at the site of infection but sometimes also at a distance from the invading pathogen. Even though PR-1 was identified over 20 years ago, and despite its frequent use as a marker for systemic resistance in plants, the mode of action of PR-1 is not well understood (van Loon *et al.*, 2006; Gibbs *et al.*, 2008). It is remarkable that this seems to hold true for many members of the SCP/TAPS protein family, which often appear to have important roles in health and disease but for which the biochemical mode of action is not clear.

The VAPs are among the most abundant proteins released by parasitic nematodes, which suggests that these proteins do have an important role in parasitism. Unfortunately, attempts to knock down VAPs in animal-parasitic nematodes with RNAi have not been successful. Recently, Lozano (unpublished data) has been able to knock down VAPs in the potato cyst nematode *G. rostochiensis*. A specific dsRNA treatment of the infective-stage juveniles significantly reduced the infectivity of the nematodes on potato plants, which demonstrates that VAPs are indeed required for parasitism. However, the function and the molecular targets of nematode VAPs in host plants remain elusive at present.

The sequence conservation in animal-parasitic and plant-parasitic nematode VAPs is relatively high, so they are expected to have similar biochemical activities and possibly similar effects on host cells. Unfortunately, there is no conclusive data available on the biochemical activities of nematode VAPs. However, the relatively well-characterized VAPs from the animal-parasitic hookworms are none the less linked to modulation of mammalian immune cells. The VAP homologue NIF, for example, inhibits neutrophils and blocks the release of ROS from these cells (Moyle *et al.*, 1994). Another *Ancylostoma* SCP/TAPS protein interferes with the extracellular integrin receptors, which inhibits platelet aggregation (Del Valle *et al.*, 2003). Similarly, a large family of VAP homologues in the trematode *Schistosoma mansoni* is also associated with modulating immune responses in the host during the infection process (Chalmers *et al.*, 2008). It is tempting to speculate that VAPs in both animal- and plant-parasitic nematodes are modulators of immunity. However, further investigations into the extracellular targets of VAPs on host cells and a systematic analysis of host defence expression and cytokine profiling following exposure to VAPs are required to classify them conclusively as immune modulators.

2.6 Conclusions and Future Directions

For decades the scientific focus in the field of plant–nematode interactions has centred mainly on host invasion and feeding site formation in susceptible plants. These aspects of parasitism are indeed extremely important for the survival of the nematode inside the host. Currently there is a growing awareness that suppression of host plant immunity may also be essential for a nematode to enable host invasion and feeding. Our field is, therefore, now slowly shifting more towards understanding the role of immune modulation by nematodes in plants. In this chapter we have reviewed current insights in the molecular and cellular aspects of nematode survival in plants, with an emphasis on plant innate immunity. Most of these insights stem from studies with bacterial plant pathogens, but they none the less reveal possible entries for nematodes to attack the immune system of the plant. Ongoing investigations on the role of nematode effectors in parasitism will reveal to what extent these parasites have exploited the same vulnerabilities in host innate immunity as other plant microbes. Animal parasitologists have been ahead of plant nematologists by recognizing that immune modulation is a critical issue for the survival of the parasite. We have briefly entered the world of animal parasites in several sections of this chapter to explore potential overlaps in the mechanisms of immune evasion and suppression. We hope that this chapter will contribute to further comparative analyses of immune modulation by animal- and plant-parasitic nematodes. There seems to be sufficient overlap to accelerate the advances in both fields.

2.7 Acknowledgements

For help in producing the figures, the authors are grateful to Natalia Pineros Arenas (Fig. 2.1), Wiebe Postma (Fig. 2.2c), Anna Tomczak (Fig. 2.2b) and Hein Overmars (Fig. 2.2a). The authors are financially supported by the European Commission's Integrated Project BIOEXPLOIT (FOOD CT-2005-513959) and the Dutch Centre for BioSystems Genomics (CBSG II).

2.8 References

Abad, P., Gouzy, J., Aury, J.M. *et al.* (2008) Genome sequence of the metazoan plant-parasitic nematode *Meloidogyne incognita*. *Nature Biotechnology* 26, 909–915.

Abbas, A.K. and Lichtman, A.H. (2005) *Cellular and Molecular Immunology*. Elsevier and Saunders, Philadelphia, Pennsylvania.

Atkinson, H.J., Urwin, P.E. and McPherson, M.J. (2003) Engineering plants for nematode resistance. *Annual Review of Phytopathology* 41, 615–639.

Badley, J.E., Grieve, R.B., Rockey, J.H. and Glickman, L.T. (1987) Immune-mediated adherence of eosinophils to *Toxocara canis* infective larvae: the role of excretory–secretory antigens. *Parasite Immunology* 9, 133–143.

Baldwin, J.G., Nadler, S.A. and Adams, B.J. (2004) Evolution of plant parasitism among nematodes. *Annual Review of Phytopathology* 42, 83–105.

Bednarek, P. and Osbourn, A. (2009) Plant–microbe interactions: chemical diversity in plant defense. *Science* 324, 746–748.

Bellafiore, S., Shen, Z., Rosso, M.N., Abad, P., Shih, P. and Briggs, S.P. (2008) Direct identification of the *Meloidogyne incognita* secretome reveals proteins with host cell reprogramming potential. *PLoS Pathogens* 4, e1000192.

Bent, A.F. and Mackey, D. (2008) Elicitors, effectors, and R genes: the new paradigm and a lifetime supply of questions. *Annual Review of Phytopathology* 46, 399–436.

Besson-Bard, A., Pugin, A. and Wendehenne, D. (2008) New insights into nitric oxide signaling in plants. *Annual Review of Plant Biology* 59, 21–39.

Bhuiyan, N.H., Selvaraj, G., Wei, Y. and King, J. (2009) Role of lignification in plant defense. *Plant Signaling and Behavior* 4, 158–159.

Birch, P.R.J., Rehmany, A.P., Pritchard, L., Kamoun, S. and Beynon, J.L. (2006) Trafficking arms: oomycete effectors enter host plant cells. *Trends in Microbiology* 14, 8–11.

Bird, A.F. (1968) Changes associated with parasitism in nematodes. III. Ultrastructure of the egg shell, larval cuticle, and contents of the subventral esophageal glands in *Meloidogyne javanica*, with some observations on hatching. *Journal of Parasitology* 54, 475–489.

Bird, A.F. (1974) Plant response to root-knot nematode. *Annual Review of Phytopathology* 12, 69–85.

Bird, A.F. (1983) Changes in the dimensions of the esophageal glands in root-knot nematodes during the onset of parasitism. *International Journal of Parasitology* 54, 879–890.

Blaxter, M.L., De Ley, P., Garey, J.R. *et al.* (1998) A molecular evolutionary framework for the phylum Nematoda. *Nature* 392, 71–75.

Boller, T. and Felix, G. (2009) A renaissance of elicitors: perception of microbe-associated molecular patterns and danger signals by pattern-recognition receptors. *Annual Review of Plant Biology* 60, 379–407.

Boller, T. and He, S.Y. (2009) Innate immunity in plants: an arms race between pattern recognition receptors in plants and effectors in microbial pathogens. *Science* 324, 742–743.

Bowles, D.J., Gurr, S.J., Scollan, C., Atkinson, H.J. and Hammond-Kosack, K.E. (1991) Local and systemic changes in plant gene expression following root infection by cyst nematodes. In: Iacobellis, N.S., Collmer, A., Hutcheson, S.W. *et al.* (eds) *Biochemistry and Molecular Biology of Plant–Pathogen Interactions*. Kluwer Academic, Dordrecht, The Netherlands, pp. 225–236.

Brivio, M.F., Mastore, M. and Moro, M. (2004) The role of *Steinernema feltiae* body-surface lipids in host–parasite immunological interactions. *Molecular and Biochemical Parasitology* 135, 111–121.

Brivio, M.F., Moro, M. and Mastore, M. (2006) Down-regulation of antibacterial peptide synthesis in an insect model induced by the body-surface of an entomoparasite (*Steinernema feltiae*). *Developmental and Comparative Immunology* 30, 627–638.

Caillaud, M.C., Dubreuil, G., Quentin, M., Perfus-Barbeoch, L., Lecomte, P., de Almeida Engler, J., Abad, P., Rosso, M.N. and Favery, B. (2008) Root-knot nematodes manipulate plant cell functions during a compatible interaction. *Journal of Plant Physiology* 165, 104–113.

Cantacessi, C., Campbell, B.E., Visser, A. *et al.* (2009) A portrait of the 'SCP/TAPS' proteins of eukaryotes – developing a framework for fundamental research and biotechnological outcomes. *Biotechnology Advances* 27, 376–388.

Carpita, N.C. and Gibeaut, D.M. (1993) Structural models of primary cell walls in flowering plants: consistency of molecular structure with the physical properties of the walls during growth. *Plant Journal* 3, 1–30.

Chalmers, I.W., McArdle, A.J., Coulson, R.M.R., Wagner, M.A., Schmid, R., Hirai, H. and Hoffmann, K.F. (2008) Developmentally regulated expression, alternative splicing and distinct sub-groupings in members of the *Schistosoma mansoni* venom allergen-like (SmVAL) gene family. *BMC Genomics* 9, 89.

Chisholm, S.T., Coaker, G., Day, B. and Staskawicz, B.J. (2006) Host–microbe interactions: shaping the evolution of the plant immune response. *Cell* 124, 803–814.

Choi, D., Cho, H.T. and Lee, Y. (2006) Expansins: expanding importance in plant growth and development. *Physiologia Plantarum* 126, 511–518.

Cooper, W.R., Jia, L. and Goggin, L. (2005) Effects of jasmonate-induced defenses on root-knot nematode infection of resistant and susceptible tomato cultivars. *Journal of Chemical Ecology* 31, 1953–1967.

Cosgrove, D.J., Li, L.C., Cho, H.T., Hoffmann-Benning, S., Moore, R.C. and Blecker, D. (2002) The growing world of expansins. *Plant Cell Physiolology* 43, 1436–1444.

Craig, A., Ewan, R., Mesmar, J., Gudipati, V. and Sadanandom, A. (2009) E3 ubiquitin ligases and plant innate immunity. *Journal of Experimental Botany* 60, 1123–1132.

Crombie, W.M.L., Crombie, L., Green, J.B. and Lucas, J.A. (1986) Pathogenicity of 'take-all' fungus to oats: its relationship to the concentration and detoxification of the four avenacins. *Phytochemistry* 25, 2075–2083.

Dangl, J.L. and Jones, J.D.G. (2001) Plant pathogens and integrated defence responses to infection. *Nature* 411, 826–833.

Dautova, M., Overmars, H.A., Schots, A., Gommers, F.J., Bakker, J. and Smant, G. (2001) A novel cell degrading enzyme in the plant-parasitic nematode *Meloidogyne incognita*. *Phytopathology* 91, S21. (Abstract.)

Davis, E.L., Allen, R. and Hussey, R.S. (1994) Developmental expression of esophageal gland antigens and their detection in stylet secretions of *Meloidogyne incognita*. *Fundamental and Applied Nematology* 17, 255–262.

Davis, E.L., Hussey, R.S. and Baum, T.J. (2004) Getting to the roots of parasitism by nematodes. *Trends in Parasitology* 20, 134–141.

De Boer, J.M., McDermott, J.P., Davis, E.L., Hussey, R.S., Popeijus, H., Smant, G. and Baum, T.J. (2002a) Cloning of a putative pectate lyase gene expressed in the subventral esophageal glands of *Heterodera glycines*. *Journal of Nematology* 34, 9–11.

De Boer, J.M., McDermott, J.P., Wang, X.H., Maier, T., Qu, F., Hussey, R.S., Davis, E.L. and Baum, T.J. (2002b) The use of DNA microarrays for the developmental expression analysis of cDNAs from the oesophageal gland cell region of *Heterodera glycines*. *Mololecular Plant Patholology* 3, 261–270.

De Meutter, J., Tytgat, T., Van der Schueren, E., Smant, G., Schots, A., Coomans, A., Van Montagu, M. and Gheysen, G. (1998) Cloning of two endoglucanase genes from *Heterodera schachtii*. *Mededelingen van de Faculteit van de Landbouwwetenschappen. University Gent* 63, 619–623.

Del Valle, A., Jones, B.F., Harrison, L.M., Chadderdon, R.C. and Cappello, M. (2003) Isolation and molecular cloning of a secreted hookworm platelet inhibitor from adult *Ancylostoma caninum*. *Molecular and Biochemical Parasitology* 129,167–177.

Dieterich, C. and Sommer, R.J. (2009) How to become a parasite – lessons from the genomes of nematodes. *Trends in Genetics* 25, 203–209.

Ding, X., Shields, J., Allen, R. and Hussey, R.S. (1998) A secretory cellulose-binding protein cDNA cloned from the root-knot nematode (*Meloidogyne incognita*). *Molecular Plant–Microbe Interactions* 11, 952–959.

Doyle, E.A. and Lambert, K.N. (2002) Cloning and characterization of an esophageal-gland-specific pectate lyase from the root-knot nematode *Meloidogyne javanica. Molecular Plant–Microbe Interactions* 15, 549–556.

Doyle, E.A. and Lambert, K.N. (2003) *Meloidogyne javanica* chorismate mutase 1 alters plant cell development. *Molecular Plant–Microbe Interactions* 16, 123–131.

Dubreuil, G., Magliano, M., Deleury, E., Abad, P. and Rosso, M.N. (2007) Transcriptome analysis of root-knot nematode functions induced in the early stages of parasitism. *New Phytologist* 176, 426–436.

Elliger, C.A., Waiss, A.C. Jr, Dutton, H.L. and Rose, M.F. (1988) α-Tomatine and resistance of tomato cultivars toward the nematode, *Meloidogyne incognita. Journal of Chemical Ecology* 14, 1253–1259.

Endo, B.Y. (1975) Pathogenesis of nematode-infected plants. *Annual Review of Phytopathology* 13, 213–238.

Endo, B.Y. (1987) Ultrastructure of esophageal gland secretory granules in juveniles of *Heterodera glycines. Journal of Nematology* 19, 469–483.

Engler, J. de A., De Vleesschauwer, V., Burssens, S., Celenza, J.L. Jr, Inze, D., Van Montagu, M., Engler, G. and Gheysen, G. (1999) Molecular markers and cell cycle inhibitors show the importance of cell cycle progression in nematode-induced galls and syncytia. *Plant Cell* 11, 793–807.

Eulgem, T. and Somssich, I.E. (2007) Networks of WRKY transcription factors in defense signaling. *Current Opinions in Plant Biology* 10, 366–371.

Feys, B.J. and Parker, J.E. (2000) Interplay of signaling pathways in plant disease resistance. *Trends in Genetics* 16, 449–455.

Fioretti, L., Warry, A., Porter, A., Haydock, P. and Curtis, R. (2001) Isolation and localisation of an annexin gene (*gp-nex*) from the potato cyst nematode, *Globodera pallida. Nematology* 3, 45–54.

Fosu-Nyarko, J., Jones, M.G.K. and Wang, Z. (2009) Functional characterization of transcripts expressed in early-stage *Meloidogyne javanica*-induced giant cells isolated by laser microdissection. *Molecular Plant Pathology* 10, 237–248.

Fuller, V.L., Lilley, C.J. and Urwin, P.E. (2008) Nematode resistance. *New Phytologist* 180, 27–44.

Gao, B., Allen, R., Maier, T., Davis, E.L., Baum, T.J. and Hussey, R.S. (2003) The parasitome of the phytonematode *Heterodera glycines. Molecular Plant–Microbe Interactions* 16, 720–726.

Gao, X., Starr, J., Göbel, C., Engelberth, J., Feussner, I., Tumlinson, J. and Kolomiets, M. (2008) Maize 9-lipoxygenase ZmLOX3 controls development, root-specific expression of defense genes, and resistance to root-knot nematodes. *Molecular Plant–Microbe Interactions* 21, 98–109.

Garcia-Brugger, A., Lamotte, O., Vandelle, E., Bourque, S., Lecourieux, D., Poinssot, B., Wendehenne, D. and Pugin, A. (2006) Early signaling events induced by elicitors of plant defenses. *Molecular Plant–Microbe Interactions* 19, 711–724.

Gheysen, G. and Fenoll, C. (2002) Gene expression in nematode feeding sites. *Annual Review of Phytopathology* 40, 191–219.

Gheysen, G. and Mitchum, M.G. (2009) Molecular insights in the susceptible plant response to nematode infection. In: Berg, R.H. and Taylor, C.G. (eds) *Cell Biology of Plant Nematode Parasitism. Plant Cell Monographs, Vol. 15*. Springer, New York/Heidelberg, pp. 45–81.

Gheysen, G., De Almeida-Engler, J. and van Montagu, M. (1997) Cell cycle regulation in nematode feeding sites. In: Iacobellis, N.S., Collmer, A., Hutcheson, S.W. *et al.* (eds) *Cellular and Molecular Aspects of Plant–Nematode Interactions*. Kluwer Academic Publishers, Dordrecht, The Netherlands, pp. 120–132.

Gibbs, G.M., Roelants, K. and O'Bryan, M.K. (2008) The CAP superfamily: cysteine-rich secretory proteins, antigen 5, and pathogenesis-related 1 proteins – roles in reproduction, cancer, and immune defense. *Endocrine Reviews* 29, 865–897.

Goellner, M., Smant, G., De Boer, J.M., Baum, T.J. and Davis, E.L. (2000) Isolation of beta-1,4-endoglucanase genes from *Globodera tabacum* and their expression during parasitism. *Journal of Nematology* 32, 154–165.

Gomez-Gomez, L. and Boller, T. (2000) FLS2: an LRR receptor-like kinase involved in the perception of the bacterial elicitor flagellin in *Arabidopsis*. *Molecular Cell* 5, 1003–1011.

Gomez-Gomez, L. and Boller, T. (2002) Flagellin perception: a paradigm for innate immunity. *Trends in Plant Science* 7, 251–256.

Goverse, A., Overmars, H., Engelbertink, J., Schots, A., Bakker, J. and Helder, J. (2000a) Both induction and morphogenesis of cyst nematode feeding cells are mediated by auxin. *Molecular Plant–Microbe Interactions* 13, 1121–1129.

Goverse, A., De Almeida-Engler, J., Verhees, J., Van der Krol, S., Helder, J. and Gheysen, G. (2000b) Cell cycle activation by plant parasitic nematodes. *Plant Molecular Biology* 43, 747–761.

Grundler, F.-M.W., Sobczak, M. and Lange, S. (1997) Defence responses of *Arabidopsis thaliana* during invasion and feeding site induction by the plant-parasitic nematode *Heterodera glycines*. *Physiological and Molecular Plant Pathology* 50, 419–429.

Haegeman, A., Jacob, J., Vanholme, B., Kyndt, T. and Gheysen, G. (2008) A family of GHF5 endo-1,4-beta-glucanases in the migratory plant-parasitic nematode *Radopholus similis*. *Plant Pathology* 57, 581–590.

Haegeman, A., Vanholme, B. and Gheysen, G. (2009a) Characterization of a putative endoxylanase in the migratory plant-parasitic nematode *Radopholus similis*. *Molecular Plant Pathology* 10, 389–401.

Haegeman, A., Jacob, J., Vanholme, B., Kyndt, T., Mitreva, M. and Gheysen, G. (2009b) Expressed sequence tags of the peanut pod nematode *Ditylenchus africanus*: the first transcriptome analysis of an anguinid nematode. *Molecular and Biochemical Parasitology* 167, 32–40.

Hansen, E., Harper, G., McPherson, M.J. and Atkinson, H.J. (1996) Differential expression patterns of the wound-inducible transgene *wun1-uidA* in potato roots following infection with either cyst or root knot nematodes. *Physiological and Molecular Plant Pathology* 48, 161–170.

Hardham, A.R., Jones, D.A. and Takemoto, D. (2007) Cytoskeleton and cell wall function in penetration resistance. *Current Opinion in Plant Biology* 10, 342–348.

Heath, M.C. (2000) Hypersensitive response-related death. *Plant Molecular Biology* 44, 321–334.

Hewezi, T., Howe, P., Maier, T.R., Hussey, R.S., Mitchum, M.G., Davis, E.L. and Baum, T.J. (2008) Cellulose binding protein from the parasitic nematode *Heterodera schachtii* interacts with *Arabidopsis* pectin methylesterase: cooperative cell wall modification during parasitism. *Plant Cell* 20, 3080–3093.

Hofius, D., Tsitsigiannis, D.I., Jones, J.D.G. and Mundy, J. (2007) Inducible cell death in plant immunity. *Seminars in Cancer Biology* 17, 166–187.

Hominick, W.M., Forrest, J.M.S. and Evans, A.A.F. (1985) Diapause in *Globodera rostochiensis* and variability in hatching trials. *Nematologica* 31, 159–170.

Huang, J.S. and Barker, K.R. (1991) Glyceollin I in soybean–cyst nematode interactions: spatial and temporal distribution in roots of resistant and susceptible soybeans. *Plant Physiology* 96, 1302–1307.

Hussey, R.S. (1987) Secretions of esophageal glands of Tylenchida nematodes. In: Veech, J.A. and Dickson, D.W. (eds) *Vistas on Nematology*. Society of Nematologists, Hyattsville, Maryland, pp. 221–228.

Hussey, R.S. and Mims, C.W. (1990) Ultrastructure of esophageal glands and their secretory granules in the root-knot nematode *Meloidogyne incognita*. *Protoplasma* 156, 9–18.

Hussey, R.S., Mims, C.W. and Westcott, S.W. III. (1992) Immunocytochemical localization of callose in root cortical cells parasitized by the ring nematode *Criconemella xenoplax. Protoplasma* 171, 1–6.

Ithal, N., Recknor, J., Nettleton, D., Maier, T., Baum, T.J. and Mitchum, M.G. (2007a) Developmental transcript profiling of cyst nematode feeding cells in soybean roots. *Molecular Plant–Microbe Interactions* 20, 510–525.

Ithal, N., Recknor, J., Nettleton, D., Hearne, L., Maier, T., Baum, T.J. and Mitchum, M.G. (2007b) Parallel genome-wide expression profiling of host and pathogen during soybean cyst nematode infection of soybean. *Molecular Plant–Microbe Interactions* 20, 293–305.

Jammes, F., Lecomte, P., De Almeida-Engler, J., Bitton, F., Martin-Magniette, M.L., Renou, J.P., Abad, P. and Favery, B. (2005) Genome-wide expression profiling of the host response to root-knot nematode infection in *Arabidopsis. Plant Journal* 44, 447–458.

Jasmer, D.P., Goverse, A. and Smant, G. (2003) Parasitic nematode interactions with mammals and plants. *Annual Review of Phytopathology* 41, 245–270.

Jaubert, S., Laffaire, J.B., Abad, P. and Rosso, M.N. (2002) A polygalacturonase of animal origin isolated from the root-knot nematode *Meloidogyne incognita. FEBS Letters* 522, 109–112.

Jaubert, S., Milac, A.L., Petrescu, A.J., De Almeida-Engler, J., Abad, P. and Rosso, M.N. (2005) *In planta* secretion of a calreticulin by migratory and sedentary stages of root-knot nematode. *Molecular Plant–Microbe Interactions* 18, 1277–1284.

Jia, X.Y., He, L.H., Jing, R.L. and Li, R.Z. (2009) Calreticulin: conserved protein and diverse functions in plants. *Physiologia Plantarum* 136, 127–138.

Jolivet, K., Grenier, E., Bouchet, J.P., Esquibet, M., Kerlan, M.C., Caromel, B., Mugniery, D. and Lefebvre, V. (2007) Identification of plant genes regulated in resistant potato *Solanum sparsipilum* during the early stages of infection by *Globodera pallida. Genome* 50, 422–427.

Jones, J.D.G. and Dangl, J.L. (2006) The plant immune system. *Nature* 444, 323–329.

Jones, J.T., Robertson, L., Perry, R.N. and Robertson, W.M. (1997) Changes in gene expression during stimulation and hatching of the potato cyst nematode *Globodera rostochiensis. Parasitology* 114, 309–315.

Jones, J.T., Furlanetto, C., Bakker, E., Banks, B., Blok, V., Chen, Q., Phillips, M. and Prior, A. (2003) Characterisation of a chorismate mutase from the potato cyst nematode *Globodera pallida. Molecular Plant Pathology* 4, 43–50.

Jones, J.T., Reavy, B., Smant, G. and Prior, A.E. (2004) Glutathione peroxidases of the potato cyst nematode *Globodera rostochiensis. Gene* 324, 47–54.

Jones, M.G.K. (1981) Host cell responses to endoparasitic nematode attack: structure and function of giant cells and syncytia. *Annals Applied Biology* 97, 353–372.

Jones, M.G.K. and Northcote, D.H. (1972) Nematode-induced syncytium – a multinucleate transfer cell. *Journal of Cell Science* 10, 789–809.

Kende, H., Bradford, K.J., Brummell, D.A. *et al.* (2004) Nomenclature for members of the expansin superfamily of genes and proteins. *Plant Molecular Biology* 55, 311–314.

Khan, R., Alkharouf, N., Beard, H., MacDonald, M., Chouikha, I., Meyer, S., Grefenstette, J., Knap, H. and Matthews, B. (2004) Microarray analysis of gene expression in soybean roots susceptible to the soybean cyst nematode two days post invasion. *Journal of Nematology* 36, 241–248.

Kikuchi, T., Jones, J.T., Aikawa, T., Kosaka, H. and Ogura, N. (2004) A family of glycosyl hydrolase family 45 cellulases from the pine wood nematode *Bursaphelenchus xylophilus. FEBS Letters* 572, 201–205.

Kikuchi, T., Li, H.M., Karim, N., Kennedy, M.W., Moens, M. and Jones, J.T. (2009) Identification of putative expansin-like genes from the pine wood nematode, *Bursaphelenchus xylophilus* and evolution of the expansin gene family within the Nematoda. *Nematology* 11, 355–364.

Klink, V.P. and Matthews, B.F. (2008) The use of laser capture microdissection to study the infection of *Glycine max* (soybean) by *Heterodera glycines* (soybean cyst nematode). *Plant Signaling and Behavior* 3, 105–107.

Klink, V.P., Alkharouf, N., MacDonald, M. and Matthews, B. (2005) Laser capture microdissection (LCM) and expression analyses of *Glycine max* (soybean) syncytiucontaining root regions formed by the plant pathogen *Heterodera glycines* (soybean cyst nematode). *Plant Molecular Biology* 59, 965–979.

Klink, V.P., Overall, C.C., Alkharouf, N.W., MacDonald, M.H. and Matthews, B.F. (2007a) A time-course comparative microarray analysis of an incompatible and compatible response by *Glycine max* (soybean) to *Heterodera glycines* (soybean cyst nematode) infection. *Planta* 226, 1423–1447.

Klink, V.P., Overall, C.C., Alkharouf, N.W., MacDonald, M.H. and Matthews, B.F. (2007b) Laser capture microdissection (LCM) and comparative microarray expression analysis of syncytial cells isolated from incompatible and compatible soybean (*Glycine max*) roots infected by the soybean cyst nematode (*Heterodera glycines*). *Planta* 226, 1389–1409.

Koiwa, H., Bressan, R.A. and Hasegawa, P.M. (1997) Regulation of protease inhibitors and plant defense. *Trends in Plant Science* 2, 379–384.

Kudla, U., Qin, L., Milac, A. *et al.* (2005) Origin, distribution and 3D-modeling of Gr-EXPB1, an expansin from the potato cyst nematode *Globodera rostochiensis*. *FEBS Letters* 579, 2451–2457.

Kudla, U., Milac, A.L., Qin, L. *et al.* (2007) Structural and functional characterization of a novel, host penetration-related pectate lyase from the potato cyst nematode *Globodera rostochiensis*. *Molecular Plant Pathology* 8, 293–305.

Kyndt, T., Haegeman, A. and Gheysen, G. (2008) Evolution of GHF5 endoglucanase gene structure in plant-parasitic nematodes: no evidence for an early domain shuffling event. *BMC Evolutionary Biology* 8, 305–321.

Lambert, K.N., Allen, K.D. and Sussex, I.M. (1999) Cloning and characterization of an esophageal-gland-specific chorismate mutase from the phytoparasitic nematode *Meloidogyne javanica*. *Molecular Plant–Microbe Interactions* 12, 328–336.

Lecourieux, D., Ranjeva, R. and Pugin, A. (2006) Calcium in plant defence-signalling pathways: Tansley review. *New Phytologist* 171, 249–269.

Ledger, T.N., Jaubert, S., Bosselut, N., Abad, P. and Rosso, M.N. (2006) Characterization of a new β-1,4-endoglucanase gene from the root-knot nematode *Meloidogyne incognita* and evolutionary scheme for phytonematode family 5 glycosyl hydrolases. *Gene* 382, 121–128.

Lee, D.L. (1972) The structure of the helminth cuticle. *Advances in Parasitology* 10, 347–379.

Lindermayr, C. and Durner, J. (2009) *S*-Nitrosylation in plants: pattern and function. *Journal of Proteomics* 73, 1–9.

Loake, G. and Grant, M. (2007) Salicylic acid in plant defence – the players and protagonists. *Current Opinion in Plant Biology* 10, 466–472.

Loukas, A. and Maizels, R.M. (2000) Helminth C-type lectins and host–parasite interactions. *Parasitology Today* 16, 333–339.

Ma, W. and Berkowitz, G.A. (2007) The grateful dead: calcium and cell death in plant innate immunity. *Cellular Microbiology* 9, 2571–2585.

Maor, R. and Shirasu, K. (2005) The arms race continues: battle strategies between plants and fungal pathogens. *Current Opinion in Microbiology* 8, 399–404.

Marathe, R. and Dinesh-Kumar, S.P. (2003) Plant defense: one post, multiple guards?! *Molecular Cell* 11, 284–286.

Martin, G.B., Bogdanove, A.J. and Sessa, G. (2003) Understanding the functions of plant disease resistance proteins. *Annual Review of Plant Biology* 54, 23–61.

Mastore, M. and Brivio, M.F. (2008) Cuticular surface lipids are responsible for disguise properties of an entomoparasite against host cellular responses. *Developmental and Comparative Immunology* 32, 1050–1062.

McDowell, J.M. and Dangl, J.L. (2000) Signal transduction in the plant immune response. *Trends in Biochemical Science* 25, 79–82.

Melillo, M.T., Leonetti, P., Bongiovanni, M., Castagnone-Sereno, P. and Bleve-Zacheo, T. (2006) Modulation of reactive oxygen species activities and H_2O_2 accumulation during compatible and incompatible tomato–root-knot nematode interactions. *New Phytologist* 170, 501–512.

Milligan, S.B., Bodeau, J., Yaghoobi, J., Kaloshian, I., Zabel, P. and Williamson, V.M. (1998) The root knot nematode resistance gene *Mi* from tomato is a member of the leucine zipper, nucleotide binding, leucine-rich repeat family of plant genes. *Plant Cell* 10, 1307–1319.

Mitreva-Dautova, M., Roze, E., Overmars, H., De Graaff, L., Schots, A., Helder, J., Goverse, A., Bakker, J. and Smant, G. (2006) A symbiont-independent endo-1,4-β-xylanase from the plant-parasitic nematode *Meloidogyne incognita*. *Molecular Plant–Microbe Interactions* 19, 521–529.

Morgan, W. and Kamoun, S. (2007) RXLR effectors of plant pathogenic oomycetes. *Current Opinion in Microbiology* 10, 332–338.

Moyle, M., Foster, D.L., McGrath, D.E. *et al.* (1994) A hookworm glycoprotein that inhibits neutrophil function is a ligand of the integrin CD11b/CD18. *Journal of Biology Chemistry* 269, 10008–10015.

Mukhopadhyay, D. and Riezman, H. (2007) Proteasome-independent functions of ubiquitin in endocytosis and signaling. *Science* 315, 201–205.

Mur, L.A.J., Kenton, P., Lloyd, A.J., Ougham, H. and Prats, E. (2008) The hypersensitive response; the centenary is upon us but how much do we know? *Journal of Experimental Botany* 59, 501–520.

Nurnberger, T., Brunner, F., Kemmerling, B. and Piater, L. (2004) Innate immunity in plants and animals: striking similarities and obvious differences. *Immunological Reviews* 198, 249–266.

Pareja-Jaime, Y., Roncero, M.I.G. and Ruiz-Roldán, M.C. (2008) Tomatinase from *Fusarium oxysporum* f. sp. *lycopersici* is required for full virulence on tomato plants. *Molecular Plant–Microbe Interactions* 21, 728–736.

Pedley, K.F. and Martin, G.B. (2005) Role of mitogen-activated protein kinases in plant immunity. *Current Opinion in Plant Biology* 8, 541–547.

Perry, R.N., Zunke, U. and Wyss, U. (1989) Observations on the response of the dorsal and subventral oesophageal glands of *Globodera rostochiensis* to hatching stimulation. *Revue de Nématologie* 12, 91–96.

Pieterse, C.M.J. and Van Loon, L.C. (2004) NPR1: the spider in the web of induced resistance signaling pathways. *Current Opinion in Plant Biology* 7, 456–464.

Popeijus, H., Overmars, H., Jones, J., Blok, V., Goverse, A., Helder, J., Schots, A., Bakker, J. and Smant, G. (2000) Enzymology: degradation of plant cell walls by a nematode. *Nature* 406, 36–37.

Portillo, M., Lindsey, K., Casson, S., García-Casado, G., Solano, R., Fenoll, C. and Escobar, C. (2009) Isolation of RNA from laser-capture-microdissected giant cells at early differentiation stages suitable for differential transcriptome analysis. *Molecular Plant Pathology* 10, 523–535.

Prior, A., Jones, J.T., Blok, V.C., Beauchamp, J., McDermott, L., Cooper, A. and Kennedy, M.W. (2001) A surface-associated retinol- and fatty acid-binding protein (Gp-FAR-1) from the potato cyst nematode *Globodera pallida*: lipid binding activities, structural analysis and expression pattern. *Biochemical Journal* 356, 387–394.

Puthoff, D.P., Ehrenfried, M.L., Vinyard, B.T. and Tucker, M.L. (2007) GeneChip profiling of transcriptional responses to soybean cyst nematode, *Heterodera glycines*, colonization of soybean roots. *Journal of Experimental Botany* 58, 3407–3418.

Qin, L., Kudla, U., Roze, E.H.A., Goverse, A., Popeijus, H., Nieuwland, J., Overmars, H., Jones, J.T., Schots, A., Smant, G., Bakker, J. and Helder, J. (2004) Plant degradation: a nematode expansin acting on plants. *Nature* 427, 30.

Ramsay, K., Wang, Z. and Jones, M.G.K. (2004) Using laser capture microdissection to study gene expression in early stages of giant cells induced by root-knot nematodes. *Molecular Plant Pathology* 5, 587–592.

Raski, D.J., Jones, N.O. and Roggen, D.R. (1969) On the morphology and ultrastructure of the esophageal region of *Trichodorus allius* Jensen. *Proceedings of the Heminthological Society of Washington* 36, 106–118.

Rehman, S., Butterbach, P., Popeijus, H., Overmars, H., Davis, E.L., Jones, J.T., Goverse, A., Bakker, J. and Smant, G. (2009a) Identification and characterization of the most abundant cellulases in stylet secretions from *Globodera rostochiensis*. *Phytopathology* 99, 194–202.

Rehman, S., Postma, W., Tytgat, T. *et al.* (2009b) A secreted SPRY domain-containing protein (SPRYSEC) from the plant-parasitic nematode *Globodera rostochiensis* interacts with a CC-NB-LRR protein from a susceptible tomato. *Molecular Plant–Microbe Interactions* 22, 330–340.

Rice, S.L., Leadbeater, B.S.C. and Stone, A.R. (1985) Changes in cell structure in roots of resistant potatoes parasitized by potato cyst-nematodes. I. Potatoes with resistance gene *H1* derived from *Solanum tuberosum* ssp. *andigena*. *Physiological Plant Pathology* 27, 219–234.

Robertson, L., Robertson, W.M. and Jones, J.T. (1999) Direct analysis of the secretions of the potato cyst nematode *Globodera rostochiensis*. *Parasitology* 119, 167–176.

Robertson, L., Robertson, W.M., Sobczak, M., Helder, J., Tetaud, E., Ariyanayagam, M.R., Ferguson, M.A.J., Fairlamb, A. and Jones, J.T. (2000) Cloning, expression and functional characterisation of a peroxiredoxin from the potato cyst nematode *Globodera rostochiensis*. *Molecular and Biochemical Parasitology* 111, 41–49.

Rosebrock, T.R., Zeng, L., Brady, J.J., Abramovitch, R.B., Xiao, F. and Martin, G.B. (2007) A bacterial E3 ubiquitin ligase targets a host protein kinase to disrupt plant immunity. *Nature* 448, 370–374.

Rosso, M.N., Favery, B., Piotte, C., Arthaud, L., De Boer, J.M., Hussey, R.S., Bakker, J., Baum, T.J. and Abad, P. (1999) Isolation of a cDNA encoding a β-1,4-endoglucanase in the root-knot nematode *Meloidogyne incognita* and expression analysis during plant parasitism. *Molecular Plant–Microbe Interactions* 12, 585–591.

Sacco, M.A., Koropacka, K., Grenier, E., Jaubert, M.J., Blanchard, A., Goverse, A., Smant, G. and Moffett, P. (2009) The cyst nematode SPRYSEC protein RBP-1 elicits Gpa2- and RanGAP2-dependent plant cell death. *PLoS Pathogens* 5, e1000564.

Schwessinger, B. and Zipfel, C. (2008) News from the frontline: recent insights into PAMP-triggered immunity in plants. *Current Opinion in Plant Biology* 11, 389–395.

Sharon, E. and Spiegel, Y. (1996) Gold-conjugated reagents for the labelling of carbohydrate-recognition domains and glycoconjugates on nematode surfaces. *Journal of Nematology* 28, 124–127.

Sharon, E., Spiegel, Y., Salomon, R. and Curtis, R.H.C. (2002) Characterization of *Meloidogyne javanica* surface coat with antibodies and their effect on nematode behaviour. *Parasitology* 125, 177–185.

Sheridan, J.P., Miller, A.J. and Perry, R.N. (2004) Early responses of resistant and susceptible potato roots during invasion by the potato cyst nematode *Globodera rostochiensis*. *Journal of Experimental Botany* 55, 751–760.

Shingles, J., Lilley, C.J., Atkinson, H.J. and Urwin, P.E. (2007) *Meloidogyne incognita*: molecular and biochemical characterisation of a cathepsin L cysteine proteinase and the effect on parasitism following RNAi. *Experimental Parasitology* 115, 114–120.

Shirasu, K. (2009) The HSP90–SGT1 chaperone complex for NLR immune sensors. *Annual Review of Plant Biology* 60, 139–164.

Smant, G., Stokkermans, J., Yan, Y.T. *et al.* (1998) Endogenous cellulases in animals: isolation of β-1,4-endoglucanase genes from two species of plant-parasitic cyst nematodes. *Proceedings of the National Academy of Sciences of the USA* 95, 4906–4911.

Sobczak, M. and Golinowski, W. (2009) Structure of cyst nematode feeding sites. In: Berg, R.H. and Taylor, C.G. (eds) *Cell Biology of Plant Nematode Parasitism. Plant Cell Monographs, Vol. 15*. Springer, New York, Heidelberg, pp. 153–187.

Spiegel, Y. (1995) Carbohydrate-recognition domains on the surface of phytophagous nematodes. *Experimental Parasitology* 80, 220–227.

Spiegel, Y. and McClure, M.A. (1995) The surface coat of plant-parasitic nematodes: chemical composition, origin, and biological role – a review. *Journal of Nematology* 27, 127–134.

Spiegel, Y., Kahane, I., Cohen, L. and Sharon, E. (1997) *Meloidogyne javanica* surface proteins: characterization and lability. *Parasitology* 115, 513–519.

Swiecicka, M., Filipecki, M., Lont, D., Van Vliet, J., Qin, L., Goverse, A., Bakker, J. and Helder, J. (2009) Dynamics in the tomato root transcriptome on infection with the potato cyst nematode *Globodera rostochiensis*. *Molecular Plant Pathology* 10, 487–500.

Szakasits, D., Heinen, P., Wieczorek, K., Hofmann, J., Wagner, F., Kreil, D.P., Sykacek, P., Grundler, F.M.W. and Bohlmann, H. (2009) The transcriptome of syncytia induced by the cyst nematode *Heterodera schachtii* in *Arabidopsis* roots. *Plant Journal* 57, 771–784.

Theodoropoulos, G., Hicks, S.J., Corfield, A.P., Miller, B.G. and Carrington, S.D. (2001) The role of mucins in host–parasite interactions: part II – helminth parasites. *Trends in Parasitology* 17, 130–135.

Thomma, B.P.H.J., Penninckx, I.A.M.A., Broekaert, W.F. and Cammue, B.P.A. (2001) The complexity of disease signaling in *Arabidopsis*. *Current Opinion in Immunology* 13, 63–68.

Tomczak, A., Koropacka, K., Smant, G., Goverse, A. and Bakker, E. (2009) Resistant plant responses. In: Berg, R.H. and Taylor, C.G. (eds) *Cell Biology of Plant Nematode Parasitism. Plant Cell Monographs, Vol. 15*. Springer, New York/Heidelberg, pp. 83–113.

Turrà, D., Bellin, D., Lorito, M. and Gebhardt, C. (2009) Genotype dependent expression of specific members of potato protease inhibitor gene families in different tissues and in response to wounding and nematode infection. *Journal of Plant Physiology* 166,762–774.

Urwin, P.E., Lilley, C.J. and Atkinson, H.J. (2002) Ingestion of double-stranded RNA by preparasitic juvenile cyst nematodes leads to RNA interference. *Molecular Plant–Microbe Interactions* 15, 747–752.

Van Der Biezen, E.A. and Jones, J.D.G. (1998) Plant disease-resistance proteins and the gene-for-gene concept. *Trends in Biochemical Sciences* 23, 454–456.

Vanholme, B., Haegeman, A., Jacob, J., Cannoot, B. and Gheysen, G. (2009a) Arabinogalactan endo-1,4-beta-galactosidase: a putative plant cell wall-degrading enzyme of plant-parasitic nematodes. *Nematology* 11, 739–747.

Vanholme, B., Kast, P., Haegeman, A., Jacob, J., Grunewald, W. and Gheysen, G. (2009b) Structural and functional investigation of a secreted chorismate mutase from the plant-parasitic nematode *Heterodera schachtii* in the context of related enzymes from diverse origins. *Molecular Plant Pathology* 10, 189–200.

van Loon, L.C. (1983) The induction of pathogenesis-related proteins by pathogens and specific chemicals. *Netherlands Journal of Plant Pathology* 89, 265–273.

van Loon, L.C., Rep, M. and Pieterse, C.M.J. (2006) Significance of inducible defense-related proteins in infected plants. *Annual Review of Phytopathology* 44, 135–162.

Van Ooijen, G., Van Den Burg, H.A., Cornelissen, B.J.C. and Takken, F.L.W. (2008) Structure and function of resistance proteins in solanaceous plants. *Annual Review of Phytopathology* 46, 43–72.

Veronico, P., Giannino, D., Melillo, M.T., Leone, A., Reyes, A., Kennedy, M.W. and Bleve-Zacheo, T. (2006) A novel lipoxygenase in pea roots. Its function in wounding and biotic stress. *Plant Physiology* 141, 1045–1055.

Waetzig, G.H., Sobczak, M. and Grundler, F.M.W. (1999) Localization of hydrogen peroxide during the defence response of

Arabidopsis thaliana against the plant-parasitic nematode *Heterodera glycines. Nematology* 1, 681–686.

Wang, X., Meyers, D., Yan, Y., Baum, T., Smant, G., Hussey, R. and Davis, E. (1999) *In planta* localization of a β-1,4-endoglucanase secreted by *Heterodera glycines. Molecular Plant–Microbe Interactions* 12, 64–67.

Wendehenne, D., Durner, J. and Klessig, D.F. (2004) Nitric oxide: a new player in plant signalling and defence responses. *Current Opinion in Plant Biology* 7, 449–455.

Wiermer, M., Feys, B.J. and Parker, J.E. (2005) Plant immunity: the EDS1 regulatory node. *Current Opinion in Plant Biology* 8, 383–389.

Willats, W.G., McCartney, L., Mackie, W. and Knox, J.P. (2001) Pectin: cell biology and prospects for functional analysis. *Plant Molecular Biology* 47, 9–27.

Williamson, V.M. (1998) Root-knot nematode resistance genes in tomato and their potential for future use. *Annual Review of Phytopathology* 36, 277–293.

Williamson, V.M. (2000) Plant nematode resistance genes. *Current Opinion in Plant Biology* 2, 327–331.

Williamson, V.M. and Kumar, A. (2006) Nematode resistance in plants: the battle underground. *Trends in Genetics* 22, 396–403.

Wright, K.A. (1987) The nematode's cuticle – its surface and the epidermis: function, homology, analogy – a current consensus. *Journal of Parasitology* 73, 1077–1083.

Wubben, M.J.E., Jin, J. and Baum, T.J. (2008) Cyst nematode parasitism of *Arabidopsis thaliana* is inhibited by salicylic acid (SA) and elicits uncoupled SA-independent pathogenesis-related gene expression in roots. *Molecular Plant–Microbe Interactions* 21, 424–432.

Wuyts, N., Lognay, G., Verscheure, M., Marlier, M., De Waele, D. and Swennen, R. (2007) Potential physical and chemical barriers to infection by the burrowing nematode *Radopholus similis* in roots of susceptible and resistant banana (*Musa* spp.). *Plant Pathology* 56, 878–890.

Wyss, U. (1992) Observations on the feeding behaviour of *Heterodera schachtii* throughout development, including events during moulting. *Fundamental and Applied Nematology* 15, 75–89.

Wyss, U. and Zunke, U. (1986) Observations on the behaviour of second stage juveniles of *Heterodera schachtii* inside host roots. *Revue de Nématologie* 9, 153–165.

Zipfel, C. (2008) Pattern-recognition receptors in plant innate immunity. *Current Opinion in Immunology* 20, 10–16.

Zipfel, C. (2009) Early molecular events in PAMP-triggered immunity. *Current Opinion in Plant Biology* 12, 414–420.

3 Survival of Animal-parasitic Nematodes inside the Animal Host

Richard Grencis[1] and William Harnett[2]

[1]Faculty of Life Sciences, University of Manchester, Manchester, UK;
[2]Strathclyde Institute of Pharmacy and Biomedical Sciences, Glasgow, UK

3.1 Introduction

The worldwide success of animal parasitism by nematodes is testament to the remarkable adaptive capacity of this invertebrate phylum. This is particularly so for parasitic nematodes of vertebrates, hosts that have evolved highly sophisticated defence mechanisms to control infection by potential prokaryotic and eukaryotic pathogens. The mechanisms of host immunity are central to our understanding of parasitic nematode survival and have been a rich area of investigation since the 1930s. The rapid increase in our understanding of the immune system genes, molecules, cells and networks that operate during infectious challenge that has occurred over the last 25 years, together with new techniques of genetic manipulation and emerging genomic data on nematodes, has enabled us to begin to investigate in precise detail the way in which nematodes evade immune system-mediated destruction, and this will be the main focus of this chapter. The animal-parasitic nematodes can be split into two basic groups: those that reside predominately in the gastrointestinal (GI) tract and those that reside in extra-intestinal tissues, the tissue-dwelling nematodes, predominately the filariae.

3.2 Gastrointestinal-dwelling Nematodes

Gastrointestinal-dwelling nematodes are ubiquitous parasites that infect numerous animal phyla. The intestinal tract is the site that the adult stages of the parasites occupy and the site at which eggs or larval stages of the parasites are produced. For most, but not all, species of gut-dwelling nematodes these life cycle stages are shed into the gut lumen and ultimately are voided with the faeces into the external environment, where they continue their life cycle after development into embryonated eggs, which are subsequently ingested, or into infective larval stages, of which the third-stage larva (L3) is the most common. The embryonated eggs or L3 are either ingested or actively penetrate through the skin of the next host. For species that do the latter, the nematodes have to transverse several extra-intestinal body systems before ultimately arriving in the GI tract. Moreover, protective immune responses are mounted against such migrating larval nematodes, with roles for eosinophils and complement (Daly *et al.*, 1999; Giacomin *et al.*, 2005, 2008; Knott *et al.*, 2007). Survival strategies used by the parasites as they transiently traverse these tissues may be more akin to those used by tissue-dwelling nematodes, although this area has not received a great deal of attention.

The GI tract presents a varied array of sites for parasites to occupy, from the stomach through the small intestine and large bowel to the rectum. Within these sites, varied niches exist, from free living in the lumen through intraepithelial sites to submucosal locations. It is very clear that the different species of nematode have adapted to occupy many of these distinct niches, and these can vary within species, depending upon the life cycle stage, or can remain the same throughout nematode development. For example, the L3 and fourth-stage larvae (L4) of *Heligmosomoides bakeri* develop in the submucosa of the small intestine of mice, before emerging to live as adults in the lumen (Behnke, 1987; Behnke *et al.*, 2009). *Trichuris muris*, however, spends its entire life cycle within the epithelium of the large intestine (Cliffe and Grencis, 2004). Thus, different species present different challenges in terms of mechanisms of survival.

3.2.1 Gastrointestinal nematode infection – chronicity is the norm

For the so-called geohelminths, the parasite maximizes its reproductive potential by deposition of eggs into the external environment. Female gut-dwelling nematodes shed large numbers of eggs on a daily basis. Moreover, GI nematode infections are generally chronic in nature, with the lifespan of worms often estimated to be up to several years. Thus, longevity within the gut is an advantage and maintaining productive fecundity is important for transmission.

From human studies of GI nematode infection, prevalence of infection is very high, although intensity of infection shows an overdispersed distribution, with 70% of the worm burden contained within 15% of the population (Bundy, 1994; Chan *et al.*, 1994). Under natural infection conditions, individuals would most likely be exposed to small numbers of infective stages, repeatedly, over protracted periods of time. For most individuals, they could acquire a small number of parasites and maintain them over a prolonged period. For some individuals, this would increase after repeated infection, and such individuals would eventually have a high worm burden. The norm, however, is clearly to maintain only relatively small numbers of parasites in the gut. In this case, individuals must either maintain low numbers, reflecting limited exposure, or lose them and re-acquire a few parasites, with no protection generated by prior exposure, or generally become resistant over time after repeated exposure and expel the majority but not all of their parasites. Which of these mechanisms is responsible for the field observations is unclear and is confounded by the difficulty in accurately measuring exposure and infection rates under field conditions.

3.2.2 The immune response to gastrointestinal nematodes – can it be protective?

It is abundantly clear from numerous field and experimental studies that GI-dwelling nematodes are recognized by and stimulate both the innate and adaptive host immune response. It is also very clear, particularly from experimental model systems, that immunity can be host protective, resulting in the expulsion of parasites from the gut. Indeed, the observation of spontaneous cure that is noted in GI nematode infections of domestic stock and epidemiological studies of human geohelminth infections strongly suggests that protective immunity can be induced even if the norm is for chronic, long-lived infection. The fact that host immunity is not generally effective indicates that some kind of immunosuppression or immune evasion is operating. The mechanisms underlying the strategies that ensure prolonged survival have been most extensively studied in rodent models that exhibit chronic infections in the laboratory, such as *H. bakeri* (formerly named *Heligmosomoides polygyrus*, *Nematospiroides dubius*) (Behnke *et al.*, 1991; Behnke and Harris, 2009), and the mouse whipworm, *T. muris* (Cliffe and Grencis, 2004).

A prerequisite to understanding how GI nematodes evade host protective immunity is to determine what mechanisms of host immunity can effectively control GI nematode infections in situations where it does operate. Data from the majority of rodent systems have investigated the host protective immunity that is observed following the administration of a single bolus of moderate to high numbers of infective stages. For many GI nematode species, most notably *Nippostrongylus brasiliensis*, *Trichinella spiralis*, *T. muris* and *Strongyloides* spp., this induces a response that expels the worm burden from the intestine. In the case of *H. bakeri*, a primary high-dose infection progresses to chronic infection in most strains of inbred mouse (Behnke *et al.*, 1992). However, host protective immunity can be generated by giving an abbreviated primary infection, which is terminated before the adult stages of the parasites establish, or by infecting with irradiated L3, which fail to develop to adulthood (Hagan *et al.*, 1981; Finkelman *et al.*, 1997). In all these systems (despite the nematodes living in different intestinal niches), a consistent observation is that resistance is accompanied by a strong type 2 cytokine response controlled by $CD4^{+}$Th2 cells. Key cytokines involved are interleukin (IL-) 4 and particularly IL-13. Also, in most of the systems studied, multiple potential effector responses are elicited, including eosinophlia, intestinal mastocytosis, intestinal goblet cell hyperplasia and mucin production, elevated parasite-specific IgE, production of antimicrobials such as resistin-like molecules (RELM; produced by goblet cells) and changes in the gut epithelium, such as increased intestinal epithelial cell turnover (Urban *et al.*, 1992; Finkelman *et al.*, 1997; Else and Finkelman, 1998; Anthony *et al.*, 2007; Artis and Grencis, 2008).

Recent investigations have dissected the particular effector responses that play dominant roles in expulsion of the different nematode species from a primary infection. For example, intestinal mastocytosis plays a dominant role in expulsion of *T. spiralis* (Grencis *et al.*, 1993; Donaldson *et al.*, 1996; McDermott *et al.*, 2003) but is dispensable for expulsion of *N. brasiliensis* (Madden *et al.*, 1991) and *T. muris* (Betts and Else, 1999). Goblet cell hyperplasia is important for expulsion of *N. brasiliensis* (McKenzie *et al.*, 1998) as is the production of the antimicrobial RELM-β (Herbert *et al.*, 2009). Goblet cell hyperplasia is also important for efficient expulsion of *T. muris* through production of particular mucins (Muc-2 and Muc 5Ac) (Hasnain *et al.*, 2010), whereas RELMb is dispensible (Nair *et al.*, 2008). Epithelial cell turnover is important for displacing *T. muris* from its epithelial niche (Cliffe *et al.*, 2005). In *H. bakeri*, a primary infection is not expelled and resistance has to be induced experimentally. Artificial elevation of intestinal mastocytosis during a primary infection, however, does induce worm expulsion (Hayes *et al.*, 2004). During a secondary or challenge infection, data have also shown that destruction or trapping of the larval stages of *H. bakeri* in its submucosal niche is protective and involves alternatively activated macrophages (controlled by type 2 cytokines) (Anthony *et al.*, 2006). Type 2 cytokine control of intestinal muscle contraction is also thought to contribute to expulsion of GI nematodes (Goldhill *et al.*, 1995; Vallance and Collins, 1998; Zhao *et al.*, 2003; Horsnell *et al.*, 2007).

A consistent observation of many studies is the lack of a definitive role for antibody as an effector mechanism, despite investigation for over 40 years. Investigations have traditionally relied upon passive transfer of serum or purified antibody from previously infected animals. Recently, the availability of a variety of genetically modified mouse strains with various lesions in cells of the B cell lineage have added significantly to this area (Wojciechowski *et al.*, 2009). For *H. bakeri*, studies have clearly shown that there is a role for antibody in a primary infection, but only in suppressing egg production (McCoy *et al.*, 2008). A primary infection is not expelled in this system. During secondary infections, however, a clear role for antibody in mediating worm expulsion was demonstrated, with IgG the important class. A role for IgE or IgA or indeed interaction with immune cells and complement were dispensable. The mechanism of antibody-mediated protection was hypothesized to be via interference with worm feeding, thus reducing viability (McCoy *et al.*, 2008). Indeed, taken together with the passive transfer studies from many other systems (Pleass and Behnke, 2009), the data may well imply that there is a protective role for antibody against GI nematodes, but this is most effective against pre-adult stages of the parasites, particularly during challenge infections. Interestingly, the function of antibody via interference with feeding ability echoes early *invitro* studies in which nematodes were incubated in serum or antibody and 'plugs' were observed, particularly around the mouth regions (Sarles and Taliaferro, 1936).

Nevertheless, it is also very clear from epidemiological and experimental studies that GI nematodes can survive in the presence of high titres of parasite-specific antibodies. This raises the possibility of a deficiency in the ability of antibody to mediate its effects *in vivo*. An insight into this may again come from the *H. bakeri* system, in which the antibody produced during a primary infection is predominately by cells outside the germinal centres of secondary lymphoid organs, suggesting it is likely to be of low affinity (McCoy *et al.*, 2008). How the parasite influences this is unknown.

The above brief discussion clearly highlights that type 2 cytokine responses play a dominant role in protective immunity against primary infections by GI nematodes. Moreover, a battery of type 2 effector mechanisms is generated against the worms, some of which are redundant to host protection, although different species appear to be controlled by different effectors. The evidence suggests that antibody plays a minor role during a primary infection and may have more of a role during challenge infections. Nevertheless, intestinal nematode infections are generally chronic in nature. It is reasonable to suggest that, to achieve this, GI nematodes must evade type 2 responses. The mechanisms whereby they do this are currently an area of active investigation and only now are we beginning to define these alongside a greater genetic knowledge of the parasites themselves.

3.2.3 Immunoregulation during chronic infection – a necessary compromise?

An important consequence of chronic nematode infection that is interwoven with evasion of host protective effector mechanisms is the avoidance of excessive host pathology following prolonged chronic infection. In the intestinal tract, occupying a luminal niche is one way to avoid direct damage to the gut, although the induction of immunopathology via the activation of a Th2 response must also be controlled. Interestingly, following a primary *H. bakeri* infection, intestinal mastocytosis is downregulated along with some of the cytokines that control this response (Hayes *et al.*, 2004). Other aspects of the type 2 response remain unregulated (Wahid *et al.*, 1994). A feature of this infection is, however, an increase in the production of FoxP3 regulatory CD4 T cells (Tregs). This cell type has been extensively studied recently, with a number of subsets now identified (Belkaid and Rouse, 2005; Belkaid and Tarbell, 2009a,b). During *H. bakeri* infections, induction of FoxP3 Tregs is associated with a strong regulation of pathology that can extend to other mucosal sites, such as the lung. Elegant studies by Wilson *et al.* (2005) demonstrated that Tregs generated by *H. bakeri* could suppress sensitization to lung allergens. The major way in which Tregs control responses is thought to be via the secretion of cytokines such as IL-10 or transforming growth factor (TGF)-β. Data from the *H. bakeri* system would suggest that IL-10 does not play a major role but TGF-β may be the major cytokine involved (Wilson *et al.*, 2005). There is indeed some data to support a role for TGF-β in suppressing worm expulsion

(Doligalska *et al.*, 2006). It is well known that induction of Tregs is dependent upon activation of TGF-β via dendritic cells (Travis *et al.*, 2007) and presents an attractive route for parasite modulation of the host response (see below).

Whipworm in the mouse has adopted a different strategy of evading immune-mediated expulsion, although subversion of type 2 effector responses is again central to this. Unlike *H. bakeri*, chronic *T. muris* infection is associated with the activation of a Th1 response (Else and Grencis, 1991a,b; Else *et al.*, 1994). The net effect of this is not only a suppression of a Th2 response (Th1 and Th2 cells reciprocally regulate each other) but also the slowing down of intestinal epithelial cell turnover via the secretion of interferon (IFN)-γ and CXCL10 (Cliffe *et al.*, 2005). This provides a slow-moving extended epithelium, which is the ideal niche for parasite survival. A role for FoxP3$^+$ Treg cells in protective immunity is also suggested from recent studies (D'Elia *et al.*, 2009), and CD4 T cell-derived IL-10 is also critical to control the severe intestinal pathology that is evident in the absence of this cytokine (Schopf *et al.*, 2002).

A role for Tregs in human GI nematode infections is also receiving attention (Maizels and Yazdanbakhsh, 2008; Wammes *et al.*, 2010) and is central to our current concepts of the 'hygiene hypothesis', in which Treg induction following intestinal nematode infection generates a regulatory circuit in the infected host which can suppress responses to common allergens (Wilson and Maizels, 2006; Maizels *et al.*, 2009; Platts-Mills and Cooper, 2010). Thus, there is an emerging theme that induction of Tregs is central to both suppression of host protective immunity and immunopathogy. However, it must be remembered that Treg induction is central to numerous immune responses and is not exclusive to GI nematode infection (Shevach, 2009).

Nevertheless, there is considerable interest in identifying parasite molecules and mechanisms that are involved in induction of Tregs. An acceleration of this process is likely to occur following publication of the genomes of both experimentally used and human and large-animal-infecting GI nematode parasites within the next few years, together with efficient strategies for modulating gene expression within parasitic nematodes.

Recent studies have identified a major secretory product of *H. bakeri* that mimics the function of host TGF-β (Maizels, 2009). A number of parasitic nematodes have TGF-β analogues (McSorley *et al.*, 2010), although whether this particular immunomodulatory molecule belongs to this family remains to be defined. Host TGF-β is a multifunctional cytokine involved in many cellular pathways. In the immune system, recent work has identified an important role for TGF-β and potent antigen-presenting cells (dendritic cells). Activation of TGF-β by integrins on dendritic cells is crucial to the induction of Tregs. It could be envisaged that the parasite-derived molecule mimics the natural host molecule and strongly drives the induction of Tregs for its own advantage, although this remains to be proven (Hewitson *et al.*, 2009).

Indeed, the role of parasite-derived molecules in activation of dendritic cells (DCs) is an active area of investigation, as DCs can have such profound influences upon the subsequent adaptive immune response generated (Coquerelle and Moser, 2010). Most *in vitro* work to date, however, indicates that helminth antigens influence DCs to make Th2 responses the host protective pathway for

GI nematodes (MacDonald and Maizels, 2008). In the intestine, early events can also often occur in the gut epithelium. Epithelial-derived cytokines such as IL-33 and thymic stromal lymphopoietin (TSLP) can play important roles in modulating DC responses. However, the importance of these molecules varies between species of GI nematodes. For example, TSLP is important for the generation of a Th2 response to *T. muris* but dispensable for others, including *H. bakeri* and *N. brasiliensis* (Massacand *et al.*, 2009; Taylor *et al.*, 2009).

Cell types that are believed to be important in the regulation and induction of type 2 responses, particularly in helminths, have increased dramatically recently. Cell types now thought to be involved in this important role include basophils (Perrigoue *et al.*, 2009), nuocytes (Neill *et al.*, 2010) and a lineage-negative multi-potent progenitor (Saenz *et al.*, 2010), with a prominent role for IL-25 in the amplification of the type 2 response (Owyang *et al.*, 2006; Barlow and McKenzie, 2009). This opens up many new levels at which a nematode could interfere with the generation of a host protective response. The challenge is to identify how GI nematodes manipulate this potent protective host response to their own advantage, i.e. whether there are parasite molecules that are capable of doing this and at what stage of the immune response they operate.

Many studies have identified GI nematode-derived molecules that can alter a variety of *in vitro* and *in vivo* immune responses. However, most are undefined mixtures of excretory and secretory molecules collected after *in vitro* incubation of different life cycle stages. A notable exception is the TGF-β-like molecule from *H. bakeri* (discussed above). *Trichuris muris* has been reported to secrete a molecule that mimics the function of the host IFN-γ (Grencis and Entwistle, 1997). Although there is little sequence homology with mammalian IFN-γ, one could envisage how such a molecule could interfere with the generation of effective Th2 responses and promote a suitable epithelial niche for parasite survival. In both cases the production of recombinant molecules, determination of their structure and deletion of the gene in the nematode will ultimately define their importance in the host–parasite relationship.

3.2.4 *Trichinella*, a gut- and tissue-dwelling nematode that bucks the trend

Paradoxically, one group of parasitic nematodes that bridges both the GI-dwelling and tissue-dwelling niche is the Trichinellidae. There are now known to be multiple species of *Trichinella* that infect a wide variety of hosts, including mammals, birds and reptiles (Pozio and Darwin Murrell, 2006; Pozio, 2007). When resident in the GI stages of its life cycle, the parasite induces a potent Th2 response, which expels the parasite from the gut (as described above) (Despommier, 1977; Grencis *et al.*, 1991). However, in the laboratory, even in immunodeficient hosts, the parasite is usually lost from the intestine within 28 days (Vallance *et al.*, 1999). It could be argued that the parasite has a vested interest in not remaining for a very long time in the intestine, as to continue the life cycle the adult female worms must shed live first-stage larvae (L1) into the lymphatics and circulation. These L1 eventually move into striated muscle all around the body and modify the myocyte into a structure known as a nurse cell (Despommier, 1975, 1990, 1998; Despommier *et al.*, 1991). Dramatic changes in gene expression are induced by the parasite (Connolly *et al.*, 1996; Lindh *et al.*, 1998), although the exact mechanisms whereby this occurs remain to be defined. There are evident parallels between induction and maintenance of nurse cells by *Trichinella* and the syncytia and giant cells induced in plant hosts by the plant-parasitic cyst and root-knot nematodes, respectively (see Lozano and Smant, Chapter 2, this volume). The net result of induction of a nurse cell by *Trichinella* is an encapsulated L1, which grows and can remain infective for many months, if not years (Boonmars *et al.*, 2004; Wu *et al.*, 2005, 2008). It is clear, as mentioned above, that data from murine studies has shown that in immunocompetent animals the adaptive immune response contributes considerably to the speed of worm expulsion following a primary infection. It is possible that this is simply a mechanism to prevent excessive larvae being deposited into the tissues. Removal of the adult females from the gut is an effective way to do this, but normally the worms are eventually lost from the gut. It may be that the strong immune response plays additional benefits to stop an unsustainable build-up of L1 during *subsequent* infections by priming the host to make a rapid and effective response upon challenge, expelling the parasites before new L1 are released. Excessive larvae in the muscles can be debilitating or fatal and it is important for the parasite to mature over a period of a month in the muscle to withstand ingestion and transit through the stomach of the new host. It is notable that *T. spiralis* has a very rapid life cycle, going through four moults within 31 h post-infection. Female worms give birth to live L1 by 5 days post-infection. This is in marked contrast to closely related species such as *T. muris*, which takes over 30 days to become sexually mature. Thus, paradoxically, for *Trichinella* to survive, it may be important that the host mounts a very strong host protective response in the intestine. Survival in the muscle tissues may require evasion strategies more akin to the tissue-dwelling nematodes such as the filariae.

3.3 Filarial Nematodes

3.3.1 Adaptation to changes in environment

The filarial nematodes are also parasites of animals but are distinct from their GI relatives with respect to mode of transmission in addition to location within the host. Uniquely, the filarial nematodes are transmitted by biting/blood-sucking arthropods, meaning that they have to cope with a very different environment in the vertebrate host following transfer. Within the vertebrate host, the adult worms, depending on species, may be found in the lymphatics, the subcutaneous tissues or in body cavities. Similarly, the habitat of the larval microfilaria stage varies depending on the species, but they are primarily found circulating in the bloodstream or migrating through the skin.

The infective L3 is the life cycle stage of filarial nematodes that is transmitted from the arthropod to the vertebrate host. It has long been postulated that, in order to withstand the very different challenges of the vertebrate host environment (increased temperature, change in food source, more advanced host immune response, etc.), the arriving infective larva must make a number of adaptations. Adaptations to changes in temperature associated with moving from one host to another are covered by Devaney, Chapter 10, this volume. Changes in feeding habits in filarial nematodes have perhaps not been as thoroughly explored, but some interesting observations have been made; for example, the existence of blood feeding by young adult *Litomosoides sigmodontis* (Attout *et al.*, 2005). On the other hand, the immune response to filarial nematodes has been particularly well documented, and mechanisms for combating it and hence surviving in the animal host will constitute the main component of this section of the chapter.

As the adaptations following transfer to a new host referred to above will certainly involve changes in gene expression, a rewarding approach that has been pursued over the past decade has been to compare the gene expression profile of infective larvae pre- and post-transfer (e.g. Devaney *et al.*, 1996). Probably the most recent example of this is a study on *Brugia malayi* by Li *et al.* (2009), in which infective larvae derived from mosquitoes were compared with larvae exposed to *in vitro* culture conditions designed to mimic the mammalian host. This work showed that the former had 353 genes upregulated in comparison to the latter. These genes were considered important for establishment of infection following transmission, examples being immunomodulatory molecules such as filarial ALT proteins and cystatins and enzymes including cathepsin L-like proteins that may aid migration. On the other hand, a different set of 232 genes was upregulated in the larvae subjected to culture, and these included genes important for growth and development, such as ribosomal proteins involved in protein expression. Some of the genes considered important with respect to immune system evasion will be examined in more detail later.

Moving from the gene to the protein level, recently there have been several publications describing a proteomic analysis of excreted–secreted products (ES) of filarial nematodes, all of which utilize *B. malayi*. *Brugia malayi* has become particularly attractive with respect to the study of filarial nematodes as it is the only species for which a draft genome sequence is available (Scott and Ghedin, 2009). ES (possibly along with surface-exposed molecules) are considered the most likely helminth products to be involved in immunomodulation or evasion of the immune system. Nutman's group (Bennuru *et al.*, 2009) was able to show evidence of considerable stage specificity among the *B. malayi* ES; indeed, the majority of proteins associated with each of infective larvae, moulting infective larvae, adult males, adult females and microfilariae could be categorized in this way. A similar result was reported by Moreno and Geary (2008) when comparing male, female and microfilaria-stage parasites. A third study (Hewitson *et al.*, 2008) focused solely on adult-stage products. All three were able to find molecules known or likely to be involved in immunomodulation, such as ES-62, MIF-1, cystatin and galectin; again, some of these will be covered in more detail later.

Finally, before moving on to discuss immunomodulation and evasion of the immune response, it is worth emphasizing that RNA interference (RNAi), so powerful in assessing the role and importance of particular genes in *Caenorhabditis elegans*, has more recently been applied successfully to filarial nematodes. As an example, treatment of adult female *B. malayi* with dsRNA corresponding to cathepsin-like cysteine protease genes resulted in disruption of embryogenesis and a decrease in the number of microfilariae released (Ford *et al.*, 2009). The same experimental approach has shown the importance of this family of proteases (Lustigman *et al.*, 2004) and also the filarial serine protease inhibitor, Ov-SPI-1 (Ford *et al.*, 2005), during moulting of *Onchocerca volvulus* infective larvae. Likewise, RNAi employing dsRNA corresponding to a chitinase of the rodent filarial nematode *Acanthocheilonema viteae* was found to result in inhibition of moulting of infective larvae, inhibition of hatching of microfilaraie and the death of 50% of adult female worms (Tachu *et al.*, 2008). Thus, the potential now exists to investigate the importance of every individual gene product to survival of filarial nematodes in the animal host.

3.3.2 Immunomodulation during filarial nematode infection

A perhaps surprising feature of filarial nematodes is their longevity. It is reported, for example, that *Wuchereria bancrofti* can live in excess of a decade (Subramanian *et al.*, 2004) and, although a number of factors may well be involved, it is assumed that one major component contributing to long-term infection is suppression of the host immune system. In relation to this, analysis of the immune response during human infection generally indicates that people having patent infection (i.e. having detectable microfilariae) demonstrate reduced lymphocyte proliferative responses (to filarial nematode antigens and in some studies other antigens and/or mitogens) (Harnett and Harnett, 2008) and decreased production of IFN-γ but increased production of IL-4 and IL-10 and the IgG subclass IgG_4 (reviewed by Nutman and Kumaraswami, 2001; Hoerauf *et al.*, 2005). As alluded to earlier, IL-4 is classically a cytokine associated with Th2 cells (Nakamura *et al.*, 1997) and IL-10 is an anti-inflammatory cytokine (Moore *et al.*, 2001), and hence such an immune response may be characterized as being of a Th2, anti-inflammatory phenotype. There is probably a benefit for humans in the generation of an anti-inflammatory immunological phenotype because, although filarial nematode infection can result in severe pathology such as elephantiasis, the majority of infected individuals exhibit little evidence of an inflammatory response or overt tissue destruction/disruption (Rajan, 2005). More recently, filarial nematode infections have also been associated with Treg responses and alternatively activated macrophages (Hoerauf *et al.*, 2005), and each of these may contribute to the anti-inflammatory effect.

Although the host may have evolved regulatory responses to deal with any persistent significant antigen challenge, it is clear that filarial nematodes are truly adept at polarizing immune responses towards Th2, in particular by their effects on DCs. As alluded to earlier, these cells are the sentinels of the immune system that sample and process foreign molecules for presentation to T cells. Presentation of peptides derived from antigens is accompanied by signals generated by DC surface co-stimulatory molecules and secreted cytokines interacting with receptors on T cells, and this combination impacts on T cell polarization (Banchereau *et al.*, 2000). Thus, filarial nematode antigens must contain information that facilitates DC-driven Th2 polarization. Interestingly, relative to antigens that induce Th1 responses, the filarial nematode-derived product ES-62, which drives Th2 polarization, has been shown to have little effect on DC activation as measured by upregulation of markers such as co-stimulatory molecules (Whelan *et al.*, 2000).

The induction of Th2 responses may be a deliberate policy that nematodes employ, which reflects particular as yet not fully defined characteristics of nematode molecules (see Harnett and Harnett, 2006). However, the free-living *C. elegans* also possesses molecules that drive Th2 responses (Tawill *et al.*, 2004). Furthermore, the advantage of such a strategy to the worms is uncertain, because, as mentioned earlier, there is evidence from rodent models of GI nematode infection that Th2 immune responses are protective (Gause *et al.*, 2003), and this is also supported by some human epidemiological studies (Bradley and Jackson, 2004). However, the emerging consensus with respect to filarial nematode infection is that protective immunity may depend on components of both the Th1 and Th2 arms of the adaptive immune response (Hoerauf *et al.*, 2005). Interestingly, with respect to regulatory responses, as alluded to earlier, these may be generated in response to any chronic inflammatory insult in an attempt to prevent it getting out of hand, but with respect to filarial nematodes, recent evidence intriguingly suggests that Tregs may contribute via inhibition of protective immunity to worm persistence during infection (Taylor *et al.*, 2007).

Other possible mechanisms of evasion of immune responses relate to antibody production. Filarial nematodes have a tendency to induce considerable non-specific IgE secretion during infection, and it has been suggested that, by saturating high-affinity IgE receptor (FcεRI) sites, these antibodies may block mast cell degranulation (reviewed by Erb, 2007). However, studies on humans infected with filarial nematode parasites indicate that the relative levels of polyclonal IgE to filaria-specific IgE do not tend to reach that required for inhibition of filaria-induced mast cell or, in addition, basophil degranulation (Mitre *et al.*, 2005). The same outcome was observed when examining dust mite-specific IgE. As mentioned earlier, filarial nematode infections can also be associated with the production of high levels of IgG_4 (Hussain and Ottesen, 1988), an antibody that competes with IgE for binding to filarial nematode antigens but does not promote mast cell or basophil degranulation. This offers an alternative antibody-mediated mechanism for inhibiting potentially protective immune responses.

3.3.3 Defined filarial nematode molecules known to modulate the immune system

3.3.3.1 Cystatins

A group of immunomodulators characterized in a number of filarial nematode species is the cystatins (cysteine protease inhibitors) (reviewed by Hartmann and Lucius, 2003; Gregory and Maizels, 2008). These secreted molecules appear to interfere directly with antigen presentation, contributing to a significant degree of inhibition of both antigen-specific and polyclonal T cell proliferation. Such inhibition is also associated with filarial nematode cystatins reducing co-stimulatory molecule expression and inducing large-scale IL-10 production by macrophages. Interestingly, filarial nematode cystatins can also upregulate production of NO, and it has been speculated that this may be a further factor in inhibition of T cell responses. However, subsequent to this it was found that the use of NO synthase inhibitors did not restore proliferative responses. In addition, unlike increased production of IL-10, upregulated NO production can also be generated from cystatins of *C. elegans* and these have little effect on T cell proliferation. Hence the relevance of NO production to immunomodulation during parasitism is uncertain.

More recently, *A. viteae* recombinant cystatin has been found to inhibit inflammation associated with both ovalbumin-induced allergic airway responsiveness and dextran sulfate sodium-induced colitis in mice, revealing a therapeutic potential of this molecule in treating inflammatory diseases (Schnoeller *et al.*, 2008).

3.3.3.2 Dirofilaria immitis*-derived antigen*

Imai and Fujita (2004) have reported on a purified 15 kDa protein from the canine filarial nematode *Dirofilaria immitis* – DiAg (*Dirofilaria immitis*-derived antigen). This molecule induces polyclonal proliferation of B cells and non-specific IgE production. Interestingly, DiAg also prevents the spontaneous generation of IgG anti-insulin antibodies that develops in the non-obese diabetic (NOD) mouse, and this is associated with lack of development of Th1-dependent autoimmune diabetes in these animals. Consistent with this latter observation, DiAg increases levels of IL-4 and IL-10 when injected into BALB/c mice.

DiAg has also been shown to inhibit passive cutaneous anaphylaxis reactions in rats. This was not due to an effect on the number or viability of mast cells but, interestingly, given the dismissal of this mode of action in the human situation referred to above (Mitre *et al.*, 2005), appears to be due to non-specific saturation of FcεRI.

3.3.3.3 ES-62

ES-62 is an immunomodulatory protein with a novel post-translational modification of phosphorylcholine (PC) attachment to an *N*-type glycan, discovered in the secretions of *A. viteae* (reviewed by Harnett and Harnett, 2009; Harnett *et al.*, 2010). In the mouse, ES-62 was found to polarize T cell responses by modulating the maturation and functional responses of antigen-presenting cells

(APCs), such as DCs and macrophages. In particular, it tends to drive Th2, rather than Th1, antigen-specific responses, as indicated by the generation of IgG_1 rather than IgG_{2a} antibodies. Moreover, ES-62 inhibits pro-inflammatory cytokine production by APCs and induces production of anti-inflammatory IL-10 by B1 cells. ES-62 can also inhibit antigen receptor-mediated conventional B cell proliferation. Employment of knockout mice revealed that ES-62 subverts Toll-like receptor-4 (TLR4) signalling to mediate its anti-inflammatory effects. More recent analysis revealed that ES-62 also targets mast cells, directly preventing degranulation and release of mediators of allergy induced via ligation of FcεRI and by a mechanism involving inhibition of phospholipase D-coupled, sphingosine kinase-mediated calcium mobilization and NF-κB activation. Consistent with the TLR4 knockout mouse studies, it was found that ES-62 mediates these effects by forming a complex with TLR4, which results in the sequestration and perinuclear degradation of protein kinase C (PKC)-α, a molecule found to be critical for mast cell activation.

The PC moiety appears to be the active immunomodulatory component of ES-62 as it has been found to largely mimic ES-62's activity when conjugated to albumin or ovalbumin. Of relevance to immunomodulation in human filarial nematode infection, PC-containing molecules have been detected in the *in vitro* secretions of the major medically important human parasites (*W. bancrofti*, *B. malayi* and *O. volvulus*) and/or in the bloodstream of infected individuals (reviewed by Harnett *et al.*, 1998). The anti-inflammatory properties of ES-62 are such that it can protect mice from developing collagen-induced arthritis and it can prevent pro-inflammatory cytokine release by cultured synovial cells from rheumatoid arthritis patients. Again, PC-conjugated proteins share these properties and such PC-mediated immunomodulation is dependent on TLR4. Furthermore, ES-62 has more recently been found to be active in two mouse models of allergy: ovalbumin-induced airway hypersensitivity and immediate-type hypersensitivity to oxazolone in the skin. In the former, ES-62 was found to reduce peri-bronchial inflammation and mucosal hyperplasia, inhibit eosinophila and prevent release of the signature cytokine required for airway inflammation development, IL-4. In the latter, ES-62 targeted inflammation, as shown by a reduction in ear swelling, and this was correlated with the effects on mast cells reported earlier.

3.4 Conclusions and Future Directions

Parasitic nematodes are exposed to the full armoury of the host immune system and hence, during the evolution of host–parasite interactions, have come up with a multitude of strategies to counter this (Fig. 3.1). These strategies involve both actively suppressing effector responses of cells such as B lymphocytes and mast cells and modulating the polarity of immune responses by acting on APCs, thereby inhibiting responses that are potentially protective. Nematodes also induce an anti-inflammatory response that is so potent that it is being explored with respect to its therapeutic potential in humans. This investigation has now reached the stage where individual nematode

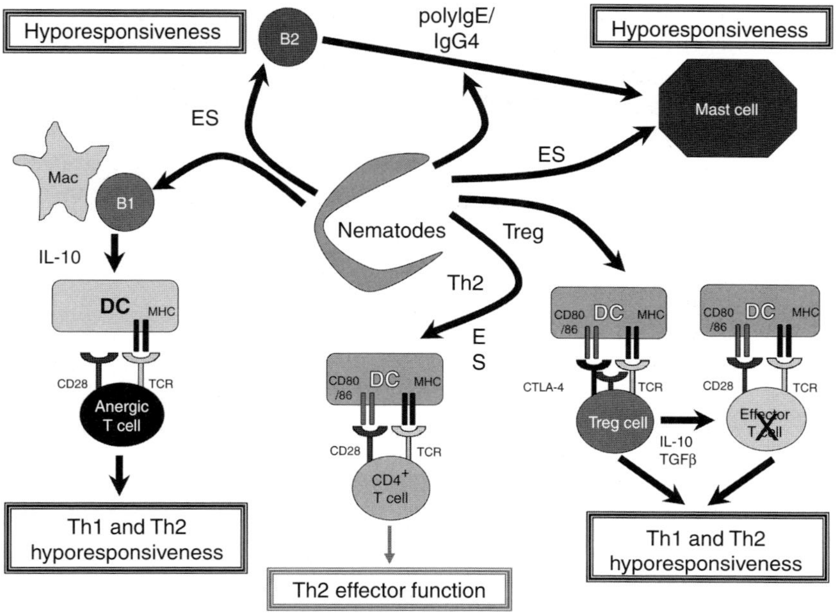

Fig. 3.1. Potential mechanisms of nematode-induced immunomodulation. Infection with nematodes can induce T cell hyporesponsiveness by directly or indirectly (through antigen-presenting cells such as DCs) suppressing T cell activation. For example, infection can result in the induction of T regulatory (Treg) cells that are refractory to antigen and can prevent maximal T effector responses of either Th1 or Th2 cells. Treg cells may also act via the release of anti-inflammatory cytokines such as IL-10 and TGF-β. Some nematode excretions–secretions (ES) contain homologues of the latter cytokine, and *T. muris* secretes a molecule that acts like IFN-γ, thereby potentially inhibiting protective Th2 responses. Moreover, worm ES products such as cystatins and ES-62 can induce other cells, such as B1 cells or macrophages, respectively, to produce IL-10. Alternatively, the worms can modulate DC maturation to polarize T cell responses towards a Th2 phenotype and hence counteract Th1-based inflammation. By contrast, it has been proposed that nematodes can antagonize Th2-driven inflammation by producing polyspecific IgE or IgG4 to block nematode antigen-induced mast cell degranulation, although the evidence in the human situation with respect to the former is generally not supportive. Finally, ES-62 has been shown to induce mast cell responsiveness directly by subverting FcεRI signalling.

molecules responsible for immunomodulation are being defined and their mechanism of action characterized. Thus, we may be approaching the exciting time when the mechanisms utilized by parasitic nematodes to survive in the animal host are successfully exploited to the benefit of that host.

3.5 References

Anthony, R.M., Urban, J.F. Jr, Alem, F., Hamed, H.A., Rozo, C.T., Boucher, J.L., Van Rooijen, N. and Gause W.C. (2006) Memory T_H2 cells induce alternatively activated macrophages to mediate protection against nematode parasites. *Nature Medicine* 12, 955–960.

Anthony, R.M., Rutitzky, L.I., Urban, J.F. Jr, Stadecker, M.J. and Gause W.C (2007) Protective immune mechanisms in helminth infection. *Nature Reviews* 7, 975–987.

Artis, D. and Grencis, R.K. (2008) The intestinal epithelium: sensors to effectors in nematode infection. *Mucosal Immunology* 1, 252–264.

Attout, T., Babayan, S., Hoerauf, A., Taylor, D.W., Kozek, W.J., Martin, C. and Bain, O. (2005) Blood-feeding in the young adult filarial worms *Litomosoides sigmodontis*. *Parasitology* 130, 421–428.

Banchereau, J., Briere, F., Caux, C., Davoust, J., Lebecque, S., Liu, Y.J., Pulendran, B. and Palucka, K. (2000) Immunobiology of dendritic cells. *Annual Review of Immunology* 18, 767–811.

Barlow, J.L. and McKenzie, A.N. (2009) IL-25: a key requirement for the regulation of type-2 immunity. *BioFactors* 35, 178–182.

Behnke, J.M. (1987) Evasion of immunity by nematode parasites causing chronic infections. *Advances in Parasitology* 26, 1–71.

Behnke, J. and Harris, P. (2009) *Heligmosomoides bakeri* or *Heligmosomoides polygyrus*? *American Journal of Tropical Medicine and Hygiene* 80, 684–685.

Behnke, J.M., Keymer, A.E. and Lewis, J.W. (1991) *Heligmosomoides polygyrus* or *Nematospiroides dubius*? *Parasitology Today* 7, 177–179.

Behnke, J.M., Barnard, C.J. and Wakelin, D. (1992) Understanding chronic nematode infections: evolutionary considerations, current hypotheses and the way forward. *International Journal for Parasitology* 22, 861–907.

Behnke, J.M., Menge, D.M. and Noyes, H. (2009) *Heligmosomoides bakeri*: a model for exploring the biology and genetics of resistance to chronic gastrointestinal nematode infections. *Parasitology* 136, 1565–1580.

Belkaid, Y. and Rouse, B.T. (2005) Natural regulatory T cells in infectious disease. *Nature Immunology* 6, 353–360.

Belkaid, Y. and Tarbell, K. (2009a) Regulatory T cells in the control of host–microorganism interactions. *Annual Review of Immunology* 27, 551–589.

Belkaid, Y. and Tarbell, K.V. (2009b) Arming Treg cells at the inflammatory site. *Immunity* 30, 322–323.

Bennuru, S., Semnani, R., Meng, Z., Ribeiro, J.M., Veenstra, T.D. and Nutman, T.B. (2009) *Brugia malayi* excreted/secreted proteins at the host/parasite interface: stage- and gender-specific proteomic profiling. *PLoS Neglected Tropical Diseases* 3, e410.

Betts, C.J. and Else, K.J. (1999) Mast cells, eosinophils and antibody-mediated cellular cytotoxicity are not critical in resistance to *Trichuris muris*. *Parasite Immunology* 21, 45–52.

Boonmars, T., Wu, Z., Nagano, I. and Takahashi, Y. (2004) Expression of apoptosis-related factors in muscles infected with *Trichinella spiralis*. *Parasitology* 128, 323–332.

Bradley, J.E. and Jackson, J.A. (2004) Immunity, immunoregulation and the ecology of trichuriasis and ascariasis. *Parasite Immunology* 26, 429–441.

Bundy, D.A. (1994) Immunoepidemiology of intestinal helminthic infections. 1. The global burden of intestinal nematode disease. *Transactions of the Royal Society of Tropical Medicine and Hygiene* 88, 259–261.

Chan, M.S., Medley, G.F., Jamison, D. and Bundy, D.A. (1994) The evaluation of potential global morbidity attributable to intestinal nematode infections. *Parasitology* 109, 373–387.

Cliffe, L.J. and Grencis, R.K. (2004) The *Trichuris muris* system: a paradigm of resistance and susceptibility to intestinal nematode infection. *Advances in Parasitology* 57, 255–307.

Cliffe, L.J., Humphreys, N.E., Lane, T.E., Potten, C.S., Booth, C. and Grencis, R.K. (2005) Accelerated intestinal epithelial cell turnover: a new mechanism of parasite expulsion. *Science* 308, 1463–1465.

Connolly, B., Trenholme, K. and Smith, D.F. (1996) Molecular cloning of a *myoD*-like gene from the parasitic nematode, *Trichinella spiralis*. *Molecular and Biochemical Parasitology* 81, 137–149.

Coquerelle, C. and Moser, M. (2010) DC subsets in positive and negative regulation of immunity. *Immunological Reviews* 234, 317–334.

Daly, C.M., Mayrhofer, G. and Dent, L.A. (1999) Trapping and immobilization of *Nippostrongylus brasiliensis* larvae at the site of inoculation in primary infections of interleukin-5 transgenic mice. *Infection and Immunity* 67, 5315–5323.

D'Elia, R., Behnke, J.M., Bradley, J.E. and Else, K.J. (2009) Regulatory T cells: a role in the control of helminth-driven intestinal pathology and worm survival. *Journal of Immunology* 182, 2340–2348.

Despommier, D. (1975) Adaptive changes in muscle fibers infected with *Trichinella spiralis*. *American Journal of Pathology* 78, 477–496.

Despommier, D. (1977) Immunity to *Trichinella spiralis*. *American Journal of Tropical Medicine and Hygiene* 26, 68–75.

Despommier, D.D. (1990) *Trichinella spiralis*: the worm that would be virus. *Parasitology Today* 6, 193–196.

Despommier, D.D. (1998) How does *Trichinella spiralis* make itself at home? *Parasitology Today* 14, 318–323.

Despommier, D., Symmans, W.F. and Dell, R. (1991) Changes in nurse cell nuclei during synchronous infection with *Trichinella spiralis*. *Journal of Parasitology* 77, 290–295.

Devaney, E., Martin, S.A. and Thompson, F.J. (1996) Stage-specific gene expression in lymphatic filarial nematodes. *Parasitology Today* 12, 418–424.

Doligalska, M., Rzepecka, J., Drela, N., Donskow, K. and Gerwel-Wronka, M. (2006) The role of TGF-β in mice infected with *Heligmosomoides polygyrus*. *Parasite Immunology* 28, 387–395.

Donaldson, L.E., Schmitt, E., Huntley, J.F., Newlands, G.F. and Grencis, R.K. (1996) A critical role for stem cell factor and c-*kit* in host protective immunity to an intestinal helminth. *International Immunology* 8, 559–567.

Else, K.J. and Finkelman, F.D. (1998) Intestinal nematode parasites, cytokines and effector mechanisms. *International Journal for Parasitology* 28, 1145–1158.

Else, K.J. and Grencis, R.K. (1991a) Cellular immune responses to the murine nematode parasite *Trichuris muris*. I. Differential cytokine production during acute or chronic infection. *Immunology* 72, 508–513.

Else, K.J. and Grencis, R.K. (1991b) Helper T-cell subsets in mouse trichuriasis. *Parasitology Today* 7, 313–316.

Else, K.J., Finkelman, F.D., Maliszewski, C.R. and Grencis, R.K. (1994) Cytokine-mediated regulation of chronic intestinal helminth infection. *Journal of Experimental Medicine* 179, 347–351.

Erb, K.J. (2007) Helminths, allergic disorders and IgE-mediated immune responses: where do we stand? *European Journal of Immunology* 37, 1170–1173.

Finkelman, F.D., Shea-Donohue, T., Goldhill, J., Sullivan, C.A., Morris, S.C., Madden, K.B., Gause, W.C. and Urban, J.F. Jr (1997) Cytokine regulation of host defense against parasitic gastrointestinal nematodes: lessons from studies with rodent models. *Annual Review of Immunology* 15, 505–533.

Ford, L., Guiliano, D.B., Oksov, Y., Debnath, A.K., Liu, J., Williams, S.A., Blaxter, M.L. and Lustigman, S. (2005) Characterization of a novel filarial serine protease inhibitor, Ov-SPI-1, from *Onchocerca volvulus*, with potential multifunctional roles during development of the parasite. *Journal of Biological Chemistry* 280, 40845–40856.

Ford, L., Zhang, J., Liu, J., Hashmi, S., Fuhrman, J.A., Oksov, Y. and Lustigman, S. (2009) Functional analysis of the cathepsin-like cysteine protease genes in adult *Brugia malayi* using RNA interference. *PLoS Neglected Tropical Diseases* 3, e377.

Gause, W.C., Urban, J.F. Jr and Stadecker, M.J. (2003) The immune response to parasitic helminths: insights from murine models. *Trends in Immunology* 24, 269–277.

Giacomin, P.R., Wang, H., Gordon, D.L., Botto, M. and Dent, L.A. (2005) Loss of complement activation and leukocyte adherence as *Nippostrongylus brasiliensis* develops within the murine host. *Infection and Immunity* 73, 7442–7449.

Giacomin, P.R., Gordon, D.L., Botto, M., Daha, M.R., Sanderson, S.D., Taylor, S.M.

and Dent, L.A. (2008) The role of complement in innate, adaptive and eosinophil-dependent immunity to the nematode *Nippostrongylus brasiliensis. Molecular Immunology* 45, 446–455.

Goldhill, J.M., Finkelman, F.D., Morris, S.C. and Shea-Donohue, T. (1995) Neural control of mouse small intestinal longitudinal muscle: interactions with inflammatory mediators. *Journal of Pharmacology and Experimental Therapeutics* 274, 72–77.

Gregory, W.F. and Maizels, R.M. (2008) Cystatins from filarial parasites: evolution, adaptation and function in the host–parasite relationship. *International Journal of Biochemistry & Cell Biology* 40, 1389–1398.

Grencis, R.K. and Entwistle, G.M. (1997) Production of an interferon-gamma homologue by an intestinal nematode: functionally significant or interesting artefact? *Parasitology* 115 (Supplement), S101–S106.

Grencis, R.K., Hultner, L. and Else, K.J. (1991) Host protective immunity to *Trichinella spiralis* in mice: activation of Th cell subsets and lymphokine secretion in mice expressing different response phenotypes. *Immunology* 74, 329–332.

Grencis, R.K., Else, K.J., Huntley, J.F. and Nishikawa, S.I. (1993) The *in vivo* role of stem cell factor (c-*kit* ligand) on mastocytosis and host protective immunity to the intestinal nematode *Trichinella spiralis* in mice. *Parasite Immunology* 15, 55–59.

Hagan, P., Behnke, J.M. and Parish, H.A. (1981) Stimulation of immunity to *Nematospiroides dubius* in mice using larvae attenuated by cobalt 60 irradiation. *Parasite Immunology* 3, 149–156.

Harnett, M.M., Melendez, A.J. and Harnett, W. (2010) The therapeutic potential of the filarial nematode-derived immunodulator, ES-62 in inflammatory disease. *Clinical and Experimental Immunology* 159, 256–267.

Harnett, W. and Harnett, M.M. (2006) Molecular basis of worm-induced immunomodulation. *Parasite Immunology* 28, 535–543.

Harnett, W. and Harnett, M.M. (2008) Lymphocyte hyporesponsiveness during filarial nematode infection. *Parasite Immunology* 30, 447–453.

Harnett, W. and Harnett, M.M. (2009) Immunomodulatory activity and therapeutic potential of the filarial nematode secreted product, ES-62. *Advances in Experimental Medicine and Biology* 666, 88–94.

Harnett, W., Bradley, J.E. and Garate, T. (1998) Molecular and immunodiagnosis of human filarial nematode infections. *Parasitology* 117 (Supplement), S59–S71.

Hartmann, S. and Lucius, R. (2003) Modulation of host immune responses by nematode cystatins. *International Journal for Parasitology* 33, 1291–1302.

Hasnain, S.Z., Wang, H., Ghia, J.E., Haq, N., Deng, Y., Velcich, A., Grencis, R.K., Thornton, D.J. and Khan, W.I. (2010) Mucin gene deficiency in mice impairs host resistance to an enteric parasitic infection. *Gastroenterology* 138, 1763–1771.

Hayes, K.S., Bancroft, A.J. and Grencis, R.K. (2004) Immune-mediated regulation of chronic intestinal nematode infection. *Immunological Reviews* 201, 75–88.

Herbert, D.R., Yang, J.Q., Hogan, S.P. *et al.* (2009) Intestinal epithelial cell secretion of RELM-β protects against gastrointestinal worm infection. *Journal of Experimental Medicine* 206, 2947–2957.

Hewitson, J.P., Harcus, Y.M., Curwen, R.S., Dowle, A.A., Atmadja, A.K., Ashton, P.D., Wilson, A. and Maizels, R.M. (2008) The secretome of the filarial parasite, *Brugia malayi*: proteomic profile of adult excretory–secretory products. *Molecular and Biochemical Parasitology* 160, 8–21.

Hewitson, J.P., Grainger, J.R. and Maizels, R.M. (2009) Helminth immunoregulation: the role of parasite secreted proteins in modulating host immunity. *Molecular and Biochemical Parasitology* 167, 1–11.

Hoerauf, A., Satoguina, J., Saeftel, M. and Specht, S. (2005) Immunomodulation by filarial nematodes. *Parasite Immunology* 27, 417–429.

Horsnell, W.G., Cutler, A.J., Hoving, J.C. *et al.* (2007) Delayed goblet cell hyperplasia, acetylcholine receptor expression, and worm expulsion in SMC-specific IL-4Rα-deficient mice. *PLoS Pathogens* 3, e1.

Hussain, R. and Ottesen, E.A. (1988) Parallel recognition of filarial antigens by IgE and IgG4 antibodies. *New England and Regional Allergy Proceedings* 9, 51–56.

Imai, S. and Fujita, K. (2004) Molecules of parasites as immunomodulatory drugs. *Current Topics in Medicinal Chemistry* 4, 539–552.

Knott, M.L., Matthaei, K.I., Giacomin, P.R., Wang, H., Foster, P.S. and Dent, L.A. (2007) Impaired resistance in early secondary *Nippostrongylus brasiliensis* infections in mice with defective eosinophilopoeisis. *International Journal for Parasitology* 37, 1367–1378.

Li, B.W., Rush, A.C., Mitreva, M., Yin, Y., Spiro, D., Ghedin, E. and Weil, G.J. (2009) Transcriptomes and pathways associated with infectivity, survival and immunogenicity in *Brugia malayi* L3. *BMC Genomics* 10, 267.

Lindh, J.G., Connolly, B., McGhie, D.L. and Smith, D.F. (1998) Identification of a developmentally regulated *Trichinella spiralis* protein that inhibits MyoD-specific protein: DNA complexes *in vitro*. *Molecular and Biochemical Parasitology* 92, 163–175.

Lustigman, S., Zhang, J., Liu, J., Oksov, Y. and Hashmi, S. (2004) RNA interference targeting cathepsin L and Z-like cysteine proteases of *Onchocerca volvulus* confirmed their essential function during L3 molting. *Molecular and Biochemical Parasitology* 138, 165–170.

MacDonald, A.S. and Maizels, R.M. (2008) Alarming dendritic cells for Th2 induction. *Journal of Experimental Medicine* 205, 13–17.

Madden, K.B., Urban, J.F. Jr, Ziltener, H.J., Schrader, J.W., Finkelman, F.D. and Katona, I.M. (1991) Antibodies to IL-3 and IL-4 suppress helminth-induced intestinal mastocytosis. *Journal of Immunology* 147, 1387–1391.

Maizels, R.M. (2009) Parasite immunomodulation and polymorphisms of the immune system. *Journal of Biology* 8, 62.

Maizels, R.M. and Yazdanbakhsh, M. (2008) T-cell regulation in helminth parasite infections: implications for inflammatory diseases. *Chemical Immunology and Allergy* 94, 112–123.

Maizels, R.M., Pearce, E.J., Artis, D., Yazdanbakhsh, M. and Wynn, T.A. (2009) Regulation of pathogenesis and immunity in helminth infections. *Journal of Experimental Medicine* 206, 2059–2066.

Massacand, J.C., Stettler, R.C., Meier, R., Humphreys, N.E., Grencis, R.K., Marsland, B.J. and Harris, N.L. (2009) Helminth products bypass the need for TSLP in Th2 immune responses by directly modulating dendritic cell function. *Proceedings of the National Academy of Sciences of the USA* 106, 13968–13973.

McCoy, K.D., Stoel, M., Stettler, R. *et al.* (2008) Polyclonal and specific antibodies mediate protective immunity against enteric helminth infection. *Cell Host & Microbe* 4, 362–373.

McDermott, J.R., Bartram, R.E., Knight, P.A., Miller, H.R., Garrod, D.R. and Grencis, R.K. (2003) Mast cells disrupt epithelial barrier function during enteric nematode infection. *Proceedings of the National Academy of Sciences of the USA* 100, 7761–7766.

McKenzie, G.J., Bancroft, A., Grencis, R.K. and McKenzie, A.N. (1998) A distinct role for interleukin-13 in Th2-cell-mediated immune responses. *Current Biology* 8, 339–342.

McSorley, H. J., Grainger, J.R., Harcus, Y., Murray, J., Nisbet, A.J., Knox, D.P. and Maizels, R.M. (2010) *daf-7*-related TGF-β homologues from *Trichostrongyloid* nematodes show contrasting life-cycle expression patterns. *Parasitology* 137, 159–171.

Mitre, E., Norwood, S. and Nutman, T.B. (2005) Saturation of immunoglobulin E (IgE) binding sites by polyclonal IgE does not explain the protective effect of helminth infections against atopy. *Infection and Immunity* 73, 4106–4111.

Moore, K.W., de Waal Malefyt, R., Coffman, R.L. and O'Garra, A. (2001) Interleukin-10 and the interleukin-10 receptor. *Annual Review of Immunology* 19, 683–765.

Moreno, Y. and Geary, T.G. (2008) Stage- and gender-specific proteomic analysis of *Brugia malayi* excretory–secretory products. *PLoS Neglected Tropical Diseases* 2, e326.

Nair, M.G., Guild, K.J., Du, Y. *et al.* (2008) Goblet cell-derived resistin-like molecule β augments $CD4^+$ T cell production of IFN-γ and infection-induced intestinal inflammation. *Journal of Immunology* 181, 4709–4715.

Nakamura, T., Kamogawa, Y., Bottomly, K. and Flavell, R.A. (1997) Polarization of IL-4- and IFN-gamma-producing CD4+ T cells following activation of naive CD4+ T cells. *Journal of Immunology* 158, 1085–1094.

Neill, D.R., Wong, S.H., Bellosi, A. *et al.* (2010) Nuocytes represent a new innate effector leukocyte that mediates type-2 immunity. *Nature* 464, 1367–1370.

Nutman, T.B. and Kumaraswami, V. (2001) Regulation of the immune response in lymphatic filariasis: perspectives on acute and chronic infection with *Wuchereria bancrofti* in South India. *Parasite Immunology* 23, 389–399.

Owyang, A.M., Zaph, C., Wilson, E.H. *et al.* (2006) Interleukin 25 regulates type 2 cytokine-dependent immunity and limits chronic inflammation in the gastrointestinal tract. *Journal of Experimental Medicine* 203, 843–849.

Perrigoue, J.G., Saenz, S.A., Siracusa, M.C. *et al.* (2009) MHC class II-dependent basophil-$CD4^+$ T cell interactions promote T_H2 cytokine-dependent immunity. *Nature Immunology* 10, 697–705.

Platts-Mills, T.A. and Cooper, P.J. (2010) Differences in asthma between rural and urban communities in South Africa and other developing countries. *Journal of Allergy and Clinical Immunology* 125, 106–107.

Pleass, R.J. and Behnke, J.M. (2009) B-cells get the T-cells but antibodies get the worms. *Trends in Parasitology* 25, 443–446.

Pozio, E. (2007) World distribution of *Trichinella* spp. infections in animals and humans. *Veterinary Parasitology* 149, 3–21.

Pozio, E. and Darwin Murrell, K. (2006) Systematics and epidemiology of *Trichinella*. *Advances in Parasitology* 63, 367–439.

Rajan, T.V. (2005) Natural course of lymphatic filariasis: insights from epidemiology, experimental human infections, and clinical observations. *American Journal of Tropical Medicine and Hygiene* 73, 995–998.

Saenz, S.A., Siracusa, M.C., Perrigoue, J.G. *et al.* (2010) IL25 elicits a multipotent progenitor cell population that promotes T_H2 cytokine responses. *Nature* 464, 1362–1366.

Sarles, M.P. and Taliaferro, W.H. (1936) The local points of defence and the passive transfer of acquired immunity to *Nippostrongylus brasiliensis* in rats. *Journal of Infectious Diseases* 59, 207–212.

Schnoeller, C., Rausch, S., Pillai, S. *et al.* (2008) A helminth immunomodulator reduces allergic and inflammatory responses by induction of IL-10-producing macrophages. *Journal of Immunology* 180, 4265–4272.

Schopf, L.R., Hoffmann, K.F., Cheever, A.W., Urban, J.F. Jr and Wynn, T.A. (2002) IL-10 is critical for host resistance and survival during gastrointestinal helminth infection. *Journal of Immunology* 168, 2383–2392.

Scott, A.L. and Ghedin, E. (2009) The genome of *Brugia malayi* – all worms are not created equal. *Parasitology International* 58, 6–11.

Shevach, E.M. (2009) Mechanisms of $Foxp3^+$ T regulatory cell-mediated suppression. *Immunity* 30, 636–645.

Subramanian, S., Stolk, W.A., Ramaiah, K.D. *et al.* (2004) The dynamics of *Wuchereria bancrofti* infection: a model-based analysis of longitudinal data from Pondicherry, India. *Parasitology* 128, 467–482.

Tachu, B., Pillai, S., Lucius, R. and Pogonka, T. (2008) Essential role of chitinase in the development of the filarial nematode *Acanthocheilonema viteae*. *Infection and Immunity* 76, 221–228.

Tawill, S., Le Goff, L., Ali, F., Blaxter, M. and Allen, J.E. (2004) Both free-living and parasitic nematodes induce a characteristic Th2 response that is dependent on the presence of intact glycans. *Infection and Immunity* 72, 398–407.

Taylor, B.C., Zaph, C., Troy, A.E., Du, Y., Guild, K.J., Comeau, M.R. and Artis, D. (2009) TSLP regulates intestinal immunity and inflammation in mouse models of helminth infection and colitis. *Journal of Experimental Medicine* 206, 655–667.

Taylor, M.D., Harris, A., Babayan, S.A., Bain, O., Culshaw, A., Allen, J.E. and Maizels, R.M. (2007) CTLA-4 and $CD4^+$ $CD25^+$ regulatory T cells inhibit protective immunity to filarial parasites *in vivo*. *Journal of Immunology* 179, 4626–4634.

Travis, M.A., Reizis, B., Melton, A.C. *et al.* (2007) Loss of integrin $\alpha_v\beta 8$ on dendritic cells causes autoimmunity and colitis in mice. *Nature* 449, 361–365.

Urban, J.F. Jr, Madden, K.B., Svetic, A., Cheever, A., Trotta, P.P., Gause, W.C., Katona, I.M. and Finkelman, F.D. (1992) The importance of Th2 cytokines in protective immunity to nematodes. *Immunological Reviews* 127, 205–220.

Vallance, B.A. and Collins, S.M. (1998) The effect of nematode infection upon intestinal smooth muscle function. *Parasite Immunology* 20, 249–253.

Vallance, B.A., Galeazzi, F., Collins, S.M. and Snider, D.P. (1999) CD4 T cells and major histocompatibility complex class II expression influence worm expulsion and increased intestinal muscle contraction during *Trichinella spiralis* infection. *Infection and Immunity* 67, 6090–6097.

Wahid, F.N., Behnke, J.M., Grencis, R.K., Else, K.J.and Ben-Smith, A.W. (1994) Immunological relationships during primary infection with *Heligmosomoides polygyrus*: Th2 cytokines and primary response phenotype. *Parasitology* 108, 461–471.

Wammes, L.J., Hamid, F., Wiria, A.E. *et al.* (2010) Regulatory T cells in human geohelminth infection suppress immune responses to BCG and *Plasmodium falciparum*. *European Journal of Immunology* 40, 437–442.

Whelan, M., Harnett, M.M., Houston, K.M., Patel, V., Harnett, W. and Rigley, K.P. (2000) A filarial nematode-secreted product signals dendritic cells to acquire a phenotype that drives development of Th2 cells. *Journal of Immunology* 164, 6453–6460.

Wilson, M.S. and Maizels, R.M. (2006) Regulatory T cells induced by parasites and the modulation of allergic responses. *Chemical Immunology and Allergy* 90, 176–195.

Wilson, M.S., Taylor, M.D., Balic, A., Finney, C.A., Lamb, J.R. and Maizels, R.M. (2005) Suppression of allergic airway inflammation by helminth-induced regulatory T cells. *Journal of Experimental Medicine* 202, 1199–1212.

Wojciechowski, W., Harris, D.P., Sprague, F. *et al.* (2009) Cytokine-producing effector B cells regulate type 2 immunity to *H. polygyrus*. *Immunity* 30, 421–433.

Wu, Z., Nagano, I., Boonmars, T. and Takahashi, Y. (2005) A spectrum of functional genes mobilized after *Trichinella spiralis* infection in skeletal muscle. *Parasitology* 130, 561–573.

Wu, Z., Sofronic-Milosavljevic, L., Nagano, I. and Takahashi, Y. (2008) *Trichinella spiralis*: nurse cell formation with emphasis on analogy to muscle cell repair. *Parasites & Vectors* 1, 27.

Zhao, A., McDermott, J., Urban, J.F. Jr, Gause, W., Madden, K.B., Yeung, K.A., Morris, S.C., Finkelman, F.D. and Shea-Donohue, T. (2003) Dependence of IL-4, IL-13, and nematode-induced alterations in murine small intestinal smooth muscle contractility on Stat6 and enteric nerves. *Journal of Immunology* 171, 948–954.

4 The Genome of *Pristionchus pacificus* and Implications for Survival Attributes

MATTHIAS HERRMANN AND RALF J. SOMMER

Max Planck Institute for Developmental Biology, Department for Evolutionary Biology, Tübingen, Germany

4.1 Introduction

Caenorhabditis elegans is one of the best-studied model organisms in modern biology. The detailed understanding of many aspects of its biology, including post-embryonic development, behaviour, dauer formation and ageing, provides a unique platform for comparative studies. In the early 1990s, a search for specifically suited nematodes to be compared to *C. elegans* was initiated, and *Pristionchus pacificus* was selected as one such comparative system, basically for two reasons. First, there are important differences in post-embryonic development, particularly vulva development, between *P. pacificus* and *C. elegans* (Sommer and Sternberg, 1996; Hong and Sommer, 2006a). Second, several techniques originally developed for *C. elegans* were successfully transferred to *P. pacificus* (Sommer *et al.*, 1996; Hong and Sommer, 2006a).

Pristionchus pacificus is a diplogastrid nematode that has been established as a model system in evolutionary developmental biology (evo-devo). Initially, *P. pacificus* was used as a convenient nematode in which to compare various developmental processes to *C. elegans* because *P. pacificus* and *C. elegans* share many technical features: a short generation time, simple laboratory culture, self-fertilization as a mode of reproduction and males arising spontaneously in laboratory cultures. These features simplify a number of forward and reverse genetic tools, as well as DNA-mediated transformation, all of which are available in *P. pacificus* (Hong and Sommer, 2006a; Schlager *et al.*, 2009). The life cycle of *P. pacificus* is similar to that of *C. elegans*: the egg develops through four larval stages before reaching the adult stage. The first-stage larva (L1) is retained in the egg so that the second-stage larva (L2) is the stage that hatches. When food supply is depleted, the *P. pacificus* L2 develops into a dauer larva instead of a third-stage larva (L3). These dauer larvae retain their cuticle from the L2 stage and are resistant to a number of abiotic factors.

Although originally established as a satellite system in evolutionary developmental biology, *P. pacificus* is currently becoming an important model in evolutionary ecology and for studying the evolution of life history traits. Several *Pristionchus* spp., including *P. pacificus*, are often found on beetles, particularly scarab beetles (Fig. 4.1) (Herrmann *et al.*, 2006a,b, 2007). In this well-defined ecological niche, *P. pacificus* has a necromenic relationship with scarab beetles: dauer larvae stay associated with the beetles and remain there until the death of their hosts, after which development is resumed and the nematodes feed on microbes on the beetle carcasses (Herrmann *et al.*, 2006a). *Pristionchus* nematodes can also be found in soil, although the proportions of nematodes found in soil and those found on beetles are not yet known. There is currently no evidence that would support a parasitic or pathogenic relationship of *Pristionchus* with beetles. However, given that not all life stages of scarab beetles have yet been studied comprehensively, one cannot completely rule out such possibilities in particular beetle life stages for certain *Pristionchus* spp.

Fig. 4.1. *Pristionchus* nematodes and their beetle associations. (a–c) Examples of beetles hosting specific *Pristionchus* spp. (a) The oriental beetle *Exomala orientalis* hosts *P. pacificus* in Japan and the USA. (b) *Geotrupes* sp. from Europe hosts *P. entomophagus*. (c) The Colorado potato beetle hosts *P. uniformis* in Europe and the USA. (d) *Melolontha* sp. from Europe caught and brought to the laboratory. (e) Beetles cut and put on plate. (f) After a certain time period (usually within days), reproducing nematodes appear on the plate, feeding on the developing microbes on the beetle carcass.

Pristionchus pacificus was one of several nematodes that had its genome sequenced in 2007 and 2008 (Dieterich *et al.*, 2008). Here, we review the implications of the *P. pacificus* genome for survival attributes. We first provide an overview of the current understanding of the association of *Pristionchus* nematodes and other members of the Diplogastridae with insects and beetles in particular. We then discuss findings of the *P. pacificus* genome in the light of the life history of this nematode.

4.2 *Pristionchus*–Beetle Interactions and Biogeography

Implications of the *P. pacificus* genome for survival attributes can best be understood in the context of the life history of *P. pacificus* and other members of the *Pristionchus* genus and the diplogasterid family. Therefore, the paragraphs below summarize current knowledge on the biodiversity, distribution and phylogeny of *Pristionchus* and the Diplogastridae.

4.2.1 Diplogastridae–insect interactions

Sudhaus and Fürst von Lieven (2003) list 28 genera in the family Diplogastridae; 14 of these genera have a well-established association with insects, and nine of those were found repeatedly on beetles. While several authors studied various aspects of the interactions of diplogastrid nematodes with insects, it was only the recent work of Sudhaus and Fürst von Lieven that started to provide a comprehensive overview of these taxa (Fürst von Lieven and Sudhaus, 2000; Sudhaus and Fürst von Lieven, 2003). A first molecular phylogeny of four genera of the Diplogastridae has been provided by Kiontke *et al.* (2007) as part of their analysis of the Rhabditidae. More recently, a molecular phylogeny of 14 insect-associated diplogastrid genera based on 12 genes provided a comprehensive phylogenetic framework for studies with *Pristionchus* (Mayer *et al.*, 2009). With the exception of the status of the genus *Koerneria*, the molecular phylogenies strongly agree with morphological trees.

4.2.2 *Pristionchus*–beetle interactions

The catalogue of Sudhaus and Fürst von Lieven (2003) lists 27 valid species in the genus *Pristionchus*, most of which have been described in the 19th century. Work initiated in 2004 searched for potential interactions of *Pristionchus* with scarab beetles. These studies indeed revealed that *Pristionchus* nematodes are often found in association with scarab beetles (Table 4.1). For example, in continental Europe *Pristionchus maupasi* is often found on the cockchafer *Melolontha melolontha*, and *Pristionchus entomophagus* associates with dung beetles of the genus *Geotrupes* (Herrmann *et al.*, 2006a). Systematic studies in Europe (Herrmann *et al.*, 2006a), North America (Herrmann *et al.*, 2006b)

Table 4.1. Schematic summary of the specificity of *Pristionchus*–beetle associations.

	Melolontha	*Geotrupes*	*Leptinotarsa*	*Exomala*	*Phyllophaga*	Region	Reference[a]
P. maupasi	×					Europe	2006a
P. lheritieri		×				Europe	2006a
P. entomophagus	×	××			×	Europe	2006a
P. uniformis			×			Europe	2006a
P. americanus					×	North America	2006b
P. marianneae					×	North America	2006b
P. aerivorus					×	North America	2006b
P. pseudaerivorus					×	North America	2006b
P. pacificus				×		Asia	2007

××: very high infestation rates (>20%); ×: infestation rates <20%; [a] the years are to references by Herrmann *et al.*

and Japan (Herrmann *et al.*, 2007), complemented by sporadic samplings in South Africa and South America, yielded a total of 25 *Pristionchus* spp., all of which are in the Tübingen stock collection as living and frozen cultures. While many of these species could be identified as species known from the literature, others are novel species that have to be described in a revision of the genus (Herrmann, unpublished). A molecular phylogeny of the available material provides a framework for studies with the genetic model system *P. pacificus* (Mayer *et al.*, 2007).

4.2.3 *Pristionchus pacificus* is a cosmopolitan species

While most *Pristionchus* spp. were only obtained in certain geographic areas, *P. pacificus* is one of a few cosmopolitan species (Zauner *et al.*, 2007). The first isolates of *P. pacificus* were from Pasadena (California, USA; Sommer *et al.*, 1996) and Port Angeles (Washington, USA; Sommer *et al.*, 1996). However, initial samplings in Europe and North America did not identify *P. pacificus* as a major component of the scarab beetle fauna (Herrmann *et al.*, 2006a,b). As we know now, this is at least in part due to a sampling bias. Studies in Japan identified the oriental beetle, *Exomala orientalis*, as one of the beetle hosts of *P. pacificus* and provided an entry point for the identification of *P. pacificus* in North America, given that *E. orientalis* had invaded the continent in the 1920s (Herrmann *et al.*, 2007). *Pristionchus pacificus* has now been found in Japan, China, the USA, Bolivia, South Africa and occasionally in Europe, Madagascar, India and Bali. In contrast to other *Pristionchus* spp., *P. pacificus* shows a wider host range, but it is currently unclear if this is the cause or a consequence of the cosmopolitan distribution.

More recent work concentrated on island biogeography and identified *P. pacificus* as a frequent *Pristionchus* sp. on the island Réunion in the Indian Ocean (Herrmann *et al.*, 2010). Interestingly, *P. pacificus* is associated with several scarab beetle species on that island and the *P. pacificus* Réunion strains show a haplotype diversity that represents a substantial amount of the haplotype diversity known from around the world.

4.3 Behaviour and Chemoattraction

Pristionchus must rely on a number of long- and short-range chemical cues to locate potential beetle hosts and to ensure host specificity. Olfaction experiments using insect sex pheromones and insect and plant semiochemicals revealed that *Pristionchus* spp. display unique chemoattraction profiles (Hong and Sommer, 2006b; Herrmann *et al.*, 2007; Hong *et al.*, 2008). This is in contrast to species of the genus *Caenorhabditis*, which have highly overlapping chemoattraction profiles. Further studies suggest that the chemoattraction of insect sex pheromones and sex attractants are involved in host recognition. *Pristionchus pacificus* is strongly attracted to the *E. orientalis* sex pheromone ((Z)-7-tetradecen-2-one) in a species-specific manner

(Herrmann *et al.*, 2007). Similarly, *P. maupasi* is strongly attracted to phenol, a sex attractant of cockchafers, whereas no other tested *Pristionchus* spp. are attracted to phenol at all (Hong *et al.*, 2008). Current studies following these observations are underway to identify the molecular mechanisms involved in these chemoattractive behaviours.

4.4 *Pristionchus*–Bacterial Interactions

The well-defined ecological niche of *Pristionchus* nematodes made the analysis of beetle and *Pristionchus*-associated bacteria straightforward. Rae *et al.* (2008) analysed bacteria from the cuticle and intestine of *P. pacificus*, *P. entomophagus* and *P. maupasi* individuals that had emerged from their associated beetles. 16S sequence analysis revealed that *Pristionchus* individuals harbour more than 40 different bacterial species from many different eubacterial taxa. More than 20 bacteria isolated from cockchafers, dung beetles and oriental beetles were established as laboratory cultures and were exposed to *P. pacificus*. A continuum of interactions from dissemination of bacteria to reduction in brood size and nematode mortality were observed (Rae *et al.*, 2008). Current studies concentrate on the interactions between *P. pacificus* and various strains of *Bacillus* (R. Rae and R.J. Sommer, unpublished observations).

Pristionchus nematodes cannot only feed on bacteria. They are also able use fungi and even other nematodes as food sources. Thus, *P. pacificus* is an omnivorous nematode. The different feeding strategies in *P. pacificus* and *C. elegans* are reflected in morphological differences between these species. While *C. elegans* has a grinder in the terminal part of the pharynx, which disrupts all bacteria under laboratory conditions, no such structure is known in *Pristionchus* (Rae *et al.*, 2008). Instead, *Pristionchus* nematodes have denticles in the buccal cavity that are used to disrupt the hyphae of fungi and/or the cuticle of other nematodes (Fürst von Lieven and Sudhaus, 2000).

4.5 From Genetics to Genomics

Pristionchus pacificus has been established as a genetic system based on its self-fertilizing hermaphroditic propagation. Hermaphrodites are modified females that produce sperm during a short period of larval development to become mature adult females. As long as no real males are around, hermaphrodites will use the self-sperm to fertilize their oocytes. Males can easily be obtained and maintained under laboratory conditions and are used in genetic experimentation and classical genetic studies (Kenning *et al.*, 2004).

Genetic approaches in *P. pacificus* have been complemented by a large-scale genomic initiative. A genetic linkage map was built using the PS312 wild-type strain from California and the polymorphic PS1843 reference strain (Srinivasan *et al.*, 2002). This genetic linkage map is integrated with the physical map (Srinivasan *et al.*, 2003) and, by now, the *P. pacificus* genome

(Dieterich *et al.*, 2008). The genome of *P. pacificus* PS312 has been sequenced by whole-genome shotgun (WGS) with approximately 9× coverage (Dieterich *et al.*, 2008; http://www.pristionchus.org). Below, we will summarize three particular features of the *P. pacificus* genome and its life cycle that are of importance in the context of nematode survival strategies.

4.5.1 Expansion of detoxification machinery

The size of the *P. pacificus* genome is around 169 Mb and is thus substantially larger than the *C. elegans* genome. Not surprisingly, therefore, the *P. pacificus* genome contains both gene predictions that are absent from *C. elegans* and expansions in gene counts relative to *C. elegans*. Both of these features might represent hallmarks of adaptations to the different ecological niches of these organisms. One of the most important questions about the ecology of this nematode is how *P. pacificus* deals with low oxygen concentrations and the toxicity of host enzymes in the context of the decaying beetle carcass. The *P. pacificus* genome contains an expansion of many gene classes that might help to provide tolerance against these challenges (Fig. 4.2). The genome encodes proteins similar to monooxygenases, dioxygenases and hydroxylases that could be involved in the rapid elimination of insect polyphenols. For example, there are 198 cytochrome P450 domains in the *P. pacificus* genome, while *C. elegans* contains only 67. Other important groups of detoxification

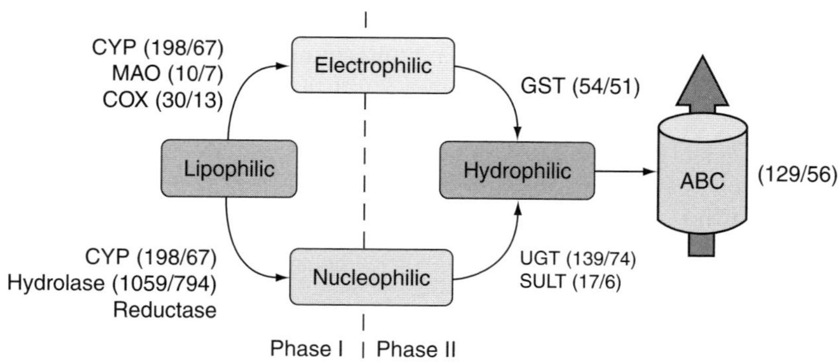

Fig. 4.2. Schematic representation of the metabolism of xenobiotics. Lipophilic xenobiotics are modified to form electrophilic or nucleophilic substances (phase I) by the activity of several enzyme groups, such as cytochrome P450 monooxygenases (CYP), monoamine oxidases (MAO), cyclooxygenases (COX), hydrolases and reductases, respectively. In phase II, hydrophilic compounds are formed by the activity of transferases (glutathione *S*-transferase, GST; UDP-glucuronosyltransferases, UGT, and sulfotransferase, SULT). ABC transporters export these modified compounds. *Pristionchus pacificus* shows an expansion of most of these enzyme groups when compared to *Caenorhabditis elegans*. Species-specific copy numbers are given in brackets (*P. pacificus/C. elegans*) next to each enzyme class. Redrawn with permission from Dieterich *et al.* (2008).

enzymes, such as the monooxygenases and sulfotransferases are also vastly expanded in *P. pacificus*, with 30 and 17 copies, in contrast to 13 and 6 copies in *C. elegans*, respectively. ABC transporters are known to play a major role in shuffling of detoxified compounds through membranes (Borst and Elferink, 2002). Interestingly, there is more than a twofold increase of ABC transporters in the *P. pacificus* genome when compared with *C. elegans*, with 129 versus 56 gene predictions, respectively.

In addition, *P. pacificus*-specific gene predictions encompass 39 EC classifications, many of which might play a role in detoxification. For example, there are several gene predictions encoding sterol esterases (EC3.1.1.13), thromboxane-A synthases (EC5.3.99.5) and polyamine oxidases (EC 1.5.3.11). While many of the gene predictions in these functional groups have been confirmed by cDNA sequencing, the exact function awaits experimental analysis.

One interesting aspect in this context is the analogy of these expansions to *Photorhabdus luminescens*. Other organisms that live in close association with, or are parasites of, insects also face the problem of host toxicity. Examples are entomopathogenic nematodes of the genus *Heterorhabditis*, which use symbiotic bacteria for host killing (see Grewal *et al.*, Chapter 7, this volume). The genome of the entomopathogenic bacterium *P. luminescens* that is a symbiont of *Heterorhabditis bacteriophora* has been sequenced and shows striking similarities to the *P. pacificus* genome (Duchaud *et al.*, 2003). The *P. luminescens* genome encodes a large number of toxins, haemolysins and proteases. Also, it encodes many triacylglycerol lipases, which are secreted enzymes that have been suggested to play a role in the conversion of the insect cadaver. Interestingly, the *P. pacificus* genome encodes 12 triacylglycerol lipases, while *C. elegans* encodes only one. One could speculate that this expansion of lipases co-evolved with the necromenic life style, in which the nematode feeds on a variety of microbes under the presence of strong competitors. Although these similarities between *P. luminescens* and *P. pacificus* are not homologous, they provide an example of how the adaptation to similar ecological niches can result in similar but analogous genome patterns.

4.5.2 Cellulases and horizontal gene transfer

Another surprising feature of the *P. pacificus* genome is the unexpected presence of glycosyl hydrolases. At the sequence level, these genes are most similar to cellulases of the GH5 family (Dieterich *et al.*, 2008). Consistent with these findings, cellulase activity has been observed in the supernatant of *P. pacificus* mixed stage cultures. While the functional analysis of these *P. pacificus* genes is still under investigation (Schuster and Sommer, unpublished results), this case represents the first example of glycosyl hydrolases/cellulases found in a non-plant-parasitic nematode.

Further bioinformatic analysis of the *P. pacificus* cellulase genes allows three major conclusions. First, *P. pacificus* genes are most closely related to GH5 family members of *Dictyostelium* and other microbes, suggesting that these genes were acquired by horizontal gene transfer (HGT) (Dieterich and Sommer, 2009). Second, in a phylogenetic tree of GH5 genes, all seven *P. pacificus* genes form a single clade. This suggests a single HGT event followed by gene duplication, rather than independent HGT into *P. pacificus*. Third, the cellulase genes of plant-parasitic nematodes, such as *Meloidogyne incognita* and *Meloidogyne hapla*, are more diverse in sequence than those from *Pristionchus* and the microbes mentioned above, indicating that *Meloidogyne* and *Pristionchus* have acquired these genes independently by HGT. It remains currently unknown how stable such HGT events are at the phylogenetic level. Therefore, future studies can reveal if different *Pristionchus* spp. share cellulase genes. Can they even be found in other diplogastrid nematodes? Next-generation sequencing might provide a powerful tool to address such questions.

4.5.3 The evolution of parasitism and the role of 'pre-adaptations'

The examples from cellulase and detoxification genes are of importance for a more general understanding of genome evolution and the acquisition of novel characters. Both the detoxification and cellulase genes in the *P. pacificus* genome are part of the versatile and complex life cycle of this organism and they are important adaptations for survival in the beetle ecosystem. In addition, however, they might also represent examples for molecular pre-adaptations, which are of theoretical importance for the understanding of the evolution of complex traits. The evolution of parasitism is a typical example for the evolution of a complex trait that requires multiple independent adaptations. Already Darwin noticed in the *Origin of Species* in 1859 that the evolution of complex traits, which require multiple independent mutations, represents a major challenge for evolutionary theory. How do independent characters evolve which have only minor or even no selective advantages on their own? This problem applies to parasitism, with the many physiological and morphological specialities required and the evolution of very specific life cycles. All of these features require genomic signatures. Thus, changes in physiology, morphology and life history traits must involve many independent mutations in the genome.

The example of the *P. pacificus* genome argues that features that are commonly regarded as typically parasitic can evolve under free-living, necromenic conditions. One might speculate, therefore, that necromenic associations represent a potential 'precursor stage' for real parasitism. This idea is consistent with the pre-adaptation hypothesis originally formulated by Osche (1956). Most recently, Poulin (2007) used the same hypothesis in his monograph on the evolutionary ecology of parasites. Pre-adaptations are adaptations to the current environment of the organism. However, such adaptations might be co-opted to a new function and might thereby facilitate the transition to a new environment (Fig. 4.3) (Dieterich and Sommer, 2009). One aspect that has been consistently considered as pre-adaptations towards parasitism is the known type of phoretic and necromenic associations of nematodes with insects and other invertebrates. For this, the ability to form dauer larvae is of

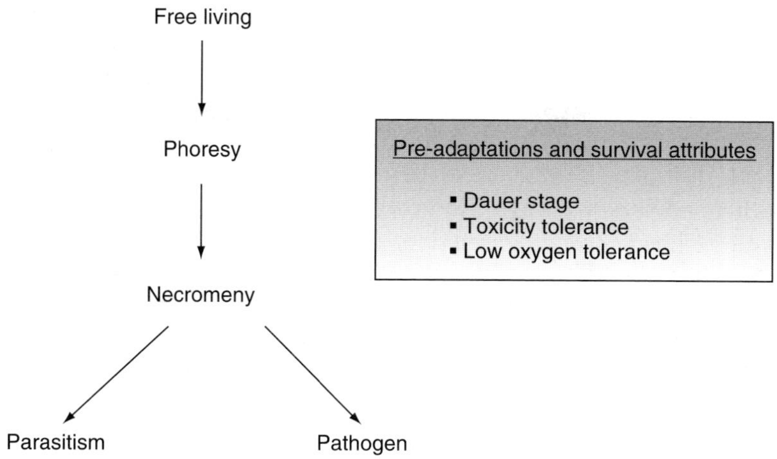

Fig. 4.3. Evolutionary trends towards parasitism in nematodes. Non-parasitic nematode species are free living in marine, freshwater and terrestrial habitats, but some species show typical associations with arthropods, other invertebrates and even vertebrates. Phoretic species associate with a host for transportation in an often unspecific manner. Necromenic nematode species associate with a host, wait for its death and then feed on the developing microbes on the host's carcass. Several studies argued that phoretic and/or necromenic associations provide important pre-adaptations for the evolution of parasitism. Pre-adaptations are adaptations to the current environment of the organism and its lifestyle. In the future, such adaptations might be co-opted to a new function and facilitate the transition to a new environment. Therefore, pre-adaptations might be helpful for an organism in the acquisition of a new niche, such as a parasitic lifestyle.

crucial importance. The recent inroads towards a genetic understanding of dauer formation in *P. pacificus* and its potential application in parasitic nematodes will be discussed in the next section.

4.6 The Analysis of *Pristionchus pacificus* Dauer Regulation Provides Inroads for the Study of Parasitism

Morphological similarities between dauer larvae of free-living nematodes and infective juveniles of parasites are striking. It has long been argued, therefore, that dauer larvae represent an important pre-adaptation towards parasitism (Fig. 4.3) (Osche, 1956; Poulin, 2007; Dieterich and Sommer, 2009). Building on the detailed knowledge of *C. elegans* dauer formation, *P. pacificus*, with its genetic and genomic toolkit, might represent a unique system to study the genetic regulation of dauer formation in a comparative context. Similarities and differences in the regulation of dauer formation between *C. elegans* and *P. pacificus* might then provide important starting points for the investigation of shared characters between dauer larvae of free-living nematodes and infective larvae of parasites.

One recent example for this is the identification of dafachronic acid (DA) as a conserved endocrine signalling system in nematode dauer and infective larvae formation (Ogawa *et al.*, 2009). In *P. pacificus*, dauer stages are formed in the absence of food, a high density of conspecifics and high temperatures, similar to *C. elegans*. Also, dauer formation requires a pheromone in *P. pacificus*. However, this pheromone is chemically distinct from the known pheromones in *C. elegans*. Two classes of genetic mutants can be distinguished in *P. pacificus*, dauer-formation-defective (*daf-d*) mutants and dauer-formation-constitutive mutants (*daf-c*). While *daf-d* mutants never form dauer stages, *daf-c* mutants enter the dauer stage even in the presence of food. Working on the paradigm of *C. elegans* dauer regulation, Ogawa *et al.* (2009) identified the nuclear hormone receptor *daf-12* as a conserved regulator of dauer formation. Loss-of-function mutations in *daf-12* are Daf-d, indicating that dauer entry requires DAF-12/nuclear hormone receptor (NHR) function. In *C. elegans*, two cholesterol derivatives, Δ4-DA and Δ7-DA, act as hormones that interact with DAF-12, making this NHR a molecular switch. In conjugation with DA, DAF-12 inhibits dauer formation; in the absence of DA, DAF-12 will induce dauer entry on its own (Motola *et al.*, 2006).

Using a group of *daf-c* mutants in *P. pacificus*, it was shown that Δ7-DA also functions in *P. pacificus*, suggesting that DA/DAF-12 represents a conserved endocrine signalling module for the regulation of dauer formation in *P. pacificus* and *C. elegans* (Ogawa *et al.*, 2009). In a next step, it was shown that Δ7-DA is also able to interfere with the generation of infective larvae in a nematode parasite. *Strongyloides papillosus* is a parasite of sheep that can be cultured in the laboratory in rabbits and has a direct (homogonic) and indirect (heterogonic) life cycle (Streit, 2008). Δ7-DA is able to block infective larva formation in both the homogonic and heterogonic cycle, providing the first evidence for a conserved mechanism in the regulation of dauer and infective larvae formation in nematodes (Ogawa *et al.*, 2009). This example highlights the importance of evolutionary comparisons and the use of genetic tools in such studies.

4.7 Conclusions and Future Directions

Although still in its beginnings, the genetic and genomic analysis of *P. pacificus* and its life history traits is a useful addition to the comprehensive study of nematode biology. By following three lines of investigation, *P. pacificus* can provide novel insights of general importance. First, the comparison between *P. pacificus* and *C. elegans* based on their unique forward genetic technologies can reveal similarities and differences in a variety of molecular processes. For example, this approach can highlight how developmental processes evolve within a conserved animal bauplan to generate developmental and morphological novelty (Sommer, 2009). Second, the well-described association with scarab beetles and the necromenic lifestyle provide an entry point for the genetic analysis of species interactions with the environment. While *Pristionchus* nematodes are not unique in their necromenic lifestyle, the ease with which these nematodes can be cultured in the laboratory and be experimentally manipulated represents unique methodological characteristics, similar to those known from *C. elegans*. Finally, *P. pacificus* might help in transferring biological insight from the primary model system *C. elegans* towards applied nematology. It has always been assumed that knowledge obtained in *C. elegans* can be useful for parasitologists and other applied scientists working with nematodes. However, given the rapid turnover of molecular processes, it is still a challenge to identify those molecular modules that can serve as successful paradigms for parasites and pathogens. The comparison of two easily accessible species, such as *P. pacificus* and *C. elegans*, can provide a framework for the identification of conserved molecular modules, which can then be tested in other organisms. This principle has been successfully applied in the analysis of the regulation of dauer and infective larvae formation in nematodes and provided the first evidence for conserved aspects in the molecular regulation of these two processes (Ogawa *et al.*, 2009).

4.8 Acknowledgements

Both authors want to express the greatest gratitude to the many entomologists and nematologists that have continuously supported our research by providing samples from all over the world. Research described in this study was and is continuously supported by the Max Planck Society.

4.9 References

Borst, P. and Elferink, R.O. (2002) Mammalian ABC transporters in health and disease. *Annual Review Biochemistry* 71, 573–592.

Dieterich, C. and Sommer, R.J. (2009) How to become a parasite – lessons from the genomes of nematodes. *Trends in Genetics* 25, 203–209.

Dieterich, C., Clifton, S.W., Schuster, L. *et al.* (2008) The *Pristionchus pacificus* genome provides a unique perspective on nematode lifestyle and parasitism. *Nature Genetics* 40, 1193–1198.

Duchaud, E., Rusniok, C., Frangeul, L. *et al.* (2003) The genome sequence of the entomopathogenic bacterium *Photorhabdus luminescens*. *Nature Biotechnology* 21, 1307–1313.

Fürst von Lieven, A. and Sudhaus, W. (2000) Comparative and functional morphology of the buccal cavity of Diplogastrina (Nematoda) and a first outline of the phylogeny of this taxon. *Journal of Zoological Systematics and Evolutionary Research* 38, 37–63.

Herrmann, M., Mayer, W.E. and Sommer, R.J. (2006a) Nematodes of the genus *Pristionchus* are closely associated with scarab beetles and the Colorado potato beetle in Western Europe. *Zoology* 109, 96–108.

Herrmann, M., Mayer, W.E. and Sommer, R.J. (2006b) Sex, bugs and Haldanes rule: the nematode genus *Pristionchus* in the United States. *Frontiers in Zoology* 3, 14.

Herrmann, M., Mayer, W.E., Hong, R.L., Kienle, S., Minasaki, R. and Sommer, R.J. (2007) The nematode *Pristionchus pacificus* (Nematoda: Diplogastridae) is associated with the Oriental beetle *Exomala orientalis* (Coleoptera: Scarabaeidae) in Japan. *Zoological Science* 24, 883–889.

Herrmann, M., Kienle, S., Rochat, J., Mayer, W.E. and Sommer, R.J. (2010) Haplotype diversity of the nematode *Pristionchus pacificus* on Réunion in the Indian Ocean suggests multiple independent invasions. *Biological Journal of the Linnean Society* 100, 170–179.

Hong, R.L. and Sommer, R.J. (2006a) *Pristionchus pacificus*: a well-rounded nematode. *BioEssays* 28, 651–659.

Hong, R.L. and Sommer, R.J. (2006b) Chemoattraction in *Pristionchus* nematodes and implication for insect host recognition. *Current Biology* 16, 2359–2365.

Hong, R.L., Svatos, A., Herrmann, M. and Sommer, R.J. (2008) The species-specific recognition of beetle cues by *Pristionchus maupasi*. *Evolution & Development* 10, 273–279.

Kenning, C., Kipping, I. and Sommer, R. J. (2004) Mutations with altered gross-morphology in the nematode *Pristionchus pacificus*. *Genesis* 40, 176–183.

Kiontke, K., Barrière, A., Kolotuev, I., Podbilewicz B., Sommer, R.J., Fitch, D. and Félix, M.-A. (2007) Evolution of development in the nematode vulva system: trends, stasis and drift. *Current Biology* 17, 1925–1937.

Mayer, W.E., Herrmann, M. and Sommer, R.J. (2007) Phylogeny of the nematode genus *Pristionchus* and implications for biodiversity, biogeography and the evolution of hermaphroditism. *BMC Evolutionary Biology* 7, 104.

Mayer, W.E., Herrmann, M. and Sommer, R.J. (2009) Molecular phylogeny of beetle associated diplogastrid nematodes suggests host switching rather than nematode–beetle coevolution. *BMC Evolutionary Biology* 9, 212.

Motola, D.L., Cummins, C.L., Rottiers, V. *et al.* (2006) Identification of ligands for DAF-12 that govern dauer formation and reproduction in *C. elegans*. *Cell* 124, 1209–1223.

Ogawa, A., Streit, A., Antebi, A. and Sommer, R.J. (2009) A conserved endocrine mechanism controls the formation of dauer and infective larvae in nematodes. *Current Biology* 19, 67–71.

Osche, G. (1956) Die Präadaptation freilebender Nematoden an den Parasitismus. *Zoologischer Anzeiger* (Supplement) 19, 391–396.

Poulin, R. (2007) *Evolutionary Ecology of Parasites*. Princeton University Press, Princeton, New Jersey.

Rae, R., Riebesell, M., Dinkelacker, I., Wang, Q., Herrmann, M., Weller, A.M., Dieterich, C. and Sommer, R.J. (2008) Isolation of naturally associated bacteria of necromenic *Pristionchus* nematodes and fitness consequences. *Journal of Experimental Biology* 211, 1927–1936.

Schlager, B., Wang, X., Braach, G. and Sommer, R.J. (2009) Molecular cloning of a dominant Roller mutant and establishment of DNA-mediated transformation in the nematode model *Pristionchus pacificus*. *Genesis* 47, 300–304.

Sommer, R.J. (2009) The future of evo-devo: model systems and evolutionary theory. *Nature Reviews Genetics* 10, 416–422.

Sommer, R.J. and Sternberg, P.W. (1996) Apoptosis limits the size of the vulval equivalence group in *Pristionchus pacificus*: a genetic analysis. *Current Biology* 6, 52–59.

Sommer, R.J., Carta, L.K., Kim, S.-Y. and Sternberg, P.W. (1996) Morphological, genetic and molecular description of *Pristionchus pacificus* sp. n. (Nematoda, Diplogastridae). *Fundamental and Applied Nematology* 19, 511–521.

Srinivasan, J., Sinz, W., Lanz, C. *et al.* (2002) A bacterial artificial chromosome-based genetic linkage map of the nematode *Pristionchus pacificus*. *Genetics* 162, 129–134.

Srinivasan, J., Sinz, W., Jesse, T., Wiggers-Perebolte, L., Jansen, K., Buntjer, J., van der Meulen, M. and Sommer, R.J. (2003) An integrated physical and genetic map of the nematode *Pristionchus pacificus*. *Molecular Genetics and Genomics* 269, 715–722.

Streit, A. (2008) Reproduction in *Strongyloides* (Nematoda): a life between sex and parthenogenesis. *Parasitology* 135, 285–294.

Sudhaus, W. and Fürst von Lieven, A. (2003) A phylogenetic classification and catalogue of the Diplogastridae (Secernentea, Nematoda). *Journal of Nematode Morphology and Systematics* 6, 43–90.

Zauner, H., Mayer, W.E., Herrmann, M., Weller, A., Erwig, M. and Sommer, R.J. (2007) Distinct patterns of genetic variation in *Pristionchus pacificus* and *Caenorhabditis elegans*, two partially selfing nematodes with cosmopolitan distribution. *Molecular Ecology* 16, 1267–1280.

5 The Dauer Phenomenon

Warwick Grant[1] and Mark Viney[2]

[1]*Genetics Department, La Trobe University, Bundoora, Victoria, Australia;*
[2]*School of Biological Sciences, University of Bristol, Bristol, UK*

5.1 Introduction

This chapter aims to discuss the dauer phenomenon in the context of nematode life history strategies and evolution, with a particular emphasis on the role that the dauer larva may have played as a pre-adaptation that facilitated the evolution of parasitism several times within the phylum.

The main feature of the dauer phenomenon is encapsulated in the name: 'dauer' is taken from the German word 'dauern', which may be translated as 'to endure', particularly in the context of nematode biology and survival. Thus, the dauer stage is a specialized, developmentally arrested and non-ageing morph, which is formed in response to specific environmental cues that signal adverse conditions and whose adaptive or evolutionary value to the nematode is thus survival under those adverse conditions. Once the adverse conditions have improved, i.e. in response to a further set of specific environmental signals, individuals emerge from dauer and resume their development to reproductive adulthood. From an ecological perspective, the dauer larva is an adaptation of nematode life history strategy to an ecology that is characterized by 'boom and bust': free-living nematodes generally inhabit patchy environments in which food, in particular, is ephemeral and highly localized (e.g. a localized bacterial bloom surrounding a dead soil invertebrate or rotten vegetation). Consistent with this notion is the observation that for at least two free-living species, *Caenorhabditis elegans* (Barrière and Félix, 2005a, 2006) and *Pristionchus pacificus* (Herrmann *et al.*, 2006), it is dauers that are almost invariably isolated from the environment. Free-living nematodes have, therefore, evolved a mechanism that permits them to survive relatively long periods of adversity between short periods of plenty. From this evolutionary perspective, the key features of dauer on which attention should be focused are:

1. The decision to make dauer. What are the environmental signals, how are they sensed, how are they transduced, and what is their relationship to the nematodes' ecology?
2. The biology of the dauer stage. What are the structural, biochemical and behavioural characteristics of dauer that promote survival under adverse conditions?
3. How the evolution of the dauer stage may have pre-adapted nematodes for the independent evolution of parasitism (of plants and animals) in several lineages.

Much of **2** above, i.e. dauer biology, and particularly molecular biology, is derived directly or indirectly from *C. elegans*, a free-living soil nematode with a global distribution that has been studied intensively as a model for developmental and cell biology for the past 40 years (Edgar and Wood, 1977). The dauer larval stage of *C. elegans* has been studied particularly closely since the early 1990s, with the observation that signal transduction pathways and patterns of gene expression that regulate dauer development in this species can also profoundly affect lifespan (Kenyon *et al.*, 1993). There are many recent, comprehensive reviews of this molecular biology, and the reader is referred to a sample of these as an introduction to this large and expanding field (Hu, 2007; Fielenbach and Antebi, 2008). We will refer only to those aspects of the molecular biology that are relevant to a more general consideration of dauer biology as a component of nematode survival and life history.

It should be stressed that extrapolation of the molecular details of dauer biology from *C. elegans* to other nematode species (Mitreva *et al.*, 2004; Elling *et al.*, 2009), or even to other wild isolates of *C. elegans*, must be done with caution (Harvey *et al.*, 2009) and that any hypothesis based on such an extrapolation must be tested. More recently, *C. elegans* and related rhabditid free-living nematodes have attracted interest as models for studies in molecular evolution (Félix and Barrière, 2005; Fitch, 2005), ecology (Kiontke and Sudhaus, 2006) and population and ecological genetics (Barrière and Félix, 2005b; Cutter *et al.*, 2006; Dolgin *et al.*, 2008). However, it should be remembered that much of the molecular biology discussed below is derived from a single species (*C. elegans*), or its near relatives, but that the Phylum Nematoda is very large and diverse, so there may not be such a thing as a 'typical' nematode.

5.2 Initiating Dauer Development

5.2.1 Environmental signals

Free-living nematodes inhabit a wide variety of ecosystems, from anaerobic marine sediments to desert soils and even the thin, frozen, desiccated soils of Antarctica. Given this variety of environments, it is risky to attempt to make general statements about nematode ecology, but one feature that many of these environments have in common is that the nematodes that inhabit them undergo cycles of 'boom and bust'. The primary reasons for this are that nematodes are small (free-living nematodes are generally in the order of millimetres or less in length) and they are essentially aquatic animals. They are, therefore, not able to disperse actively over large distances, so that once the local food source is exhausted or some other environmental variable becomes limiting, they generally have very little capacity to move on in search of better conditions. In general, therefore, nematodes are poor dispersers (Kiontke and Sudhaus, 2006): they may be carried passively in flowing water (or even blown by the wind) or, as some species show, hitch a ride with passing, more mobile animals, but they are unlikely to disperse over more than a few metres under their own power. For example, population genetic studies have shown that dispersal is probably limited to a few metres for dauer larvae of *Heterorhabditis marelatus* (Blouin *et al.*, 1999). Consequently, the ability to arrest development as a dauer larva then wait for conditions to improve is of obvious adaptive value.

Nematodes are ecdysozoans. A 'typical' nematode life cycle consists of (usually) four larval stages and an adult stage (Fig. 5.1); each stage is separated from the preceding stage by a moult, at which the collagenous outer cuticle for that stage is replaced and a new, stage-specific pattern of gene expression is initiated. Evolution of a developmentally arrested, environmentally resistant dauer stage may have been facilitated by the division of the life cycle into stages punctuated by moults: each moult represents an opportunity to arrest and/or to change the direction of development in response to environmental conditions. In most nematodes, the dauer is an alternative third-stage larva, and the dauer-specific developmental programme is initiated towards the end of the first larval stage. This timing also makes sense: the first-stage larva (L1) assesses the likelihood that it can complete its development to reproductive adult successfully *in situ*; if conditions are adverse it instead spends the following stage of development preparing for dauer by synthesizing a modified cuticle, laying by energy stores, modifying its biochemistry and initiating a body-wide programme of stress responses. It is noteworthy that the 'decision' to enter dauer is made in the first larval stage, shortly before the first larval moult, but that the dauer larva itself is an alternative third-stage larva (L3), meaning that a full second stage of development (known in *C. elegans* as L2d in order to differentiate it from a 'normal' L2) is required in order to develop to dauer (Hu, 2007). Thus, dauer is not simply a developmental arrest, but rather it is a specialized developmental stage that requires

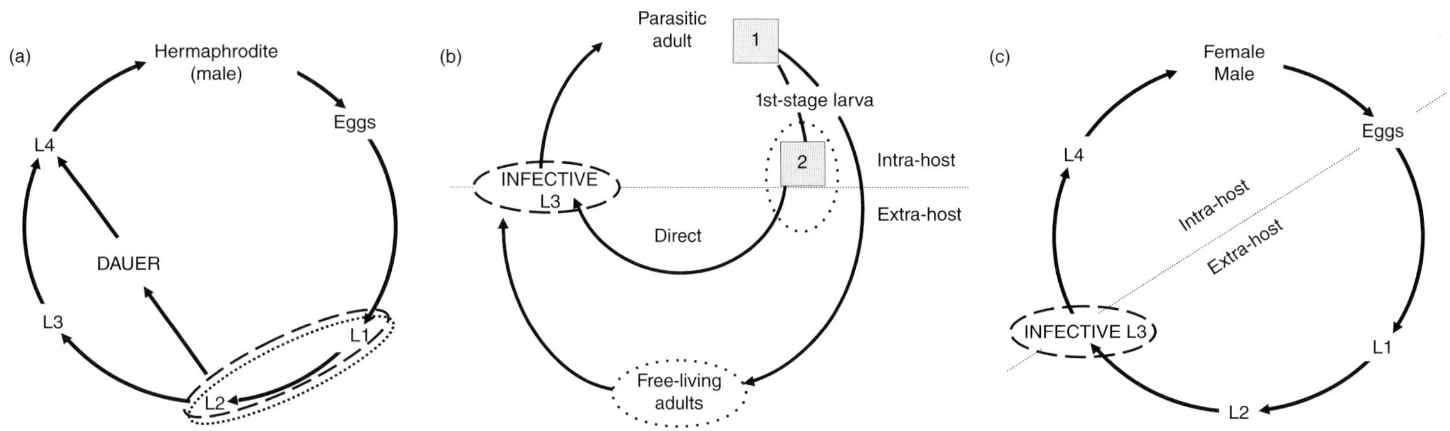

Fig. 5.1. Comparison of free-living and animal-parasitic nematode life cycles, with the dauer larva stage of free-living nematodes and the infective third-stage larvae (L3) of parasitic nematodes shown in capitals. The point in the life cycle where dafachronic acid (DA)/DAF-12 signalling occurs is in a dotted (·······) circle; DAF-7/TGF-β signalling shown is in a dashed (- - - - -) circle. (a) The life cycle of nematodes such as *Caenorhabditis elegans* and *Pristionchus pacificus*. Self-fertile hermaphrodites can also mate with males, which occur rarely in *C. elegans* but occur more frequently in *P. pacificus*. (b) The life cycle of *Strongyloides* spp., with an obligate female-only parasitic stage, which gives rise, by chromosomal sex determination (Box 1), to male and female larvae forming the free-living adult generation, in which female larvae have a developmental choice (Box 2) between direct larval development to infective larvae or development into free-living dioecious adults. Experimentally the addition of DA to larvae passed from the host as well as the progeny of the free-living adults both resulted in the development of non-infective larvae, suggesting that DA/DAF-12 signalling acts at the two places indicated. The life cycle of *Parastrongyloides* differs only in that the parasitic stage is dioecious and there can be multiple free-living generations. For clarity, larval stages (L1, L2, etc.) have been omitted. (c) The life cycle of an obligate endoparasitic nematode (e.g. *Ancylostoma* spp.) in which there is constitutive development of infective larvae. The role (if any) of DA in the development of life cycles such as these remains to be investigated.

a specific pattern of gene expression and morphological, physiological, etc., differentiation to occur over several hours before it matures. In *C. elegans* at least, additional forms of arrest are possible: sudden withdrawal of food or elevation of temperature early in the first larval stage will result in a more or less immediate developmental arrest at this stage (Munoz and Riddle, 2003; Baugh and Sternberg, 2006), and nutritional signals can also cause an arrest at the L2 stage. In L1, some specific gene expression does occur, but no differentiation (Baugh and Sternberg, 2006), so the arrested animal remains essentially a stressed but otherwise normal L1. By contrast to dauers, these arrested L1 have much more limited resistance to environmental stressors and are relatively short-lived: development must resume with a few days or the worms begin to die.

5.2.2 The chemistry of dauer induction

The environmental signals to which the L1 responds are primarily population density, temperature and the availability of food (Riddle and Albert, 1997). The population density is measured via sensing of metabolite(s) that have been termed dauer pheromone or daumone (Jeong *et al.*, 2005). In *C. elegans*, the composition of the pheromone is complex and, at the time of writing, somewhat unclear. From a strictly chemical perspective, the *C. elegans* pheromone appears to consist of a mixture of several closely related glycosylated fatty acid derivatives, ascarosides, which share a common sugar component (3,6-dideoxyhexose ascarylose) in combination with several related fatty acids (Butcher *et al.*, 2007, 2008, 2009a,b; Srinivasan *et al.*, 2008; Edison, 2009). There are at least five such compounds that have been purified by activity-guided fractionation and are therefore known to be present in natural pheromone. Each of these compounds is capable of inducing L1s to adopt a dauer fate, but they differ in their activity when applied alone and in the degree of interaction and synergy when applied in combination (Butcher *et al.*, 2008, 2009a; Edison, 2009). Thus, although it is now clear that ascarosides play a role in dauer regulation in *C. elegans*, the details of dauer chemical ecology are not yet fully elucidated.

The first of these compounds to be described, which was termed daumone and is referred to in some literature as ascaroside C7, induces dauer only at what are clearly non-physiological concentrations, with an ED_{50} of 384 μM (Jeong *et al.*, 2005). Furthermore, this compound induced a stress response in wild-type worms at this concentration and is acutely toxic if the worms are concurrently treated by mutation and/or mild detergent treatment to increase the permeability of the cuticle (Gallo and Riddle, 2009). Subsequent work showed that there are at least four additional ascarosides (C3, C5, C6 and C9) that are active in promoting dauer at 100–1000-fold lower concentrations, which is in a range in which one might expect components of the natural pheromone to act (Butcher *et al.*, 2007). One of these, ascaroside C5, has an unusual indolecarboxyl side group and is active at nanomolar concentrations (Butcher *et al.*, 2009a). The dose response to this compound is bell-shaped, i.e. it inhibits dauer formation at higher concentrations, and is very strongly temperature dependent: as with natural pheromone, worms are most sensitive to ascaroside C5 at 25°C and show virtually no response at 16°C. This is in contrast to the response to ascaroside C3, which is greater at lower compared to higher temperatures (Butcher *et al.*, 2009a). Unfortunately there are no direct chemical assays to measure ascaroside concentrations in naturally produced pheromone, so there is, as yet, no way to determine whether the concentrations of these compounds that are required to induce dauer in the laboratory (in the micromolar range) are 'physiological' in the sense that they are encountered by worms making the dauer decision in the wild. Nor is the ratio of these components in natural pheromone accurately known, although bioassay and fractionation data suggest that the ratio varies under different environmental conditions and that this variation is in accord with the differences in temperature optima for responses to ascarosides C3, C5 and C7.

Lastly, testing of the activity of daumone (ascaroside C7) alone in several assays of the dauer induction pathway suggested that, when used in isolation, this compound does not accurately mimic natural pheromone action at a molecular level. In particular, signal transduction through both the insulin and transforming growth factor (TGF)-β pathways (both of which are required for transduction of the pheromone signal, see below) differed between daumone and natural pheromone, and daumone may act via a different receptor to the other, more potent, ascarosides (Gallo and Riddle, 2009). The more potent ascarosides act through a pair of related G-protein-coupled receptor (GPCR)-coding genes, *srbc-64* and *srbc-66*, which are expressed specifically in the ASI amphid neurones of *C. elegans* (Kim *et al.*, 2009) and presumably require coupling to the GPA-3 protein that is required for dauer formation (Gallo and Riddle, 2009). This receptor expression pattern accounts also for the dauer-defective phenotype of all mutations that disrupt the ciliated dendrites of the amphid neurones, including the ASI neurones, that are exposed to the external environment and are required for normal regulation of dauer formation (Bargmann and Horvitz, 1991; Starich *et al.*, 1995). Also, daumone appears to partition differently *in vivo*, so its apparent route of entry may be primarily via accumulation in the cuticular lining of the pharynx rather than the amphidial pores (Baiga *et al.*, 2008).

Recent evidence suggests that ascarylation of the fatty acids is a detoxification mechanism in *C. elegans*, since mutations that block the pathway and cause the accumulation of toxic peroxisomal very long chain fatty acid precursors shorten lifespan and reduce fertility (Joo *et al.*, 2009). Furthermore, as discussed above, daumone is acutely toxic to worms with increased cuticular permeability. The observation that daumone is likely to be the product of a lipid detoxification pathway lends support to the contention that daumone is not a pheromone in the true sense and should be regarded as a chemical cue that is used by *C. elegans* as a surrogate measure of population density.

In addition to a role as a chemical measure of population density, ascarosides at low concentration (well below those at which dauer development is triggered) are also used by males as a chemotactic cue to aid in the search for hermaphrodites (Fig. 5.2; Srinivasan *et al.*, 2008). As is the case for dauer induction, different ascarosides differ in their potency as mate attractants in a complex fashion under laboratory conditions, but there are no data on the

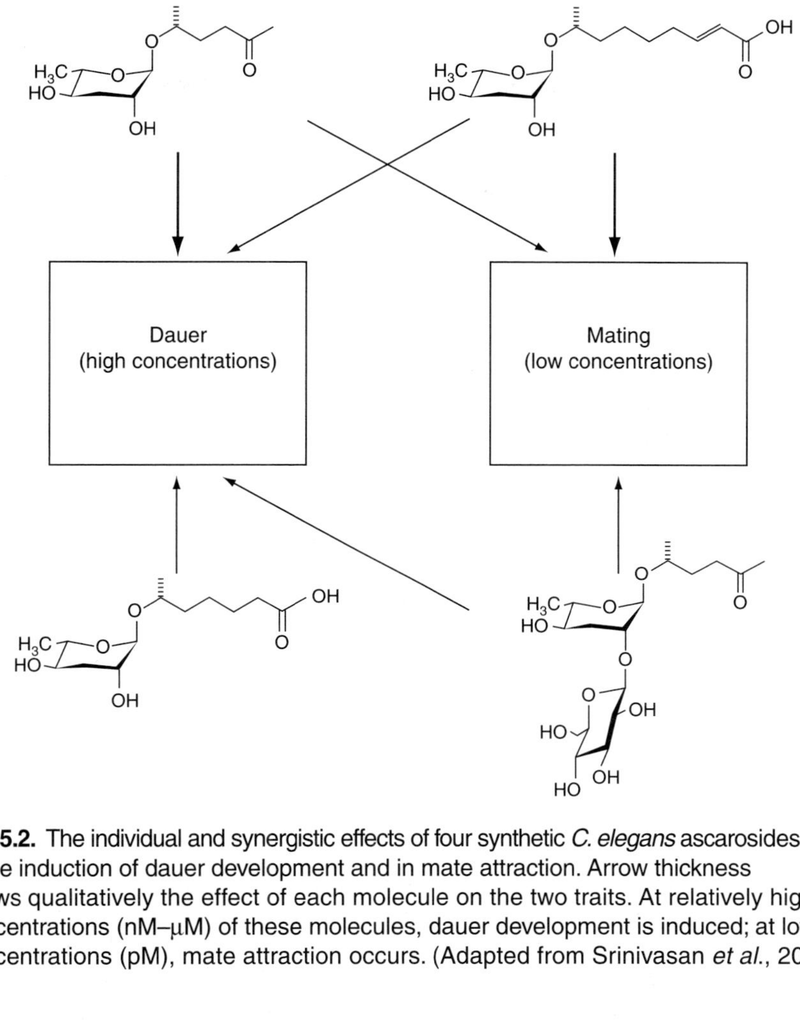

Fig. 5.2. The individual and synergistic effects of four synthetic *C. elegans* ascarosides in the induction of dauer development and in mate attraction. Arrow thickness shows qualitatively the effect of each molecule on the two traits. At relatively high concentrations (nM–μM) of these molecules, dauer development is induced; at low concentrations (pM), mate attraction occurs. (Adapted from Srinivasan *et al.*, 2008.)

role played by each family member in the wild. However, just as the accumulation of a family of constitutively produced excretory products does not meet the criteria for definition as a pheromone signal directing a change in the developmental programme, the recruitment of the same constitutively produced chemicals as a cue for the location of a potential mate does not meet the criteria for definition as a mating pheromone. What is lacking is evidence that it is possible for worms to vary pheromone production (either the overall concentration or the ratio of the components) in response to stimuli and that this variation may be under selection.

5.2.3 Sensory biology and ecology of dauer signals

Added to the chemical complexity of the ascarosides described above, the information content of these chemical cues is integrated with other environmental information, particularly temperature and the availability of food (Riddle and Albert, 1997). In *C. elegans*, dauer formation in response to dauer pheromone is potentiated by temperatures higher than the optimum growth temperature, so the pheromone dose response curve is shifted to the right as temperature moves above the optimum growth temperature (18–20°C) (Butcher *et al.*, 2007). Similar observations of changes in the balance between direct development and dauer (or indirect development, as it is termed in parasitic nematodes of the genera *Strongyloides* and *Parastrongyloides*; see below) appear to be the result of the integration of at least three environmental cues: pheromone concentration (as a measure of population density), temperature and food availability. The molecular mechanism by which this integration of qualitatively different environmental cues occurs is not known for either *C. elegans* or any other nematode species. There is increasing interest, however, in the integration of a variety of environmental cues with other life history traits in *C. elegans* (e.g. food and reproductive history with respect to lifespan; Crawford *et al.*, 2007) and it is likely that data relevant to sensory integration for dauer formation will emerge from that work.

5.2.4 Dauer signalling and the ecology of the dauer phenomenon

A critical feature that is missing from most of the *C. elegans* literature is that free-living nematodes such as *C. elegans* share their habitat with a myriad other invertebrates, including other nematodes. Most or all of these other nematode species will form dauer larvae in response to adverse conditions. Does the induction of dauer hint at any features of soil nematode ecology? Exposure of a limited range of other rhabditid species to *C. elegans* pheromone mixture extracted from growing worms has shown that there is some overlap between closely related species in pheromone response but that, in general, a given species of nematode responds only to its own chemical cue (Riddle and Albert, 1997). This implies either that the chemistries of the cues from different species do not overlap (each taxon produces a unique chemical cue) or, perhaps more intriguingly, that most (all?) soil nematodes produce chemically related cues and that the chemosensory machinery of nematodes is able to distinguish between different blends of these related compounds. The latter implies that species specificity of dauer pheromones is conferred at the sensory level rather than at the level of the composition of the cue. The application of rather basic physico-chemical criteria is consistent with the hypothesis that the chemical cues of quite distantly related nematodes may be similar: the fractionation of cue activity based on partitioning between solvents and adsorption to activated charcoal has shown that the physical chemistry of the ascarosides from *C. elegans* and the natural pheromones of *P. pacificus* and *Parastrongyloides trichosuri* are very similar (Ogawa *et al.*, 2009; Stasiuk, 2010), suggesting that, at this rather crude level, the constituents of the pheromones of these three quite unrelated species are similar. Resolution of this question has some, albeit indirect, bearing on the question of the evolution of the pheromone that is produced by worms. If all nematodes constitutively produce the same set of related compounds and species specificity is therefore conferred at the sensory level, then the evolution of the dauer induction response would require selection on the ability to distinguish between related compounds at a sensory level rather than on the ability to vary the chemical nature of the cue. This is more consistent with the pheromone compounds being waste products that are used as simple population density cues rather than true pheromones, because waste products are more likely to be shared between species with similar metabolisms. Also consistent with this are: (i) the remarkable expansion of the GPCR protein-coding gene families of nematodes (Robertson and Thomas, 2006), since it is GPCRs that appear to act as the pheromone receptor (Kim *et al.*, 2009); and (ii) the likelihood that different neurones are required for the dauer as opposed to the mate-attractive responses to ascaroside (Srinivasan *et al.*, 2008).

The chemosensory capabilities of *C. elegans* have been investigated in detail and are well characterized (Bargmann, 2006). The cellular component of their chemosensory apparatus is quite simple in comparison to mammals: there is a small number of chemosensory neurones, and the number, position, biological function and chemical repertoire of each neurone is invariant. Where there are available data, each neurone appears to express a specific complement of multiple chemoreceptor proteins of differing specificity, so each chemosensory neurone is clearly multi-modal with respect to the number of compounds it can detect. In nematodes, therefore, chemosensory complexity is achieved by endowing a few cells with a multitude of chemical sensitivities rather than a system of many cells with limited, non-overlapping specificities. Furthermore, the signals from different neurones, each with their own repertoire of chemical sensitivity, can be integrated such that the behavioural output to one compound (for example, a chemoattractant) mediated by one neurone can be modulated by a completely unrelated compound that may elicit a quite different response (perhaps a chemorepellant) mediated by a different neurone (Bargmann, 2006). Chemical cues can even modulate learning in *C. elegans* (Zhang *et al.*, 2005).

Unfortunately there are few comparative data on chemoreception in other species of nematode, so it is not possible to test directly the hypothesis that pheromone specificity is conferred primarily at the sensory level. The 'wiring diagram' and neuroanatomy of the chemosensory neurones are well conserved between at least three taxonomically disparate nematodes (*C. elegans*, *Strongyloides stercoralis* and *Ascaris suum*; Stretton *et al.*, 1991,1992; Johnson *et al.*, 1996; Fine *et al.*, 1997), and the dauer-related behaviours mediated by specific neurones may also be conserved between *C. elegans* and *S. stercoralis* (Ashton *et al.*, 1998, 2007), but there appears to be considerable divergence in the pattern of neurotransmitter, and especially neuropeptide, expression between the nervous systems of *C. elegans* and at least *A. suum* (Stretton *et al.*, 1991). The extent to which this neurotransmitter divergence translates into different responses to chemical cues is not known. Consequently, it is impossible to judge the extent to which the apparent conservation of structure may be countered by divergence of neurotransmitter or receptor use. Perhaps the greatest potential of all for divergence between nematodes in chemosensory ability and specificity lies in the diversity of predicted olfactory and gustatory receptors encoded in nematode genomes (in a nematode context, olfactory refers to volatile compounds whereas gustatory refers to water-soluble compounds): there are perhaps as many as 1000 putative olfactory GPCRs expressed in the chemosensory neurones of *C. elegans*, and each neurone also expresses a cell-specific set of signal transduction molecules downstream of these receptors (Bargmann, 2006; Robertson and Thomas, 2006). Given that there are four pairs of chemosensory neurones in *C. elegans* that are known to regulate dauer entry and/or exit, there is ample potential for complexity and hence the potential, at least, for fine discrimination between closely related molecules, such as may be produced by several nematode species sharing a single habitat. Perhaps one simple test of this hypothesis is to swap GPCRs that transduce pheromone sensation between species (e.g. between *C. elegans* and *P. trichosuri*) to test whether it is the pheromone or the receptor that determines the specificity of the response.

It seems reasonable, therefore, to propose that different species of nematode are likely to produce chemically related cues but that each species is able to discriminate between its own 'cue cocktail' and the cocktails of other species with whom they share their environment. What is the adaptive significance of this specificity? It implies, clearly, that nematodes in the soil do not generally take the population density of other nematode species into account when determining whether to enter dauer or not. This, in turn, implies that free-living nematode species occupy distinct, generally non-overlapping ecological niches in the soil and perhaps do not often compete directly for the same resources. If they were in direct competition, then the inability to detect the presence and population density of competitors, and to factor that information into the dauer decision, would surely be maladaptive. This hypothesis predicts, then, not only that from a nematode's perspective the soil is a patchy environment but also that free-living nematodes may be far more specialized in their preference for particular niches than previously suspected. Continuing this line of reasoning, selection may actually act to favour within-kin signalling, such that there is private signalling about population density. However, the role of these pheromones in mate attraction, combined with the advantage of avoiding inbreeding, may counter this force by causing selection on pheromone signalling to act species-wide. There may, therefore, be different, contrasting selection pressures acting on nematode pheromone signalling.

Data in support of the idea of very fine niche specificity comes from studies on the ecology of nematodes of the genus *Pristionchus* and from anecdotal reports of how best to hunt for caenorhabditid nematodes in the wild. *Pristionchus* spp. nematodes are found almost exclusively as dauer larvae in association with soil-dwelling beetles in quite species-specific relationships, i.e. each species of *Pristionchus* has a specific beetle species (or small group of related beetle species) with which it is associated (Herrmann *et al.*, 2006, 2007; see Herrmann and Sommer, Chapter 4, this volume). There are two ecologically and evolutionarily important characteristics of this relationship relevant to dauer biology. First, the specificity is conferred by the nematode: the nematode uses beetle pheromones to detect and then locate the appropriate beetle species (Hong *et al.*, 2008a). There is evidence that host switching can occur but that, in general, the associations are stable (Hong *et al.*, 2008a; Mayer *et al.*, 2009). Second, the adaptive value to the nematodes of the association is necromeny. The worms find a beetle of the appropriate species, get on board, then wait until the beetle dies. Bacteria associated with the dauers then colonize the cadaver and the nematodes exit dauer to feed on the bacteria (Rae *et al.*, 2008). In this privileged niche it is likely that the only population density trigger of relevance is that of conspecifics descended from other dauers that may have hitched a ride on the same beetle and thus colonized the same cadaver at the outset. Other competitors, who would not have the same advantage of being on the spot when the beetle died, would be disadvantaged and presumably are not relevant in the population density equation. Phoretic relationships of this kind may be a general feature of soil-dwelling free-living nematodes: there is anecdotal evidence that caenorhabditid nematodes are often found in association with snails and slugs, although this association (if it exists) is perhaps less specific and certainly less well documented.

Necromeny by dauers is very close to, and may be the evolutionary precursor of, the symbiotic relationship between entomopathogenic bacteria and the dauers of some entomopathogenic nematode parasites, where the nematodes do the host finding and then actively infect the insect with a specific bacterial pathogen (Forst *et al.*, 1997). It is the bacteria that then rapidly kill the insect host and use the cadaver as a food source, and the nematodes simply feed on the resultant bacterial bloom. A well-studied example of this is the relationship between gamma-Proteobacteria of the genus *Xenorhabdus* and nematodes of the genus *Steinernema*, as characterized by the association between *Xenorhabdus nematophila* and *Steinernema carpocapsae* (Richards and Goodrich-Blair, 2009). This and other observations (see below) have given rise to the hypothesis that dauer larvae are the likely evolutionary precursors of the infective larvae of parasitic nematodes.

5.3 Genetic Variation in Dauer Switching

The discussion of initiation of dauer so far has stressed the importance of environmental signals that contain information about the likelihood of reproductive success following the alternatives of development to reproductive maturity immediately or arresting development as dauer larvae in response to an adverse environment and therefore delaying reproduction. There is clearly an adaptive value in such developmental plasticity but do all members of a species behave in the same way? The widespread existence of dauer as a developmental option among free-living nematodes and its importance in necromeny and in at least some forms of nematode parasitism (see Perry and Moens, Chapter 1, this volume) suggest that dauer switching can itself evolve. Evolution requires phenotypic variation for the trait in question and that at least some component of this variation should be heritable. Again using *C. elegans* as a model, recent work has shown that there is heritable variation in the response to dauer larva-inducing conditions (i.e. food or dauer pheromone concentration). Thus, for *C. elegans* at least, there is very clear evidence that different strains or genotypes of worms isolated from their natural environment (i.e. not laboratory strains) will respond in different ways to the same quantum of signal: there are some 'anxious' genotypes, opting for the relative safety of dauer at the first sign of adverse conditions (i.e. which form dauer larvae with very low levels of dauer-inducing signal) and there are 'optimistic' genotypes, which form dauers only in response to very strong signals (Viney *et al*., 2003). Each of these strategies has, presumably, evolved to maximize fitness under certain sets of environmental conditions, which will probably include the extent to which the ecological niche occupied by that genotype is variable or patchy. The very broad range of sensitivities to dauer pheromone to which this heritable variation in dauer larva formation contributes also suggests the possibility that dauer larva formation could evolve to no longer exist, or could evolve to become an obligatory rather than a facultative part of a nematode life cycle. This has relevance to the evolution of nematode parasitism (see below). The ability to modify the degree of developmental plasticity in response to quantitatively different signalling is an important component of the evolution of developmentally plastic responses. One approach to try and understand the adaptive value of different plastic responses (i.e. 'anxious' or 'optimistic' strategies) is to determine how these traits trade off with other important life history traits. For *C. elegans*, different plasticities of dauer development are correlated with (and perhaps causally related to) the population growth rate of different genotypes under limited food conditions (Harvey *et al*., 2008). This suggests that 'anxious' or 'optimistic' decisions about dauer formation and the rates of reproduction and larval growth are all components of a coordinated strategic response of worms to changing environments.

These data may also shed light on the mechanisms by which environmental signals are sensed and on the genes that are under selection to produce the differences in plasticity of dauer development in responsiveness to variable, dauer-inducing environmental conditions. Classical mutational studies in *C. elegans* and *P. pacificus* have uncovered signal transduction pathways that converge on a conserved nuclear hormone, dafachronic acid, that is the ligand for the conserved DAF-12 nuclear hormone receptor protein (Ogawa *et al*., 2009). However, these transduction pathways are downstream of the primary dauer signals of temperature, food availability and population density; they transduce but do not receive the primary signals (Hu, 2007). Quantitative genetic analysis of *C. elegans* lines with different sensitivities of dauer larva formation (i.e. 'anxious' or 'optimistic' above) has permitted the mapping of a number of quantitative trait loci (QTL), and, to date, none of these map genetically to regions of the *C. elegans* genome in which known *daf* genes, defined by mutation to dauer-defective phenotypes, occur (Harvey *et al*., 2008). The conclusion is that, as interesting as the insulin, TGF-β and other signalling pathways may be from a basic biology perspective, these pathways are not major contributors to the detectable natural variation in dauer formation in response to variable environmental signals. In other words, if one considers dauer formation as a quantitative trait (as it clearly is, given the continuous variation in response to inducing conditions), then the insulin and TGF-β pathways are unlikely to play a role in the natural plasticity of dauer development. This is an important finding, since it implies that the evolutionary response of free-living nematodes to variation in the environment is mediated by selection on yet to be discovered genes, at least in the case of *C. elegans*.

This may also be of significance from an ecological perspective. As discussed above, the dauer pheromones of diverse species may be chemically similar, so that species specificity of response to pheromone may be conferred at the level of pheromone sensing rather than pheromone production. If the *C. elegans* pheromone were a cocktail of constitutively produced excretory products rather than a true pheromone, then there would probably be little flexibility to vary the nature of the chemical signal via natural selection for species specificity: toxic metabolites must be excreted. In this case, the processes that detect the metabolites and transduce this information may be under greater selection pressure than the production of the metabolites. The experiments described above have revealed genetic variation for dauer larvae formation following exposure to a constant 'quantum' pheromone, which may be controlled by genes that affect signal perception or the transduction and integration of this information, and which are not part of the currently known *daf* gene network or of the metabolic pathways leading to pheromone production. As also noted above, some, at least, of these genes may code for GPCRs. It will be of interest to determine whether the putative ascaroside receptors defined in molecular studies are amongst the QTL defined in the genetic studies and whether these are then conserved among diverse nematodes or, as the 'signal discrimination' hypothesis might require, these will vary between nematode species that respond to different pheromone or chemical cue cocktails.

Some data relevant to the question of how pheromone specificity may be conferred comes from studies on the chemoattraction between *P. pacificus* and its beetle associates. Natural variation in the sensitivity of geographically distinct isolates of *P. pacificus* to the beetle pheromone used in host finding by the nematodes is mediated by the EGL-4 protein kinase, which is a component of a cGMP signalling cascade downstream of a putative GPCR (Hong *et al.*, 2008b). This demonstrates, in principle, that variation in sensitivity to an environmental cue can be mediated at the level of cue reception/signal transduction.

5.4 The Biology of the Dauer Stage

The biology of *C. elegans* dauers has been reviewed from several perspectives and is quite well understood. We will not reproduce those reviews here, but refer the reader to a selection of recent reviews and somewhat older monographs as a starting point (Riddle and Albert, 1997; Burnell *et al.*, 2005). From the perspective of the 'dauer phenomenon', the salient features of dauer biology are that the dauer is structurally, behaviourally and metabolically adapted to survive stress, and to survive it for a considerable period of time. For example, the non-dauer lifespan of *C. elegans* under laboratory conditions is 2–3 weeks, whereas dauers under the same laboratory conditions can survive for several months (Cassada and Russell, 1975). There are increases of a similar magnitude in the resistance of dauers to environmental stressors such as heat, harsh detergents, oxidizing agents and desiccation (Hu, 2007). Common themes of the biology of dauer larvae include the induction of stress response pathways, shifts in metabolism away from carbohydrates to fats, modification of the cuticle to increase mechanical and chemical resistance and the downregulation of biosynthetic processes (Burnell *et al.*, 2005). The variation in the transcriptional signature of differential gene expression in dauers between species (Mitreva *et al.*, 2004; Elling *et al.*, 2007) therefore most likely represents different molecular means to the same end: a long-lived, stress-resistant developmental stage. It should also be remembered that dauer is not the only stress response available to nematodes: there are other developmentally arrested, 'resistance' states in *C. elegans* as a response to stress, and the most stress resistant of all nematodes (Antarctic species such as *Panagrolaimus davidi* and *Plectus murrayi*) arrest development and survive extremes of cold and desiccation at various stages (Wharton, 2003). Clearly, then, while dauer is important to many nematodes, there is a wide range of responses to stress that are expressed at a variety of developmental stages. The important characteristics are that all require that the nematodes monitor and then respond to their environment.

In *C. elegans*, the regulation of dauer gene expression is primarily under the control of two transcription factors, DAF-16/FOXO and DAF-12, which are the end points of the insulin/insulin-like growth factor (IGF) and nuclear hormone dafachronic acid signalling pathways, respectively (Hu, 2007). DAF-16 and DAF-12 also play a pivotal role in stress responses in stages other than dauers and in the regulation of other developmental processes related to developmental timing and lifespan, emphasizing that dauer biology is based on existing molecular and biochemical processes, most of which are conserved widely among metazoans. DAF-12 has been shown directly to play a similarly crucial role in the regulation of dauer entry in *P. pacificus* (Ogawa *et al.*, 2009), and presumably does so by direct regulation of dauer-specific changes in gene expression in this species. In general, very little is known of the molecular detail of dauer-specific patterns of gene expression or their regulation in other nematodes. As noted above, comparison of expressed sequence tag profiles between dauers and non-dauers from a handful of species suggests that the identity of differentially expressed genes differs between species, although there are some general trends: biosynthetic processes are downregulated and metabolism changes (Mitreva *et al.*, 2004; Elling *et al.*, 2007; Evans *et al.*, 2008; Thompson *et al.*, 2008). Of interest in the context of the discussion above concerning the importance of signal reception in regulating dauer biology, one of the most prominent functional classes of genes showing differential expression in dauers is signal transduction genes. For example, 19% of differentially regulated transcripts in dauers of the Antarctic species *P. murrayi* were classified as 'involved in environmental information processing' (Adhikari *et al.*, 2009).

Dauer larvae also exhibit dauer-specific behaviours. Nematodes are small, essentially aquatic animals that have very limited dispersal capacity unless assisted. An adaptationist view postulates that a dauer that exhibits behaviours that are likely to facilitate movement to other, perhaps more favourable, locations will be at an advantage over nematodes that do not, but data are lacking for all but a few species. Behaviours that have been described to a limited extent are nictation, negative geotaxis and chemotaxis. Nictation is where the dauer (or infective larva for parasitic species) stands upright on its tail and waves its body around, and is clearly an adaptation for phoresis in free-living nematodes and for host finding and infection in parasites. In some species, nictation is accompanied by negative geotaxis, where the dauer climbs up physical features in order to nictate. Although data in the literature are sparse, species for which there are anecdotal descriptions of phoretic or necromenic behaviour (such as *C. elegans*) show both behaviours: most *C. elegans* researchers are familiar with the sight of nictating dauers that have climbed up fungal hyphae on old agar plates in the laboratory. Furthermore, nictation is not a constitutive behaviour but is elicited in response to mechanical and other stimuli. In the absence of stimulation, *C. elegans* dauers are lethargic and generally immobile but nictate vigorously when disturbed. Chemotaxis is clearly another dauer behaviour of likely adaptive significance. Chemotaxis for host finding is well documented in the infective larvae of plant-parasitic and entomopathogenic nematodes (Perry and Maule, 2004; Christen *et al.*, 2007; Ramos-Rodriguez *et al.*, 2007; Curtis, 2008) and, among free-living nematodes, has been described in nematodes in phoretic association with beetles (Hong and Sommer, 2006; Hong *et al.*, 2008b). The adaptive value of the relationship is supported by the observation that there are at least two species of nematode that are attracted to and have a phoretic association with a single species of cockchafer. However, the two species utilize different chemical signals from the cockchafers: *Pristionchus maupasi* are attracted to phenol (a sex attractant in beetles) while *Diplogasteroides magnus*, which is also associated with these beetles, are not. The chemoattractant utilized by *D. magnus* is not known (Hong *et al.*, 2008a). This example also illustrates the likely ecological importance of sensory diversity between species in a different facet of dauer biology: two distantly related species of nematode responding to different signals produced by a single species of beetle which both nematodes utilize as a phoretic host.

5.5 Dauer as a Pre-adaptation for the Evolution of Parasitism in Nematodes

5.5.1 Dauer biology and parasitism

The current nematode molecular phylogenies show clearly that parasitic nematodes are polyphyletic, implying that parasitism arose independently several times in the course of nematode evolution (Blaxter *et al.*, 1998; Dorris *et al.*, 1999). Not only have parasites of plants and animals evolved separately (are polyphyletic), but there are four groups of animal parasites and three groups of plant parasites, each of which are phylogenetically interspersed with non-parasitic relatives (Dorris *et al.*, 1999). Despite multiple origins, parasites share the common nematode life history strategy of dividing their life cycle into an adult stage that is (usually) preceded by four pre-adult larval stages. Also, with few exceptions, the adult stage is parasitic, even though there is considerably more variation in the parasitic status of the larval stages. Parasite life history is characterized by developmental arrest at the point where the nematodes make the transition from outside to inside a host, and this arrest always coincides with a transition between larval stages. In other words, the point at which the parasite infects its host is coordinated with arrest at the completion of a stage of larval development. For animal-parasitic nematodes, the most common transition point is at the completion of the second larval stage, i.e. from developmental arrest at the onset of the third larval stage in one environment (outside the host) to an L3 that has resumed its development in the new environment (inside the host). More than one transition may exist for some animal parasites with more complex life cycles, but for all described animal parasites there is a period of early development in one environment, followed by arrest until the transition is made to a new, parasitic, environment. For those species with so-called direct life cycles, the first environment in which the first and second larval stages are completed is a free-living environment (e.g. in a faecal deposit for a gastrointestinal nematode) and the second environment is the host, where the parasite completes the third and fourth larval stages and reaches adulthood. For parasites with indirect life cycles, first- and second-stage development may be completed in an intermediate host (very often an arthropod), followed by arrest and transition to the definitive host, where third- and fourth-stage development occurs and the parasite reaches adulthood. In all cases, the developmentally arrested stage that makes the transition to the definitive host is called the infective larva.

The infective larvae of most direct life cycle parasites bear a very strong resemblance to the dauer larvae of free-living nematodes (Hotez *et al.*, 1993). Most obviously of course, both are arrested at the start of a third larval stage. In many cases, animal-parasitic infective larvae are essentially an arrested third-stage larva that retains the second-stage cuticle, which is extensively modified to resemble the dauer cuticle in ultrastructure and physical properties and is usually termed a sheath (Hotez *et al.*, 1993). It is important to note that, despite retention of the second-stage cuticle, the infective larvae of directly developing parasites have completed second-stage development and possess a fully formed third-stage cuticle under the sheath: the point of developmental arrest is at the initiation of the moult from L2 to L3. This moult usually occurs concomitantly with the process of infection and is termed exsheathment. Other forms of infective larva exist that bear a less obvious relationship to dauer larvae. For example, the infective stage of filarial nematodes has already moulted from L2 to L3 in its dipteran intermediate host and has migrated to the anatomical site from which it will infect the vertebrate definitive host, but, none the less, it has arrested its development at the third larval stage. Another variation on the theme is the infective stage of

Strongyloides spp., in which the L2 to L3 moult has occurred, but the arrested infective L3 possesses a specialized surface coat that is lost in the process of skin penetration (Grove *et al.*, 1984, 1987). In both of these cases, the arrest is immediately following the L2 to L3 moult rather than immediately preceding it. It is the arrested L3 that ties dauers and all these infective stages together.

There are other biological or phenotypic characteristics that are conserved between dauer larvae and infective larvae, such as stress resistance (e.g. to elevated temperature or desiccation), behaviours (e.g. nictation, negative geotaxis), cessation of feeding (occlusion of the pharynx), often extremely long survival in the arrested state (months to years), downregulation of biosynthetic processes and the accumulation of fat (presumably signalling a shift in energy metabolism associated with long survival without feeding; Hotez *et al.*, 1993). These shared biological characteristics gave rise to and support the hypothesis, known as the dauer hypothesis, that the infective larva of parasites evolved from a dauer larva precursor. Perhaps a more reasonable modification of this hypothesis is that it is not the existence of a dauer stage per se that is the critical pre-adaptation for the evolution of parasitism and accounts for the apparent frequency with which the step from free-living to parasitic life histories has been made, but rather it is the existence of a mechanism for developmental arrest and the associated phenotypic plasticity (Hawdon and Schad, 1991).

Nematode parasites do not reproduce themselves inside their hosts, so each adult parasite in the host started its development elsewhere, then had to make the transition from that environment into the host. In order to infect the host, the infective larva and the host must first meet, and it is the probability of this meeting that drives parasite life history. In directly developing parasites, the encounter with the host is left largely to chance: a very small and not particularly vagile nematode must encounter a host in order to progress from one life cycle stage to the next. The probability of this occurring is very low, so there must be very large numbers of infective larvae produced by each adult and/or infective larvae must be able to survive for long periods. Even in parasites with intermediate hosts, where the intermediate host is usually more vagile and may even actively seek out the host (e.g. mosquitoes or biting flies as intermediate hosts for filarial parasites), it is still likely to be advantageous to the parasite to be able to arrest its development and thus prolong the period during which it can find a host. It is this combination of developmental arrest and prolonged survival that is most likely the critical pre-adaptation rather than any other aspect of dauer larva biology.

This hypothesis is supported further by the existence of what appear to be extant transitional stages in the evolution of obligate parasites from free-living ancestors. An example of the first such step may be the necromenic relationship between *P. pacificus* and its beetle associates described above. The next evolutionary step may have been the development of a second association between the nematode and an entomopathogenic bacterium, so that rather than having to wait for the beetle to die of old age or by misadventure, the nematode infects the beetle with a pathogenic bacterium that is, first, the agent of the beetle's death and, second, the food source for the expanding nematode population (Goodrich-Blair, 2007). In this growth phase of their life history, the nematodes are indistinguishable from a free-living species under similar circumstances of abundant food and low population density. When the bacterial food source is exhausted and the nematode population density is high, the nematodes enter dauer and wait for another suitable insect, with their pathogenic bacterial symbiont sequestered safely in a specialized elaboration of the pharynx. It is important to note that these nematodes are not obligate parasites in the sense that they do not absolutely require an insect host for maturation and reproduction: they can be grown successfully in the laboratory on their symbiont bacteria (grown in broth), essentially as though they were free-living nematodes without recourse to an insect 'host' (Ciche, 2007). They should not, perhaps, be regarded as true parasites at all.

The next transitional step may be to a facultative parasite in which a complete free-living cycle is still present, but in which there is an optional parasitic stage that is truly parasitic, i.e. is completely dependent on the host for survival and reproduction during the parasitic phase of its life cycle. The facultative feature of this life history is the existence of a 'choice' between the two life cycles that is determined by a condition-sensitive switch. Such a facultative species would remain capable of free-living development under favourable conditions and enter parasitic development in response to adverse conditions. Species of the genera *Strongyloides* and *Parastrongyloides*, particularly *Parastrongyloides*, are extant species with these life histories (Viney, 1996; Grant *et al.*, 2006). *Parastrongyloides trichosuri* can be maintained indefinitely in the laboratory in culture as a free-living nematode feeding on a range bacterial species. The switch from this free-living life cycle to a parasitic life cycle is determined by the same cues that determine the dauer switch (i.e. population density, as measured by a 'pheromone', temperature and the availability of food). Parasites of the genus *Strongyloides* have the same free living to parasite switch (best characterized in *Strongyloides ratti* and *S. stercoralis*), but in most species in the genus the switch operates for only a single free-living generation, then must be 'reset' by passage through a parasitic life cycle (Viney, 1996). For both genera, once in the parasitic life cycle, they are true obligate parasites that can only reach reproductive adulthood in the intestine of an appropriate host (Australian brushtail possums for *P. trichosuri*, rats for *S. ratti* and humans for *S. stercoralis*).

The biological analogy between the dauer switch in free-living species, as exemplified by *C. elegans*, and these facultative parasites extends to the neuronal control of the developmental switch. A combination of electron microscope reconstruction of the amphid wiring diagram and laser ablation of specific amphid neurones in *S. stercoralis* has shown that the analogous neurones sense and transduce the environmental signals for dauer/infective larval entry and exist in both *S. stercoralis* and *C. elegans* (Ashton and Schad, 1996).

5.5.2 Dauer molecular biology and parasite evolution

Is the biological analogy between dauer larvae and the infective larvae of parasites described above underpinned by an equally precise analogy at the molecular level? This question has been investigated directly and indirectly in several species, and the answer appears to be equivocal (Viney, 2009). A direct genetic investigation of the role of dauer signal transduction in a parasitic species has been carried out in *S. ratti*. Mutagenesis of *S. ratti* followed by screening for resistance to the anthelmintic compound ivermectin resulted in infective larvae that appeared to be unable to exit dauer (Viney *et al.*, 2002). Resistance to ivermectin results in amphid defects that consequently confer a dauer-defective phenotype in *C. elegans*, and the failure of infective larvae screened for resistance to ivermectin to exit dauer is consistent with this view. Stepping a little away from parasitic species to the necromenic species, screens for dauer mutations in *P. pacificus* have yielded mutations in *Pp-daf-12* which confer a dauer-defective phenotype, as do loss-of-function mutations in *Ce-daf-12* (Ogawa *et al.*, 2009).

There have been several reports of pharmacological tests of the role of parasite orthologues of the *C. elegans* dauer signal transduction pathways, or the *daf* genes. Although, as discussed in the previous section, the evolutionarily important genetic variation upon which selection may have acted during parasite evolution is likely not to involve the *daf* genes, there is evidence that the *daf* signalling pathways are components of infective larval biology. The endogenous ligand for Ce-DAF-12 is now known to be a family of related, novel sterol derivatives known as dafachronic acids, the most active of which are Δ4 and Δ7 dafachronic acids (Motola *et al.*, 2006). Application of Δ4 or Δ7 dafachronic acids to *C. elegans* promotes reproductive development, blocking dauer development and promoting recovery from dauer (Motola *et al.*, 2006). Application of these compounds to *P. pacificus* has the same effect, and in *S. stercoralis* or *Strongyloides papillosis*, application of Δ4 or Δ7 dafachronic acids inhibits infective larval development and promotes free-living development (Ogawa *et al.*, 2009; Wang *et al.*, 2009). In *S. stercoralis*, and in the hookworm *Ancylostoma caninum*, these compounds also promote the recovery of infective larvae from dauer, as assayed by a resumption of feeding following exposure (Wang *et al.*, 2009). This therefore suggests that there is conservation of DAF-12-like molecules and their dafachronic acid-like ligands in controlling the development of free-living dauer larvae and of parasitic infective larvae.

By contrast to the apparent conservation of dafachronic acid signalling, analysis of the expression of *daf-7* (which codes for TGF-β like molecules) in different life cycle stages of *C. elegans* and parasitic nematodes suggests that its role may have changed during the evolution of parasitism (Crook *et al.*, 2005). In *C. elegans*, the peak expression of *daf-7* is in the L1 stage, consistent with TGF-β's role in transduction of the environmental signals underlying the dauer decision. In a range of parasitic nematodes (*S. ratti*, *S. stercoralis*, *P. trichosuri* and *A. caninum*), peak expression of the presumed *daf-7* orthologue is in the infective larval stage (i.e. the dauer equivalent). This divergent expression pattern is consistent with the idea that in parasitic nematodes this molecule is involved in environmental sensation, but in exiting the infective larva stage to resume development inside the host rather than in the induction of infective larva development. Further analysis of *Pt-daf-7* function by transgenesis of *C. elegans daf-7* mutants showed that *Pt-daf-7* cannot rescue the *C. elegans* Daf-c phenotype conferred by loss-of-function mutations in *Ce-daf-7* and that 'forcing' the expression of a Pt-DAF-7 ligand in wild-type *C. elegans* actually appears to interfere with wild-type DAF-7 signalling (Grant and Crook, unpublished). Furthermore, the *Pt-daf-7* promoter drives expression of green fluorescent protein (GFP) in many additional neurones and, at later stages of development, in transgenic *C. elegans* compared to a *Ce-daf-7–gfp* reporter. These data are most easily interpreted as evidence that, despite the high level of protein sequence conservation, DAF-7 signalling differs significantly between *P. trichosuri* and *C. elegans*.

Insulin/IGF signalling has also been investigated in parasitic species using pharmacological methods and by transgenesis (Tissenbaum *et al.*, 2000; Massey *et al.*, 2006; Junio *et al.*, 2008). Compounds known to stimulate insulin signalling promote the resumption of hookworm feeding (Tissenbaum *et al.*, 2000), as discussed above for *Ac-daf-12* and dafachronic acids, and a compound known to inhibit signalling through the insulin pathway (and which modulates dauer formation in *C. elegans*) appears to prevent free-living development in *P. trichosuri* (Stasiuk, 2010).

There are several indirect molecular tests of insulin signalling and parasite *daf* orthologues, primarily of candidate orthologues of *daf-16*, which utilize transformation rescue of loss-of-function mutations in *C. elegans* (Massey *et al.*, 2006; Hu *et al.*, 2010; Stasiuk, 2010). The studies differ somewhat in choice of *daf-16* allele to rescue and in the *Ce-daf-16* promoter used to drive expression of the transgene, the isoform of DAF-16 encoded by the parasite sequence and the method by which rescue is assessed, so direct comparison is difficult. However, the results of three studies, using *daf-16* sequences cloned from *S. stercoralis*, *Haemonchus contortus* and *P. trichosuri* (Massey *et al.*, 2006; Hu *et al.*, 2010; Stasiuk, 2010), indicate that these parasite DAF-16 transcription factors can function to at least partially rescue most, but perhaps not all, of the phenotypes associated with *daf-16* mutations in *C. elegans*. Similar experiments with another insulin/IGF signalling component, *age-1*, from *P. trichosuri* showed a similar partial rescue (Stasiuk, 2010). Thus, insulin/IGF signalling components from at least some parasite species are sufficiently well conserved and function sufficiently well that they can rescue mutations in the insulin/IGF pathway of *C. elegans*. However, although these data are consistent with a role for insulin/IGF signalling in infective larval development in parasites, they are not evidence that these parasite *daf-16* genes play this role in the species from which they were isolated nor, more importantly, is it evidence that insulin/IGF signalling is a necessary component of the hypothesized dauer switch in parasites of the genera *Strongyloides* and *Parastrongyloides*.

More direct evidence for a role for *daf-16* and insulin/IGF signalling in infective larva formation has been provided by expression in transgenic *S. stercoralis* of a transgene encoding a *Ss*-DAF-16::GFP translational fusion (Castelletto *et al.*, 2009). The recombinant protein was localized in tissues that are remodelled in infective larval development compared with free-living development, and expression of mutant versions of the transgene resulted in larvae with a range of developmental defects in these and other tissues. These experiments demonstrate clearly that DAF-16 is required for infective larval development. Further transgenesis utilizing 'dominant-repressor' and 'dominant-negative' forms of DAF-16 resulted in abnormal larvae that arrested development. Some features of these larvae were consistent with a partial switch to free-living development. These data are, however, difficult to interpret: these transgenic larvae are the progeny of free-living females, and under natural conditions *S. stercoralis* undergoes only a single free-living generation. In other words, it is possible that the progeny of this single free-living cycle are obligately committed to parasitic development and are not competent to complete a second free-living cycle. In other words, the 'dauer switch' that operates in the first generation may not operate in the second. In support of this interpretation, exposure of second-generation L1 progeny of *S. stercoralis* to exogenous dafachronic acid also induced only a partial switch to free-living development and resulted in abnormal, arrested 'free-living-like' larvae (Wang *et al.*, 2009). Consequently, although the transgenesis and dafachronic acid experiments may test the involvement of insulin/IGF and dafachronic acid signalling in regulating the obligate development of infective L3s, they may not directly test the function of the 'dauer switch' in the choice between free-living and parasitic life cycles.

Some resolution of this uncertainty may be provided by similar experiments in which dafachronic acid was applied to second-generation *S. papillosis* L1s that would also normally be expected to undergo obligate development to infective L3 (Ogawa *et al.*, 2009). In this species, dafachronic acid was able to bring about a complete fate transformation of 'second-generation' female larvae to fertile free-living worms, in contrast to the partial transformation discussed above for *S. stercoralis*. This is conclusive evidence that dafachronic acid signalling is necessary and sufficient to drive a dauer-like switch between free-living and parasitic development in this species, and, by extension, dafachronic acid seems likely to play the same 'dauer-switch' regulation role in other *Strongyloides* spp. The nature of the differences between *Strongyloides* spp. and the question of whether insulin/IGF signalling also plays a 'dauer switch' role or is simply required for the execution of the developmental programme remain open.

5.6 Conclusions and Future Directions

The dauer larva is central to the life history strategy of many, perhaps all, free-living nematodes. The dauer provides nematodes with the ability to survive harsh adverse conditions for long periods between short bursts of feverish reproduction and has most likely been the major factor in the ability of nematodes to colonize more ecological niches and evolve greater species richness than perhaps any other metazoan taxon. Furthermore, the dauer may have provided free-living nematodes with a crucial adaptation that facilitated the repeated evolution of parasitic life histories in this phylum. The regulation of dauer development has been studied extensively at a molecular level in one free-living species, *C. elegans*, and comparative studies between *C. elegans* and either other free-living species or parasitic species are beginning to appear. The picture that is emerging is a now familiar Evo-Devo theme of common, conserved components utilized sometimes in similar and other times in strikingly different ways to produce the same biological result: a survival stage par excellence.

5.7 Acknowledgements

WG would like to thank Dr Matt Crook and Dr Susan Stasiuk for permission to use some unpublished data and for fruitful discussions on the biology of the dauer switch in *P. trichosuri*. We would like to thank Nick Roberts for help with Fig. 5.2.

5.8 References

Adhikari, B.N., Wall, D.H. and Adams, B.J. (2009) Desiccation survival in an Antarctic nematode: molecular analysis using expressed sequenced tags. *BMC Genomics* 10, 69.

Ashton, F.T. and Schad, G.A. (1996) Amphids in *Strongyloides stercoralis* and other parasitic nematodes. *Parasitology Today* 12, 187–194.

Ashton, F.T., Bhopale, V.M., Holt, D., Smith, G. and Schad, G.A. (1998) Developmental switching in the parasitic nematode *Strongyloides stercoralis* is controlled by the ASF and ASI amphidial neurons. *Journal of Parasitology* 84, 691–695.

Ashton, F.T., Zhu, X., Boston, R., Lok, J.B. and Schad, G.A. (2007) *Strongyloides stercoralis*: amphidial neuron pair ASJ triggers significant resumption of development by infective larvae under host-mimicking *in vitro* conditions. *Experimental Parasitology* 115, 92–97.

Baiga, T.J., Guo, H., Xing, Y., O'Doherty, G.A., Dillin, A., Austin, M.B., Noel, J.P. and La Clair, J.J. (2008) Metabolite induction of *Caenorhabditis elegans* dauer larvae arises via transport in the pharynx. *ACS Chemical Biology* 3, 294–304.

Bargmann, C.I. (2006) Chemosensation in *C. elegans*. In: The *C. elegans* Research Community (ed.) *WormBook* (25.10.2006), doi/10.1895/wormbook.1.123.1, http://www.wormbook.org.

Bargmann, C.I.H. and Horvitz, H.R. (1991) Control of larval development by chemosensory neurons in *Caenorhabditis elegans*. *Science* 251, 1243–1246.

Barrière, A. and Félix, M.A. (2005a) High local genetic diversity and low outcrossing rate in *Caenorhabditis elegans* natural populations. *Current Biology* 15, 1176–1184.

Barrière, A. and Félix, M.A. (2005b) Natural variation and population genetics of *Caenorhabditis elegans*. In: The *C. elegans* Research Community (ed.) *Wormbook* (26.12.2005), doi/10.1895/wormbook.1.43.1, http://www.wormbook.org.

Barrière, A. and Félix, M.A. (2006) Isolation of *C. elegans* and related nematodes. In: The *C. elegans* Research Community (ed.) *WormBook* (01.12.2007), doi/10.1895/wormbook.1.115.1, http://www.wormbook.org.

Baugh, L.R. and Sternberg, P.W. (2006). DAF-16/FOXO regulates transcription of *cki-1*/Cip/Kip and repression of *lin-4* during *C. elegans* L1 arrest. *Current Biology* 16, 780–785.

Blaxter, M.L., De Ley, P., Garey, J.R., Liu, L.X., Scheldeman, P., Vierstaete, A., Vanfleteren, J.R., Mackey, L.Y., Dorris, M., Frisse, L.M., Vida, J.T. and Thomas, W.K. (1998) A molecular evolutionary framework for the Phylum Nematoda. *Nature* 392, 71–75.

Blouin, M.S., Liu, J. and Berry, R.E. (1999) Life cycle variation and the genetic structure of nematode populations. *Heredity* 83, 253–259.

Burnell, A.M., Houthoofd, K., O'Hanlon, K. and Vanfleteren, J.R. (2005) Alternate metabolism during the dauer stage of the nematode *Caenorhabditis elegans. Experimental Gerontology* 40, 850–856.

Butcher, R.A., Fujita, M., Schroeder, F.C. and Clardy, J. (2007) Small-molecule pheromones that control dauer development in *Caenorhabditis elegans. Nature Chemical Biology* 3, 420–422.

Butcher, R.A., Ragains, J.R., Kim, E. and Clardy, J. (2008) A potent dauer pheromone component in *Caenorhabditis elegans* that acts synergistically with other components. *Proceedings of the National Academy of Sciences of the USA* 105, 14288–14292.

Butcher, R.A., Ragains, J.R. and Clardy, J. (2009a) An indole-containing dauer pheromone component with unusual dauer inhibitory activity at higher concentrations. *Organic Letters* 11, 3100–3103.

Butcher, R.A., Ragains, J.R., Li, W., Ruvkun, G., Clardy, J. and Mak, H.Y. (2009b) Biosynthesis of the *Caenorhabditis elegans* dauer pheromone. *Proceedings of the National Academy of Sciences of the USA* 106, 1875–1879.

Cassada, R.C., and Russell, R.L. (1975) The dauer larva, a post-embryonic developmental variant of the nematode *C. elegans. Developmental Biology* 46, 326–342.

Castelletto, M.L., Massey, H.C. Jr. and Lok, J.B. (2009) Morphogenesis of *Strongyloides stercoralis* infective larvae requires the DAF-16 ortholog FKTF-1. *PLoS Pathology* 5, e1000370.

Christen, J.M., Campbell, J.F., Lewis, E.E., Shapiro-Ilan, D.I. and Ramaswamy, S.B. (2007) Responses of the entomopathogenic nematode, *Steinernema riobrave* to its insect hosts, *Galleria mellonella* and *Tenebrio molitor. Parasitology* 134, 889–898.

Ciche, T. (2007) The biology and genome of *Heterorhabditis bacteriophora*. In: The *C. elegans* Research Community (ed.) *WormBook* (20.02.2007), doi/10.1895/wormbook.1.135.1, http://www.wormbook.org.

Crawford, D., Libina, N. and Kenyon, C. (2007) *Caenorhabditis elegans* integrates food and reproductive signals in lifespan determination. *Aging Cell* 6, 715–721.

Crook, M., Thompson, F.J., Grant, W.N. and Viney, M.E. (2005) *daf-7* and the development of *Strongyloides ratti* and *Parastrongyloides trichosuri. Molecular and Biochemical Parasitology* 139, 213–223.

Curtis, R.H. (2008) Plant–nematode interactions: environmental signals detected by the nematode's chemosensory organs control changes in the surface cuticle and behaviour. *Parasite* 15, 310–316.

Cutter, A.D., Félix, M.A., Barrière, A. and Charlesworth, D. (2006) Patterns of nucleotide polymorphism distinguish temperate and tropical wild isolates of *Caenorhabditis briggsae. Genetics* 173, 2021–2031.

Dolgin, E.S., Félix, M.A. and Cutter, A.D. (2008) Hakuna Nematoda: genetic and phenotypic diversity in African isolates of *Caenorhabditis elegans* and *C. briggsae. Heredity* 100, 304–315.

Dorris, M., De Ley, P. and Blaxter, M.L. (1999). Molecular analysis of nematode diversity and the evolution of parasitism. *Parasitology Today* 15, 188–193.

Edgar, R.S. and Wood, W.B. (1977) The nematode *Caenorhabditis elegans*: a new organism for intensive biological study. *Science* 198, 1285–1286.

Edison, A.S. (2009) *Caenorhabditis elegans* pheromones regulate multiple complex behaviors. *Current Opinion in Neurobiology* 19, 378–388.

Elling, A.A., Mitreva, M., Recknor, J., Gai, X., Martin, J., Maier, T.R., Mcdermott, J.P., Hewezi, T., McCarter, K.B.D., Davis, E.L., Hussey, R.S., Nettleton, D., McCarter, J.P. and Baum, T.J. (2007) Divergent evolution of arrested development in the dauer stage of *Caenorhabditis elegans* and the infective stage of *Heterodera glycines. Genome Biology* 8, R211.

Elling, A.A., Mitreva, M., Gai, X., Martin, J., Recknor, J., Davis, E.L., Hussey, R.S., Nettleton, D., McCarter, J.P. and Baum, T.J. (2009) Sequence mining and transcript profiling to explore cyst nematode parasitism. *BMC Genomics* 10, 58.

Evans, H., Mello, L.V., Fang, Y., Wit, E., Thompson, F.J., Viney, M.E. and Paterson, S. (2008) Microarray analysis of gender- and parasite-specific gene transcription in *Strongyloides ratti. International Journal for Parasitology* 38, 1329–1341.

Félix, M.-A. and Barrière, A. (2005) Evolvability of cell specification mechanisms. *Journal of Experimental Zoology Part B: Molecular and Developmental Evolution* 304B, 536–547.

Fielenbach, N. and Antebi, A. (2008) *C. elegans* dauer formation and the molecular basis of plasticity. *Genes & Development* 22, 2149–2165.

Fine, A.E., Ashton, F.T., Bhopale, V.M. and Schad, G.A. (1997) Sensory neuroanatomy of a skin penetrating nematode parasite *Strongyloides stercoralis*. II.Labial and cephalic neurons. *Journal of Comparative Neurology* 389, 212–223.

Fitch, D.H. (2005) Evolution: an ecological context for *C. elegans. Current Biology* 15, R655–R658.

Forst, S., Dowds, B., Boemare, N. and Stackebrandt, E. (1997) *Xenorhabdus* and *Photorhabdus* spp.: bugs that kill bugs. *Annual Review of Microbiology* 51, 47–72.

Gallo, M. and Riddle, D.L. (2009) Effects of a *Caenorhabditis elegans* dauer pheromone ascaroside on physiology and signal transduction pathways. *Journal of Chemical Ecology* 35, 272–279.

Goodrich-Blair, H. (2007) They've got a ticket to ride: *Xenorhabdus nematophila–Steinernema carpocapsae* symbiosis. *Current Opinion in Microbiology* 10, 225–230.

Grant, W.N., Stasiuk, S., Newton-Howes, J., Ralston, M., Bisset, S.A., Heath, D.D. and Shoemaker, C.B. (2006) *Parastrongyloides trichosuri*, a nematode parasite of mammals that is uniquely suited to genetic analysis. *International Journal for Parasitology* 36, 453–466.

Grove, D.I., Northern, C., Warwick, A. and Lovegrove, F.T. (1984) Loss of surface coat by *Strongyloides ratti* infective larvae during skin penetration: evidence using larvae radiolabelled with gallium. *Journal of Parasitology* 70, 689–693.

Grove, D.I., Warton, A., Northern, C. and Papadimitriou, J.M. (1987) Electron microscopical studies of *Strongyloides ratti* infective larvae: loss of the surface coat during skin penetration. *Journal of Parasitology* 73, 1030–1034.

Harvey, S.C., Shorto, A. and Viney, M.E. (2008) Quantitative genetic analysis of life-history traits of *Caenorhabditis elegans* in stressful environments. *BMC Evolutionary Biology* 8, 15.

Harvey, S.C., Barker, G.L., Shorto, A. and Viney, M.E. (2009) Natural variation in gene expression in the early development of dauer larvae of *Caenorhabditis elegans. BMC Genomics* 10, 325.

Hawdon, J.M. and Schad, G.A. (1991) Developmental adaptations in nematodes. In: Toft, C.A., Aeschlimann, A. and Bolis, L. (eds) *Parasitism: Coexistence or Conflict?* Oxford University Press, Oxford, UK, pp. 274–298.

Herrmann, M., Mayer, W.E. and Sommer, R.J. (2006) Nematodes of the genus *Pristionchus* are closely associated with scarab beetles and the Colorado potato beetle in Western Europe. *Zoology (Jena)* 109, 96–108.

Herrmann, M., Mayer, W.E., Hong, R.L., Kienle, S., Minasaki, R. and Sommer, R.J. (2007) The nematode *Pristionchus pacificus* (Nematoda: Diplogastridae) is associated with the oriental beetle *Exomala orientalis* (Coleoptera: Scarabaeidae) in Japan. *Zoological Science* 24, 883–889.

Hong, R.L. and Sommer, R.J. (2006) Chemoattraction in *Pristionchus* nematodes and implications for insect recognition. *Current Biology* 16, 2359–2365.

Hong, R.L., Svatos, A., Herrmann, M. and Sommer, R.J. (2008a) Species-specific recognition of beetle cues by the nematode

Pristionchus maupasi. Evolution & Devopment 10, 273–279.

Hong, R.L., Witte, H. and Sommer, R.J. (2008b) Natural variation in *Pristionchus pacificus* insect pheromone attraction involves the protein kinase EGL-4. *Proceedings of the National Academy of Sciences of the USA* 105, 7779–7784.

Hotez, P., Hawdon, J. and Schad, G.A. (1993) Hookworm larval infectivity, arrest and amphiparatenesis: the *Caenorhabditis elegans* Daf-c paradigm. *Parasitology Today* 9, 23–26.

Hu, M., Lok, J.B., Ranjit, N., Massey, H.C. Jr, Sternberg, P.W. and Gasser, R.B. (2010) Structural and functional characterisation of the fork head transcription factor-encoding gene, *Hc-daf-16*, from the parasitic nematode *Haemonchus contortus* (Strongylida). *International Journal for Parasitology* 40, 405–415.

Hu, P.J. (2007) Dauer. In: The *C. elegans* Research Community (ed.) *WormBook* (08.08.2007), doi/10.1895/wormbook.1.144.1, http://www.wormbook.org

Jeong, P.Y., Jung, M., Yim, Y.H., Kim, H., Park, M., Hong, E., Lee, W., Kim, Y.H., Kim, K. and Paik, Y.K. (2005) Chemical structure and biological activity of the *Caenorhabditis elegans* dauer-inducing pheromone. *Nature* 433, 541–545.

Johnson, C.D., Reinitz, C.A., Sithigorngul, P. and Stretton, A. (1996) Neuronal localisation of serotonin in the nematode *Ascaris suum. Journal of Comparative Neurology* 367, 352–360.

Joo, H.J., Yim, Y.H., Jeong, P.Y., Jin, Y.X., Lee, J.E., Kim, H., Jeong, S.K., Chitwood, D.J. and Paik, Y.K. (2009) *Caenorhabditis elegans* utilizes dauer pheromone biosynthesis to dispose of toxic peroxisomal fatty acids for cellular homoeostasis. *Biochemical Journal* 422, 61–71.

Junio, A.B., Li, X., Massey, H.C. Jr, Nolan, T.J., Lamitina, S.T., Sundaram, M.V. and Lok, J.B. (2008) *Strongyloides stercoralis*: cell- and tissue-specific transgene expression and co-transformation with vector constructs incorporating a common multifunctional 3′ UTR. *Experimental Parasitology* 118, 253–265.

Kenyon, C., Chang, J., Gensch, E., Rudner, A. and Tabtiang, R. (1993) A *C. elegans* mutant that lives twice as long as wild type. *Nature* 366, 461–464.

Kim, K., Sato, K., Shibuya, M., Zeiger, D.M., Butcher, R.A., Ragains, J.R., Clardy, J., Touhara, K. and Sengupta, P. (2009) Two chemoreceptors mediate developmental effects of dauer pheromone in *C. elegans. Science* 326, 994–998.

Kiontke, K. and Sudhaus, W. (2006) Ecology of *Caenorhabditis* species. In: The *C. elegans* Research Community (ed.) *WormBook* (09.01.2006), doi/10.1895/wormbook.1.37.1, http://www.wormbook.org

Massey, J., Holman, C., Bhopale, M.K., Li, X., Castelletto, M. and Lok, J.B. (2006) The fork head transcription factor FKTF-1b from *Strongyloides stercoralis* restores DAF-16 developmental function to mutant *Caenorhabditis elegans. International Journal for Parasitology* 36, 347–352.

Mayer, W.E., Herrmann, M. and Sommer, R.J. (2009) Molecular phylogeny of beetle associated diplogastrid nematodes suggests host switching rather than nematode–beetle coevolution. *BMC Evolutionary Biology* 9, 212.

Mitreva, M., McCarter, J.P., Martin, J., Dante, M., Wylie, T., Chiapelli, B., Pape, D., Clifton, S.W., Nutman, T.B. and Waterston, R.H. (2004) Comparative genomics of gene expression in the parasitic and free-living nematodes *Strongyloides stercoralis* and *Caenorhabditis elegans. Genome Research* 14, 209–220.

Motola, D.L., Cummins, C.L., Rottiers, V., Sharma, K.K., Li, T., Li, Y., Suino-Powell, K., Xu, H.E., Auchus, R.J., Antebi, A. and Mangelsdorf, D.J. (2006) Identification of ligands for DAF-12 that govern dauer formation and reproduction in *C. elegans. Cell* 124, 1209–1223.

Munoz, M.J. and Riddle, D.L. (2003) Positive selection of *Caenorhabditis elegans* mutants with increased stress resistance and longevity. *Genetics* 163, 171–180.

Ogawa, A., Streit, A., Antebi, A. and Sommer, R.J. (2009) A conserved endocrine mechanism controls the formation of dauer and infective larvae in nematodes. *Current Biology* 19, 67–71.

Perry, R.N. and Maule, A.G. (2004) Physiological and biochemical basis of nematode behaviour. In: Gaugler, R. and Bilgrami, A. (eds) *Nematode Behaviour*. CAB International, Wallingford, UK, pp. 197–238.

Rae, R., Riebesell, M., Dinkelacker, I., Wang, Q., Herrmann, M., Weller, A.M., Dieterich, C. and Sommer, R.J. (2008) Isolation of naturally associated bacteria of necromenic *Pristionchus* nematodes and fitness consequences. *Journal of Experimental Biology* 211, 1927–1936.

Ramos-Rodriguez, O., Campbell, J.F., Christen, J.M., Shapiro-Ilan, D.I., Lewis, E.E. and Ramaswamy, S.B. (2007) Attraction behaviour of three entomopathogenic nematode species towards infected and uninfected hosts. *Parasitology* 134, 729–738.

Richards, G.R. and Goodrich-Blair, H. (2009) Masters of conquest and pillage: *Xenorhabdus nematophila* global regulators control transitions from virulence to nutrient acquisition. *Cellular Microbiology* 11, 1025–1033.

Riddle, D.L. and Albert, P.S. (1997) Genetic and environmental regulation of dauer larva development. In: Riddle, D.L., Blumenthal, T., Meyer, B.J. and Priess, J.R. (eds) *C. elegans II*. Cold Spring Harbour Press, Cold Spring Harbour, New York, pp. 739–768.

Robertson, H.M. and Thomas, J.H. (2006) The putative chemoreceptor families of *C. elegans*. In: The *C. elegans* Research Community(ed.)*WormBook*(06.01.2007), doi/10.1895/wormbook.1.66.1, http://www.wormbook.org

Srinivasan, J., Kaplan, F., Ajredini, R., Zachariah, C., Alborn, H.T., Teal, P.E., Malik, R.U., Edison, A.S., Sternberg, P.W. and Schroeder, F.C. (2008) A blend of small molecules regulates both mating and development in *Caenorhabditis elegans*. *Nature* 454, 1115–1118.

Starich, T.A., Herman, R.K., Kari, C.K., Yeh, W.-H., Schackwitz, W.S., Schuyler, M.W., Collet, J., Thomas, J.H. and Riddle, D.L. (1995) Mutations affecting the chemosensory neurons of *Caenorhabditis elegans*. *Genetics* 139, 171–188.

Stasiuk, S. (2010) Functional analysis of the insulin/IGF signalling pathway and the infective larva developmental switch in *Parastrongyloides trichosuri*. PhD thesis, Massey University, New Zealand.

Stretton, A.O.W., Cowden, C., Sithigorngul, P. and Davis, R.E. (1991) Neuropeptides in the nematode *Ascaris suum*. *Parasitology* 102, S107–S116.

Stretton, A., Donmoyer, J., Davis, R., Meade, J., Cowden, C. and Sithigorngul, P. (1992) Motor behavior and motor nervous system function in the nematode *Ascaris suum*. *Journal of Parasitology* 78, 206–214.

Thompson, F.J., Barker, G.L., Hughes, L. and Viney, M.E. (2008) Genes important in the parasitic life of the nematode *Strongyloides ratti*. *Molecular and Biochemical Parasitology* 158, 112–119.

Tissenbaum, H.A., Hawdon, J., Perregaux, M., Hotez, P., Guarente, L. and Ruvkun, G. (2000) A common muscarinic pathway for diapause recovery in the distantly related nematode species *Caenorhabditis elegans* and *Ancylostoma caninum*. *Proceedings of the National Academy of Sciences of the USA* 97, 460–465.

Viney, M.E. (1996) Developmental switching in the parasitic nematode *Strongyloides ratti*. *Proceedings of the Royal Society of London B – Biological Sciences* 263, 201–208.

Viney, M.E. (2009) How did parasitic worms evolve? *Bioessays* 31, 496–499.

Viney, M.E., Green, L.D., Brooks, J.A. and Grant, W.N. (2002) Chemical mutagenesis of the parasitic nematode *Strongyloides ratti* to isolate ivermectin resistant mutants. *International Journal for Parasitology* 32, 1677–1682.

Viney, M.E., Gardner, M.P. and Jackson, J.A. (2003) Variation in *Caenorhabditis elegans* dauer larva formation. *Development, Growth and Differentiation* 45, 389–396.

Wang, Z., Zhou, X.E., Motola, D.L., Gao, X., Suino-Powell, K., Conneely, A., Ogata, C.,

Sharma, K.K., Auchus, R.J., Lok, J.B., Hawdon, J.M., Kliewer, S.A., Xu, H.E. and Mangelsdorf, D.J. (2009) Identification of the nuclear receptor DAF-12 as a therapeutic target in parasitic nematodes. *Proceedings of the National Academy of Sciences of the USA* 106, 9138–9143.

Wharton, D.A. (2003) The environmental physiology of Antarctic terrestrial nematodes: a review. *Journal of Comparative Physiology B* 173, 621–628.

Zhang, Y., Lu, H. and Bargmann, C.I. (2005) Pathogenic bacteria induce aversive olfactory learning in *Caenorhabditis elegans. Nature* 438, 179–184.

6 Gene Induction and Desiccation Stress in Nematodes

Ann M. Burnell[1] and Alan Tunnacliffe[2]

[1]*Department of Biology, National University of Ireland Maynooth, Maynooth, Co. Kildare, Ireland;* [2]*Institute of Biotechnology, Department of Chemical Engineering and Biotechnology, University of Cambridge, Cambridge, UK*

6.1 Introduction

Water accounts for 70% of the weight a living cell (Alberts *et al.*, 2008) and water is the medium in which the majority of cellular reactions take place. By means of hydrophobic interactions, water mediates the folding of proteins and the assembly of phospholipids into biological membranes; it also stabilizes the native structures of proteins, nucleic acids and lipids (Tanford, 1978; Chaplin, 2006). Lack or scarcity of water is a severe stress for organisms. Most animals die if they lose more than 15–20% of their body water (Barrett, 1991), and loss of more than 20–50% of their water content is lethal to most higher plants (Kranner *et al.*, 2002). Nevertheless, some organisms have evolved mechanisms that enable them to survive conditions of extreme desiccation in which there is no continuous aqueous phase in the cytoplasm and where the hydration shell of biomolecules is lost (Clegg, 1979, 2001; Barrett, 1991). This ability to survive extreme desiccation is known as anhydrobiosis. During anhydrobiosis, metabolism and life processes come to a halt, but they resume again on rehydration. For example, a viable culture of the nematode *Panagrolaimus* sp. PS443 was isolated from dry soil that had been stored in the laboratory for 8 years (Aroian *et al.*, 1993), and embryonated egg cysts of the brine shrimp *Artemia salina* can survive in an anhydrobiotic state for 15 years (Clegg, 1967). Plant seeds have been shown to survive for up to 300 years in herbaria and seed collections (Godwin, 1968), and lotus and date palm seeds have germinated after 1300 and 2000 years, respectively (Shen-Miller *et al.*, 1995; Sallon *et al.*, 2008).

Since all biological kingdoms contain some anhydrobiotic taxa, it seems likely that anhydrobiosis is a phylogenetically ancient adaptation for life in habitats subject to periodic desiccation. Several aquatic invertebrate phyla contain anhydrobiotic animals whose members typically occupy aquatic habitats that are prone to temporary water loss. These include rotifers, tardigrades, microcrustaceans, nematodes and the chironomid insect *Polypedilum vanderplanki* (reviewed by Watanabe, 2006). Free-living anhydrobiotic nematodes occur in habitats susceptible to desiccation, but a capacity to undergo anhydrobiosis has also been important in the evolution of parasitic nematodes, since many nematode parasites have anhydrobiotic infective stages or cysts (reviewed by Perry, 1999; see Perry and Moens, Chapter 1, this volume). Nematodes are essentially aquatic animals which need to be fully hydrated and covered in a water film for normal activity. Thus, a large majority of nematode species are highly sensitive to desiccation and osmotic stress and are not anhydrobiotic.

In this chapter we summarize the effects of water loss on the structure of cells, membranes, organelles and biomolecules, and we describe the means used by desiccation-resistant and anhydrobiotic nematodes to combat the effects of dehydration. Our knowledge of these mechanisms has been obtained largely from data on genes induced in response to desiccation stress in nematodes, gene knockout studies in *Caenorhabditis elegans*, *in vitro* studies of recombinant proteins from anhydrobiotic nematodes and by extrapolation from other anhydrobiotic systems, principally resurrection plants, seeds and yeast. Our summary shows that current knowledge of the molecular basis of nematode anhydrobiosis is patchy and that many of the steps in the anhydrobiotic signalling and response pathways of nematodes remain to be elucidated.

6.2 The Effects of Water Loss on Living Systems

Life has evolved in aqueous environments and therefore it comes as no surprise that the structure and function of cells and their components are dependent on the presence of water. For instance, the bilayers of cell membranes assemble as they do because of the amphiphilic nature of phospholipids: the hydrophobic fatty acid tails of the two leaflets prefer each other's company to that of water, and therefore are positioned internally, while the hydrophilic head groups are orientated at the external face of each leaflet, in contact with water molecules. This is an example of the hydrophobic effect (Tanford, 1980), where the thermodynamics of interactions between water and other molecules defines the biological structure and hence its function. Another example would be the globular, soluble proteins within the cell that fold around a core of hydrophobic residues, stabilizing the overall three-dimensional structure (Creighton, 1993).

Equally unsurprisingly, as stated by Parsegian *et al.* (1986), '[i]t is almost a truism that changes in the activity of solvent water will determine the state of organization of biological molecules.' Removal of water does not necessarily mean that biological structures disintegrate completely, but in the absence of protection systems they are profoundly changed. Thus, as they lose water, membranes suffer phase changes from the liquid crystal state to a much less fluid gel, and can further undergo a lamellar to inverted hexagonal (H_{II}) phase change and fusion events with other, normally separate membranes (Crowe and Crowe, 1984, 1986). Similarly, although proteins do not lose structure altogether as they dehydrate, they can become partially denatured such that they do not spontaneously adopt their correct conformation on rehydration, and they become prone to aggregation (Goyal *et al.*, 2005a; Choe and Strange, 2008).

There are consequences of desiccation for other cell components, e.g. nucleic acids, and for cell compartments, such as the mitochondria, some of which we will touch on below. The cell itself will experience morphological and ultrastructural changes: its volume will decrease, its surface will buckle and fold, and there is a danger of physical collapse. Such problems also translate to the whole organism level but, as might be expected, successful anhydrobiosis is usually associated with their minimization. For example, in a study on *Panagrolaimus* spp., Shannon *et al.* (2005) found that survival of desiccation was correlated with robustness of structure; decreasing desiccation tolerance in different species was linked with an increasing tendency of the nematodes to experience whole body collapse during drying (see Fig. 6.1).

In addition to a profound effect on the structure and function of cells and their components, loss of water leads to diminution and eventually cessation of metabolism, equivalent, in biochemical terms, to death. However, anhydrobiotic organisms are not dead in the dry state but merely experiencing a state of suspended animation: a third type of biological organization, distinguishable from life and death, according to Clegg (1973, 2001). This is the 'reversible standstill' of metabolism referred to by Keilin (1959) in his seminal review of anhydrobiosis; 'reversible' since, after the return of water, life processes resume once more. While it is widely accepted that dried organisms are ametabolic, on theoretical grounds and since many experimental studies have been performed, it is not clear whether metabolic shutdown is controlled in those organisms that can survive desiccation or whether it is simply a consequence of the lack of water. The imposition of environmental stress is often associated with quiescence and has, in some instances, been shown to be under molecular control. For example, in the yeast *Saccharomyces cerevisiae*, this is achieved under osmotic stress conditions by direct action of the Hog1 osmostress-responsive kinase on regulators of the cell cycle (Escote *et al.*, 2004) and components of the DNA replication complex (Yaakov *et al.*, 2009). Equally, however, it is well known that cells and organisms grown in

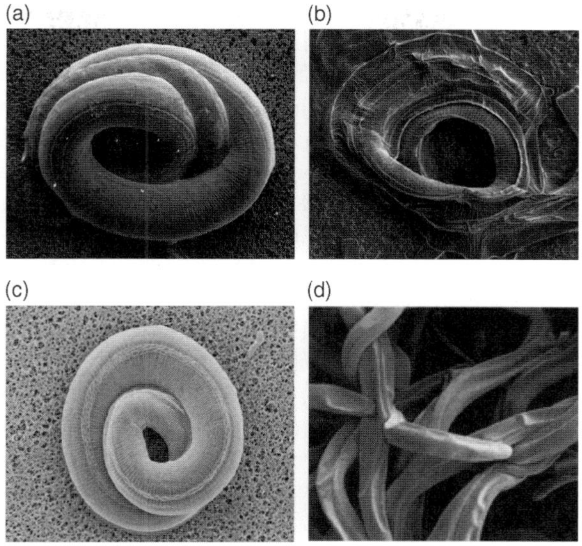

Fig. 6.1. Scanning electron micrographs of (a) *Panagrolaimus superbus* exposed *directly* to silica gel for 48h; (b) *Panagrolaimus* PS1732 preconditioned at 98% relative humidity (RH) for 78h prior to exposure to silica gel for 48h; (c) *Aphelenchus avenae* preconditioned at 98% RH for 72h, followed by exposure to silica gel for 24h; and (d) *Heterorhabditis megidis* infective juveniles following exposure to 98% RH for 96h. (Figs 6.1a and 6.1b are reproduced with permission from Shannon *et al.* (2005) *The Journal of Experimental Biology* 208(12), 2433–2445. © The Company of Biologists Ltd.).

culture, e.g. mammalian cells, can be frozen and stored in liquid nitrogen for long periods with maintenance of viability. Cells in this condition have been rendered ametabolic, but there has been no controlled shutdown since the frozen state was imposed, abruptly, by the researcher. Nevertheless, if cryoprotectants are included and freezing regimes are optimized, high viability rates are readily achievable. Similarly, the loss of water by evaporation will necessarily result in metabolic arrest: no control of metabolism need operate and, providing the cell or organism in question has evolved innate mechanisms to protect it from irreparable damage, rehydration might be all that is required to restart life processes.

So what are the protection systems which operate in anhydrobiotic organisms, including nematodes? Non-reducing disaccharides, trehalose (in animals and fungi) and sucrose (in plants), which accumulate in diverse anhydrobiotes, are thought to be involved and have been proposed to function as either water-replacement molecules or vitrification agents (see Tunnacliffe and Lapinski (2003) for a fuller description of these hypotheses and appropriate references to the literature). Other hypotheses, which are not incompatible with the above, have also been proposed (e.g. the hydration forces explanation; Bryant *et al.*, 2001; and see Teramoto *et al.*, 2008 for a review), and it seems likely that trehalose and similar molecules have multiple stabilizing functions within anhydrobiotes. However, non-reducing disaccharides are not an absolute requirement for anhydrobiosis, being absent from bdelloid rotifers (Lapinski and Tunnacliffe, 2003; Caprioli *et al.*, 2004) and some tardigrades (Hengherr *et al.*, 2008); indeed, some nematodes are also able to undergo anhydrobiosis while accumulating only very low levels of carbohydrates, namely *Plectus* sp. (Hendriksen, 1983) and *Ditylenchus* (= *Orrina*) *phyllobia* (Robinson *et al.*, 1984); the yeast *S. cerevisiae* remains desiccation tolerant when its capacity to synthesize trehalose has been removed by mutation (Ratnakumar and Tunnacliffe, 2006).

Therefore, even where the accumulation of trehalose or sucrose has been correlated with desiccation tolerance, it is highly likely that other molecules also contribute, e.g. small heat shock proteins in brine shrimp cysts (Liang *et al.*, 1997), extracellular polysaccharides and water stress proteins in the cyanobacterium *Nostoc commune* (Potts, 1999; Helm *et al.*, 2000) or amphiphilic molecules in plant embryos (Hoekstra *et al.*, 2001). Hydrophilic proteins might also play a role in drought and desiccation tolerance of plants (Ingram and Bartels, 1996), bacteria (Battista *et al.*, 2001) and anhydrobiotic nematodes (Solomon *et al.*, 2000; Browne *et al.*, 2002, 2004; Goyal *et al.*, 2003). Gene discovery programmes to determine key genes important in anhydrobiosis are well advanced in a number of organisms, and it is anticipated that the 'anhydrobiotic gene set' will become increasingly well defined. This chapter will cover some of these adaptations as they relate to nematodes.

6.3 Protein Homeostasis

The ability to detect, stabilize, repair or remove damaged proteins and maintain a functional proteome, i.e. protein homeostasis, is essential for the continued viability of the cell. This is of particular importance in a stress situation, when pressure on homeostatic mechanisms is substantially increased. Some small molecules, often called organic osmolytes or compatible solutes, which accumulate in many organisms in response to water stress, can behave as 'chemical chaperones' (Welch and Brown, 1996) and act to reduce protein unfolding. In the context of nematode anhydrobiosis, where trehalose is produced we would expect this disaccharide to behave as a chemical chaperone during the first dehydration stage of anhydrobiosis, when bulk water is being lost by evaporation but macromolecules are still fully hydrated (see Tunnacliffe and Lapinski (2003) for a discussion of the different stabilizing functions of trehalose at different stages of desiccation). Under these conditions, the cell interior will experience changes in intracellular pH, viscosity and solute concentrations, which will tend to destabilize protein conformation (Somero, 1986). Trehalose and similar chemical chaperones can counteract this tendency by preferential exclusion (Arakawa and Timasheff, 1982; Timasheff, 1992) from the protein surface due to excluded volume, preferential interaction and osmophobic effects (Bolen and Baskakov, 2001; Bolen, 2004). The disaccharide is partially excluded from a protein's hydration shell, thereby decreasing entropy in the system to a degree proportional to the surface area of the protein. The surface area of a protein is usually smaller in its native conformation than when it unfolds; therefore, in effect, trehalose forces the protein to fold correctly. It is worth noting that this is a thermodynamic stabilizing effect of trehalose, in contrast to the effect of glass formation as the dry state is attained (i.e. the vitrification hypothesis), where a kinetic stabilization is achieved by the reduction in reaction rate of degradative chemistry. Indeed, although glass formation is not a mechanism that confers anhydrobiosis by itself, the formation of intracellular glasses seems indispensable for survival of the dry state, at least in plant propagules. Both sugars and proteins contribute to glass formation, with such 'bioglasses' showing a more dense hydrogen bonding network than sugar glasses alone (Buitink and Leprince, 2004).

As well as small molecule involvement, we might expect that anhydrobiosis will require upregulation or activation of molecular chaperones (Ellis and Hartl, 2003), together with mechanisms for the removal of damaged proteins that cannot be repaired, i.e. the various ubiquitin–proteasome and autophagy–lysosome pathways (Knecht *et al.*, 2009). Prevention of widespread damage to the proteome must also extend to these repair and removal systems, of course, since the molecules involved are themselves proteins. Large-scale studies on the need for protein homeostasis under osmotic stress conditions have been performed in *C. elegans*. Although *C. elegans* is not desiccation tolerant, its response to osmotic stress could highlight important indicators of how nematodes deal with dehydration. In genome-wide screens, Strange's group has shown that 73% of genes involved in the osmostress response in *C. elegans* are concerned with protein homeostasis (Lamitina *et al.*, 2006) and that, of 40 genes essential for survival of hypertonic stress, half encode proteins governing the ubiquitin–proteasome, endosomal sorting and lysosome degradation pathways (Choe and Strange, 2008). Furthermore, using polyglutamine-containing marker proteins, Choe and Strange (2008) demonstrated that osmotic stress increases the tendency of proteins to aggregate as a result of cell shrinkage (for more on the osmotic stress responses of nematodes see Wharton and Perry, Chapter 11, this volume). Given these results, it is highly likely that the extreme water loss experienced by nematodes and other organisms undergoing desiccation will impose an even greater tendency on proteins to aggregate.

Nevertheless, although there are still relatively few studies on the anhydrobiotic genes of nematodes, the information to hand suggests that dehydration does not result in marked upregulation of many of the classical protein homeostasis genes. In the desiccation-resistant entomopathogenic species *Steinernema feltiae*, genes encoding ubiquitin and Hsp40 analogues were induced by dehydration among 81 genes identified (Gal *et al.*, 2003), but in a related nematode, *Steinernema carpocapsae*, these or other protein homeostasis genes were not found in a set of 41 dehydration-induced expressed sequence tags (ESTs) (Tyson *et al.*, 2007). Somvanshi *et al.* (2008) studied the regulation of four genes associated with stress tolerance in five entomopathogenic nematodes with varying resistance to desiccation. Interestingly, all four genes, including an analogue of the co-chaperone *hsp40*, showed the greatest degree of upregulation in the least desiccation-resistant nematode species; this suggests that this molecular chaperone gene, at least, is already transcribed at a high, 'stress-ready' level, or that other mechanisms are in place for maintenance of proteome functionality in the more tolerant species. A similar message emerges from a global study on the Antarctic nematode *Plectus murrayi*, by Adhikari *et al.* (2009): although several molecular chaperone genes were identified in a data set of ESTs, quantitative PCR experiments showed that genes for Hsp70 and Hsp90 were not upregulated by desiccation. Finally, in a small EST set from the well-characterized anhydrobiote *Aphelenchus avenae*, no canonical molecular chaperone genes or protein degradation genes were identified, although two chaperone-related genes were shown to be upregulated by desiccation and other stresses (Reardon *et al.*, 2010).

It is striking that few genes involved in disposal of damaged proteins, i.e. genes encoding components of the proteasomal and autophagy pathways, were identified in these studies, but this might simply reflect the limited amount of data available at the present time. Another caveat is that transcriptomics tells us nothing about post-transcriptional regulation. Only a single investigation of the effects of desiccation on the nematode proteome per se has been reported (Chen *et al.*, 2006), and again this is on *S. feltiae*; out of more than 400 protein spots identified by 2D-gel electrophoresis, ten showed changes in abundance on desiccation, one of which was the chaperonin Hsp60. Interestingly, these authors also found that an uncharacterized, cyclin-like F-box protein was also upregulated by desiccation; these proteins form part of the SCF complex, which recruits protein substrates for ubiquitination (Patten *et al.*, 1998). This could therefore indicate increased flux through the proteasome in *S. feltiae* during evaporative water loss.

The evidence for involvement of classical protein homeostasis systems is also rather tentative in other anhydrobiotes. For example, one study in tardigrades shows upregulation of one of three isoforms of Hsp70 during anhydrobiosis, although it seems to be most highly expressed during recovery from the dry state (Schill *et al.*, 2004). A follow-up report came to a similar conclusion at the protein level (Jönsson and Schill, 2007). In the desiccation-resistant larvae of the Antarctic midge *Belgica antarctica*, expression of genes encoding small heat shock proteins (sHsps) as well as Hsp70 and Hsp90 was increased by dehydration, although this varied between genes, depending on rate of water loss (Lopez-Martinez *et al.*, 2009). The Arctic springtail *Megaphorura arctica* survives sub-zero temperatures by undergoing cryoprotective dehydration (Worland, 1996; Clark *et al.*, 2009). Microarray and qPCR data showed that an sHsp gene (*hsp20*), but not *hsp70*, was upregulated in *M. arctica* in response to sub-zero temperatures and that the highest level of *hsp20* expression was recorded in the springtail during recovery from the desiccated state. By contrast, two springtail species from temperate regions (*Orchesella cincta* and *Folsomia candida*) showed elevated levels of Hsp70 protein under drought conditions (Bahrndorff *et al.*, 2009). In yeast, Hsp70 expression is not thought to enhance desiccation tolerance (Guzhova *et al.*, 2008), but in plant seeds upregulation of some Hsps, particularly sHsps, is correlated with entry into anhydrobiosis (Ingram and Bartels, 1996; Hoekstra *et al.*, 2001). Several molecular chaperone and proteasome-pathway-related genes are also implicated in dehydrating or rehydrating resurrection plants (O'Mahony and Oliver, 1999; Iturriaga *et al.*, 2006; Oliver *et al.*, 2009).

Therefore, despite a rather preliminary level of understanding of the role of protein homeostasis systems in anhydrobiosis, particularly in invertebrates, it seems likely that both small molecule and protein-based systems will be involved at some level in many, perhaps most, species. Indeed, recent evidence suggests a synergistic link between both systems, since trehalose seems able to stimulate autophagy, at least in mammalian cells (Sarkar *et al.*, 2007). However, in most desiccation-tolerant organisms, an additional class of protectants exists, which has been proposed to reduce some of the consequences of desiccation damage in proteins. These are highly hydrophilic proteins, the best characterized of which are the LEA (late embryogenesis abundant) proteins, originally described in plant seeds but now known to occur in invertebrates and microorganisms also (see several recent reviews: Rorat, 2006; Tunnacliffe and Wise, 2007; Battaglia *et al.*, 2008; Shih *et al.*, 2008; Tunnacliffe *et al.*, 2010).

LEA proteins have been known for many years to accumulate in maturing plant seeds as they acquire desiccation tolerance, but their discovery in invertebrates, including nematodes, suggests that they are important for anhydrobiosis in both animals and plants. Most LEA proteins are highly hydrophilic, small (10–30 kDa) and largely disordered, although exceptions have been described, and can be categorized on the basis of protein sequence motifs and peptide composition into various groups. In the plant *Arabidopsis thaliana*, for example, nine different groups have recently been proposed (Bies-Ethève *et al.*, 2008; Hundertmark and Hincha, 2008). However, in nematodes and other invertebrates, with only a few exceptions to date (Förster *et al.*, 2009; Sharon *et al.*, 2009), almost all LEA proteins characterized belong to group 3 (Pfam LEA_4). Expression of LEA proteins has been associated with tolerance to water stress caused by desiccation, salt or low temperature in many studies (see reviews above for citations). Furthermore, reduction of expression of LEA protein genes by RNA interference (RNAi) in *C. elegans* (Gal *et al.*, 2004), or by mutation in bacteria (Battista *et al.*, 2001), has been shown to decrease resistance to desiccation. Therefore, it seems likely that LEA proteins, and perhaps other highly hydrophilic proteins, play an important role in anhydrobiosis and other forms of water stress.

The precise molecular function of LEA proteins and other hydrophilins is still being evaluated, although they have been proposed to act as protein stabilizers, hydration buffers, membrane protectants, antioxidants, organic glass formers and ion sinks/chelators (Tunnacliffe and Wise, 2007; Battaglia *et al.*, 2008; Shih *et al.*, 2008). Evidence is accumulating to support some of these roles, particularly that of protein stabilization, since LEA proteins and hydrophilic proteins from various species can preserve enzyme activity *in vitro* after desiccation or freezing (Sanchez-Ballesta *et al.*, 2004; Goyal *et al.*, 2005a; Grelet *et al.*, 2005; Reyes *et al.*, 2005, 2008; Haaning *et al.*, 2008; Kovacs *et al.*, 2008; Nakayama *et al.*, 2008). One mechanism for the protection observed is the prevention of water-stress-induced aggregation of sensitive proteins. Many proteins form insoluble aggregates when dried or frozen, but aggregation is markedly reduced in the presence of LEA proteins (Goyal *et al.*, 2005a; Pouchkina-Stantcheva *et al.*, 2007; Kovacs *et al.*, 2008; Nakayama *et al.*, 2008). Probably due to their hydrophilic, unstructured nature, LEA proteins themselves are not generally susceptible to aggregation on desiccation, freezing or even boiling. This protein anti-aggregation activity can also extend *in vitro* to the protection of complex mixtures of proteins, such as the water-soluble proteomes of human and nematode cells, and to aggregation-prone targets *in vivo* (Chakrabortee *et al.*, 2007). The latter study showed that when a group 3 LEA protein from the nematode *A. avenae* was introduced into mammalian cells expressing proteins containing long polyglutamine (polyQ) or polyalanine (polyA) sequences, aggregate formation was reduced. The aggregation-prone polyQ or polyA expansion proteins are associated with a number of human neurodegenerative diseases, including Huntington's disease and oculopharyngeal muscular dystrophy. These experiments demonstrate anti-aggregation activity in the hydrated state, consistent with a potential role for LEA proteins in chilling stress and osmostress where the solid state is not reached.

The anti-aggregation activity of LEA proteins is reminiscent of that of molecular chaperones, particular the Hsps, sometimes known as 'holding' chaperones because they are capable of passively stabilizing protein species in a partially unfolded state. Holding chaperones prevent client proteins aggregating until stress has abated and refolding, by Hsp70 and Hsp60 teams or spontaneously, becomes possible or aggregates can be disposed of (e.g. Haslbeck *et al.*, 1999). However, LEA proteins can be distinguished from holding chaperones due to both their lack of structure and their relatively poor ability to prevent heat-induced protein aggregation (Goyal *et al.*, 2005a; although cf. Kovacs *et al.*, 2008). Another difference is that molecular chaperones form transient complexes with their client proteins, often through hydrophobic surfaces; such interactions are thought to be unlikely with unstructured LEA proteins, particularly since they are highly hydrophilic in nature, but no experimental evidence for or against such interactions has been reported. Finally, LEA protein genes are not usually upregulated by heat stress (Browne *et al.*, 2004; Hundertmark and Hincha, 2008), unlike many of the classical molecular chaperones.

In fact, LEA protein activity might be simpler than that of molecular chaperones: the unstructured nature of LEA proteins means they can function as entropic chains (Tompa, 2002) and exert an excluded volume effect, which, in the increasingly crowded environment of the dehydrating cytoplasm, could serve to decrease interaction between partially denatured polypeptides with the potential to aggregate. This has been called 'molecular shield' activity (Wise and Tunnacliffe, 2004; Goyal *et al.*, 2005a) and is similar to that of the entropic bristles of MAP2, tau and neurofilament side arms (Mukhopadhyay *et al.*, 2004), which act as spacers of cytoskeletal filaments. Shield proteins might also have a more general space-filling role in the dehydrating organism and help to prevent cellular collapse as water is lost (Tunnacliffe *et al.*, 2010).

In summary, protein homeostasis in desiccation-tolerant nematodes and other anhydrobiotes is likely to be achieved by a combination of chemical chaperone, molecular chaperone, molecular shield and protein degradation activities during the initial dehydration phase imposed by evaporative water loss. As the dry state is reached, different stresses begin to operate and therefore different protective mechanisms, such as water replacement and organic glass formation, might come into play. Then, during rehydration, protein repair and damage clearance systems are likely to be heavily engaged in re-establishing protein homeostasis. An important adjunct to the consideration of protein homeostasis comes from studies in the radiotolerant and highly desiccation-tolerant bacterium *Deinococcus radiodurans* and related species. Here, survival of desiccation correlates inversely with degree of protein oxidation; an unusual redox system based on high manganese and low iron concentrations seems to be critical (Fredrickson *et al.*, 2008). The implication is that stress protection and damage repair systems – at least those that are protein-based – might not be particularly unusual in anhydrobiotic organisms but that extraordinary antioxidation mechanisms are required to minimize oxidative damage to these protection and repair proteins (and presumably other cell components). It is not known whether nematodes and other anhydrobiotes also possess manganese-based antioxidant systems, but protection from oxidative stress has long been recognized as a vital component of the armoury against desiccation damage (see below).

6.4 Membrane Integrity in Anhydrobiotic Nematodes

An intact phospholipid membrane in the form of a bilayer is crucial to the survival of cells and organisms. When fully hydrated, a phospholipid bilayer can be described as a liquid crystal, i.e. its bilayer structure is conserved but the individual lipids are free to diffuse laterally through the membrane. When a hydrated lipid bilayer is desiccated, it may enter a gel phase (depending on the temperature) in which the lipid head groups and the hydrocarbon chains become tightly packed (Crowe *et al.*, 1989). Transitions between the gel and liquid crystalline phases during desiccation and rehydration are very damaging to living cells. Membrane-phase transitions can result in the aggregation of membrane proteins, the formation of non-bilayer lipid structures, fusion between adjacent bilayers and permeability changes (reviewed by Crowe and Crowe, 1984; Crowe *et al.*, 1989). Trehalose accumulation has been associated with the onset of desiccation in several anhydrobiotic nematodes including *A. avenae*, *Anguina tritici*, *Ditylenchus dipsaci*, and several species of *Panagrolaimus* (Shannon *et al.*, 2005, and references therein). The enzyme trehalose-6-phosphate synthase, which catalyses the first step in the biosynthesis of trehalose, is encoded by two genes in *A. avenae*, and the expression of both genes is upregulated in response to desiccation stress (Goyal *et al.*, 2005b). Seeds and angiosperm resurrection plants also accumulate sugars, particularly sucrose and raffinose (Amuti and Pollard, 1977; Vicré *et al.*, 2004), raffinose functioning to prevent the crystallization of sucrose (Caffrey *et al.*, 1988). Several studies have shown that sucrose and trehalose can stabilize the liquid crystalline form of phospholipid membranes and liposomes (reviewed by Crowe *et al.*, 1988).

The water replacement hypothesis proposes that trehalose and sucrose are able to insert between adjacent phosphate groups in the lipid bilayer, forming hydrogen bonds and maintaining sufficient space between the phospholipids to keep the lipid bilayer in a liquid crystalline state (Crowe *et al.*, 1984). Trehalose and sucrose may also contribute to the formation and stabilization of cytoplasmic glasses (Sun and Leopold, 1997; Crowe *et al.*, 1998). It has been possible to establish experimental conditions for aqueous sugar and phospholipid mixtures under which the sugars vitrify while the lipid membrane is still in a fluid state (Koster *et al.*, 1994, 2000). Under such circumstances, a glass forms on both sides of the membrane and, upon further desiccation, the rigidity of the glass prevents the membrane from contracting into the gel form (Bryant *et al.*, 2001). The plasma membrane contributes only 2–5% of the lipid bilayer area and mass of a eukaryotic cell; the remaining bilayer mass in a typical animal cell is derived from the endoplasmic reticulum (~50%), Golgi apparatus, and mitochondrial and nuclear membranes, and from the membranes of endosomes, peroxisomes and lysosomes (Alberts *et al.*, 2008). In severely dehydrated cells (<0.2 g H_2O/g dry weight), the membranes of cells and organelles are brought close together. It is important for anhydrobiotic survival that such membranes do not fuse. The scenario where intracellular membranes seem not to fuse with each other or undergo phase changes in desiccated anhydrobiotes has been attributed to a combination of the hydration repulsion that exists between adjacent lipid bilayers (Rand and Parsegian, 1989) and the accumulation of compatible solutes (Wolfe and Bryant, 1999; Lenne *et al.*, 2009). When hydrated phospholipid bilayers are less than 1 nm apart, it has been shown experimentally that a large repulsive force exists between the two membranes, which decreases exponentially with distance from the surface; this repulsive 'hydration force' has been attributed to non-random polarization of water molecules on the opposing surfaces (Rand and Parsegian, 1989; Bryant *et al.*, 2001). Severe desiccation produces vertical (i.e. at right angles to the lamellar surface of the bilayer) and lateral suction forces that remove water molecules from phospholipid membranes. The lateral suction force tends to compress the lipid bilayer and favours a transition from the liquid crystalline to gel phase (Wolfe and Bryant, 1999). According to the hydration force model, when trehalose and sucrose (or other compatible solutes) are present between the two lipid membranes, the osmotic pressure of these solutes would offset the suction forces caused by desiccation and thereby counteract the transition to gel phase; in addition, the volumetric effects of these solutes would also contribute to keeping the lipid bilayers apart (Wolfe and Bryant, 1999; Lenne *et al.*, 2009).

Plant seeds and pollen contain abundant amounts of polyphenolic antioxidant compounds, among them hydroquinones, such as rutin and arbutin, and flavonoids, such as quercetin. There is evidence that such amphiphilic molecules become partitioned from the cytoplasm to the lipid bilayer at the onset of desiccation in seeds (and liposomes), where they increase membrane fluidity (reviewed by Hoekstra *et al.*, 2001; Hoekstra and Golovina, 2002). A similar set of endogenous amphiphilic molecules has not been described in anhydrobiotic animals. Several studies have shown that LEA proteins may fold into an α-helical form upon desiccation (e.g. Goyal *et al.*, 2003). Some LEA α-helices may be amphipathic, with negatively charged residues forming a stripe along the axis of the helix and with a corresponding (relatively) hydrophobic stripe on the opposing helical face, which is flanked by stripes of positively charged residues (Tolleter *et al.*, 2007). Such molecules could become aligned parallel to the lipid bilayer with their negative charges in contact with bulk water, their positive charges interacting with the phosphate groups of the bilayer and the hydrophobic stripe dipping into the fatty acid tails of the membrane. Tolleter *et al.* (2007) have shown that an amphipathic LEA protein which is targeted to the inner membrane of pea seed mitochondria can protect liposome membranes from phase changes during desiccation. This discovery raises the possibility that LEA or other lipochaperone proteins (Tsvetkova *et al.*, 2002; Coucheney *et al.*, 2005; Horvath *et al.*, 2008) may also play a role in preserving membrane fluidity and bilayer stability in anhydrobiotic nematodes during desiccation. In summary, currently available data indicate that a variety of molecular mechanisms may be involved in the protection of phospholipid membranes in the anhdyrobiotic state. However, further research is required to determine the extent to which each of these mechanisms or combinations of mechanisms may be utilized by anhydrobiotic taxa at successive stages during the entry into (and exit from) anhydrobiosis.

Membrane fluidity is affected by composition of the phospholipids and the degree of fatty acid desaturation. The introduction of unsaturated bonds into the fatty acids of membrane phospholipids decreases the temperature for the transition from gel to liquid crystalline state (Los and Murata, 1998). Synthesis of transcripts encoding the fatty acid desaturase 2 (*fat-2*) gene was strongly upregulated in response to desiccation in *S. carpocapsae* (Tyson *et al.*, 2007). FAT-2 catalyses the insertion of a double bond at the Δ^{12} position of plamitoleic and linoleic acids. Lamitina and Strange (2005) found that RNAi knockdown of the fatty acid desaturase gene *fat-6* in *C. elegans* adults led to a reduction in the nematodes' ability to survive osmotic stress. Polyunsaturated fatty acids readily undergo lipid peroxidation (Gunstone, 1996); thus, an increase in fatty acid desaturation in response to desiccation in anhydrobiotic nematodes would require compensatory adaptations to repair oxidized membrane lipids.

Lipid membranes contain a complex mixture of proteins – surveys suggest that up to 30% of all predicted proteins in an organism are integral membrane proteins (Wallin and von Heijne, 1998). Some of these membrane proteins serve as cross-linkers between the cortical cytoskeleton and plasma membranes of adjacent cells. The cortical cytoskeleton is composed of a network of actin fibres (F-actin), which lie underneath the plasma membrane, providing strength and shape to the lipid bilayer. These cortical fibres form part of the cytoskeleton, a network of protein fibres which extends throughout the cytoplasm and which also contains intermediate filaments and microtubules (composed of tubulin). Exposure of mammalian cells to hyperosmotic stress induces rapid F-actin polymerization and remodelling of the actin cytoskeleton (Burg *et al.*, 2007). Microtubule reorganization has been associated with salt tolerance in seedlings of *Arabidopsis* (Wang *et al.*, 2007) and *Brassica napus* (Bagniewska-Zadworna, 2008) and with rehydration in the moss *Polytrichum formosum* (Pressel *et al.*, 2006). Thus, adjustments to the cytoskeleton may also be associated with the apparent structural stability of desiccated anhydrobiotic nematodes, but the effects of desiccation on the cytoskeleton of nematodes await investigation.

6.5 Oxidative Stress and its Effects during Desiccation and Anhydrobiosis

Aerobic organisms are constantly exposed to reactive oxygen species (ROS), which are generated in mitochondria during oxidative phosphorylation and are also generated in other subcellular compartments by a variety of oxidases, oxygenases and peroxidases. The major ROS in biological systems are superoxide ($O_2^{\bullet -}$) and hydroxyl ($^{\bullet}OH$) radicals and hydrogen peroxide (H_2O_2). The mitochondrial electron transport chain is 'leaky'; approximately 1–5% of electrons flowing through this transport chain pass directly to oxygen (generating $O_2^{\bullet -}$), instead of passing to the next electron carrier in the chain (reviewed by Cadenas and Davies, 2000; Halliwell and Gutteridge, 2007). These $O_2^{\bullet -}$ radicals are usually converted by mitochondrial superoxide dismutase (SOD) into H_2O_2, a potent oxidizing agent which can diffuse across the mitochondrial membranes to enter the cytoplasm. Hydroxyl ($^{\bullet}OH$) radicals, which are the most reactive free radical species known (Crichton *et al.*, 2002), are formed from the interaction of H_2O_2 and redox-active metal ions such as Fe^{2+} and Cu^{2+}. ROS accumulation is also triggered by cellular dehydration (Potts, 1994; Kranner and Birtic, 2005; França *et al.*, 2007). The contributing factors to the desiccation-induced build-up of ROS are likely to include impairment of the mitochondrial electron transport chain, dysfunctional enzymes, increased cytoplasmic ionic strength and the dehydration of biomolecules.

ROS accumulation in cells leads to the oxidative modification of proteins, lipids, DNA and other macromolecules, with proteins likely to be the first major class of molecules to be attacked (Du and Gebicki, 2004; Nauser *et al.*, 2005). Oxidative modifications can affect the activity of proteins and increase their susceptibility to aggregation, unfolding and degradation (Davies, 2005). The sulfur-containing amino acids, cysteine and methionine, are very sensitive ROS targets, although the side chains of arginine, lysine, proline, histidine, tryptophan and tyrosine are also prone to oxidation (Berlett and Stadtman, 1997). ROS cause substantial damage to membrane lipids because lipid peroxidation proceeds by a chain reaction mechanism in which the lipid radicals interact with oxygen, leading to the formation of other radical species that further propagate the peroxidation process (reviewed by Niki, 2009). Lipid peroxidation decreases membrane fluidity and increases membrane leakage, and, in addition, membrane proteins, enzymes and ion channels also experience oxidative damage (Halliwell and Gutteridge, 2007). ROS also damage DNA by causing base and nucleotide modifications and strand breaks.

Aerobic organisms have developed robust mechanisms to detoxify ROS, maintain redox homeostasis and repair ROS-mediated damage to macromolecules. Such systems are further enhanced in plants, which have to deal with a second source of ROS generated by the photosynthetic electron chain (Rouhier *et al.*, 2008). Cells possess several enzymes (catalase, glutathione peroxidase and peroxiredoxin) to deactivate the H_2O_2 formed in the mitochondria and cytoplasm by the activity of SOD; specialist phospholipid hydroperoxidases are targeted to membranes; malondialdehyde, an important end product of lipid peroxidation and itself a reactive molecule, is deactivated by aldehyde dehydrogenase. Catabolism via lysosomes, peroxisomes and the ubiquitin–proteasome system is the main fate of oxidatively damaged proteins (Davies, 2005; Halliwell and Gutteridge, 2007). These organelles are also responsible for the removal of unfolded and denatured proteins caused by dehydration. Choe and Strange (2008) carried out a whole genome RNAi screen in *C. elegans* and identified 40 genes that are essential for survival during acute hypertonic stress. Half of these genes encode proteins that function to detect, transport and degrade damaged proteins, including components of the ubiquitin–proteasome system, endosomal sorting complexes and lysosomes. Many enzymatic antioxidant defence systems are reported to be upregulated in response to desiccation or osmotic stress in microorganisms, plants and animals and in both desiccation-sensitive and anhydrobiotic nematodes. Some nematode examples are: in *S. feltiae*, aldehyde dehydrogenase (ALDH) (Gal *et al.*, 2003); in *S. carpocapsae*, glutaredoxin (GRx) and mitochondrial mnSOD (Tyson *et al.*, 2007); in *A. avenae*, GRx (Browne *et al.*, 2004) and glutathione peroxidase (GPx) (Reardon *et al.*, 2010); in *P. murrayi*, ALDH, glutathione *S*-transferase (GST) and GPx (Adhikari *et al.*, 2009); and in *Panagrolaimus superbus*, thioredoxin peroxidase and GST (Tyson *et al.*, 2010, unpublished data).

Enzymatic antioxidants require aqueous conditions and thus can provide protection against ROS activity only during the induction and recovery phases of anhydrobiosis. Although metabolic activity ceases in the fully anhydrobiotic state (Clegg, 2001), damage from ROS continues to accumulate; for example, yeast cells showed a greater than tenfold increase in intracellular oxidation levels after dehydration (Pereira *et al.*, 2003). When in a fully desiccated state, anhydrobiotes rely on non-enzymatic antioxidants to provide protection from ROS damage. Unlike plants, nematodes do not synthesize the potent antioxidants α-tocopherol (vitamin E) or ascorbic acid, but anhydrobiotic nematodes contain several other molecules which may function as antioxidants. The tripeptide glutathione (GSH) is the most abundant low molecular weight thiol in most biological systems – its concentration in the cytoplasm is $\sim$10 mM (Halliwell and Gutteridge, 2007). GSH functions as a scavenger of free radicals and as an electron donor to maintain protein thiols in the reduced state (Ghezzi, 2005). Schafer and Buettner (2001) have shown that the concentration of the two components of the glutathione disulfide/glutathione [GSSG]/[2GSH] redox couple can be used to provide a reliable estimate of the redox environment of the cell. Kranner *et al.* (2006) have shown that viability loss in seeds from four plant species coincides with the oxidation of GSH and accumulation of GSSG. Metallothioneins – small, cysteine-rich, metal-binding proteins – can also act as free radical scavengers (Sato and Bremner, 1993; Xue *et al.*, 2009). They are upregulated in the long-lived dauer larval stage of *C. elegans* (Wang and Kim, 2003). The polyamines spermine and spermidine are positively charged at physiological pH; thus, they can interact with negatively charged macromolecules such as nucleic acids, proteins and phospholipids. Spermine and spermidine can also act as scavengers of free radicals and aldehydes, and there is increasing interest in these molecules as protectants against stress-derived antioxidative damage in plants (reviewed by Groppa and Benavides, 2008). *Caenorhabditis elegans* also synthesizes these polyamines and requires them for normal growth and embryogenesis (Macrae *et al.*, 1998), but no data are available on a putative role for polyamines as antioxidants in free-living or anhydrobiotic nematodes. In addition to its proposed roles in protecting membranes and proteins from desiccation and contributing to organic glass formation, evidence is accumulating that trehalose may also function as an antioxidant (Benaroudj *et al.*, 2001; Herdeiro *et al.*, 2006; Luo *et al.*, 2008). Finally, LEA proteins and other natively unfolded proteins have abundant lysine and arginine residues, which are readily oxidized and could serve as a non-enzymatic antioxidant defence system in anhydrobiotic nematodes (Hara *et al.*, 2004), thus providing a protective sink to inactivate free radicals until the nematodes are rehydrated and the antioxidant enzymes begin to function again.

6.6 Stabilizing Nucleic Acids

There is sparse evidence for the effects of desiccation on DNA in nematodes and nothing at all on RNA: one preliminary report suggests that drying itself has little immediate effect on DNA integrity of fouth-stage larvae of *D. dipsaci* (Barrett and Butterworth, 1985). Although brief, this study does fit with observations in most other species; for example, in a recent analysis of dried tardigrade storage cells, Neumann *et al.* (2009) showed that initially, on drying, only a small fraction (2%) of genomic DNA was damaged sufficiently to migrate in a comet assay. However, this fraction increased to 14% after 6 weeks in the dry state, and to 24% after 10 months. Thus, it seems that DNA breaks accumulate during time in the dry state. Moreover, this occurs whether the organism is anhydrobiotic or not. In bacteria, DNA double-strand breaks become extensive over a period of days to weeks of desiccation in both *D. radiodurans*, a well-known anhydrobiote, and *Escherichia coli*, a desiccation-sensitive prokaryote. The striking difference – surely related to their respective anhydrobiotic potential – is that *D. radiodurans* is able to repair this damage, whereas *E. coli* is not (Mattimore and Battista, 1996). Furthermore, the same authors showed that the extreme radiotolerance for which *D. radiodurans* is named correlates with its desiccation tolerance. Indeed, it is known that radiation-resistant bacteria can be selected without prior exposure to radiation by screening for desiccation-tolerant organisms (Sanders and Maxcy, 1979). Therefore, it is perhaps not surprising that anhydrobiotic tardigrades and bdelloid rotifers also exhibit extreme radiotolerance associated with a highly efficient DNA repair system (Jönsson *et al.*, 2005; Gladyshev and Meselson, 2008). However, on a note of caution, these capabilities alone are probably insufficient for anhydrobiosis; directed evolution of radioresistance in *E. coli*, which results at least in part from enhanced DNA repair systems, is unlikely to give improved desiccation tolerance in this bacterium (Harris *et al.*, 2009). An intriguing exception to the above situation of damaged DNA being repaired in anhydrobiotes seems to occur in cyanobacteria. Even after years of desiccation, cyanobacterial genomes are reported to be largely intact (Jäger and Potts, 1988; Shirkey *et al.*, 2003; Billi, 2009). To our knowledge, it is not yet known whether desiccation-tolerant nematodes also exhibit radiotolerance and an efficient DNA repair capability, or whether perhaps they manage to avoid genome fragmentation on drying, like cyanobacteria. However, management of genome integrity is certain to be an important feature of their resistance to desiccation damage.

6.7 Model Nematodes for Anhydrobiosis Studies

Molecular phylogenetic approaches have separated the Phylum Nematoda into five major clades along with a sixth group, corresponding to the Order Chromadoria, which is considered to be paraphyletic (Blaxter *et al.*, 1998). Holterman *et al.* (2006) have further extended this work and have subdivided the Nematoda into 12 clades; however, in this review we use the simpler clade structure described by Blaxter and colleagues. Clade IV (Blaxter *et al.*, 1998) contains the best-described anhydrobiotic nematodes, including the free-living nematodes *A. avenae* and *Panagrolaimus* spp. (which can undergo anhydrobiosis at all stages in their life cycles) and the plant-parasitic nematodes of the Order Tylenchida, some of which can survive anhydrobiotically for >20 years (Norton, 1978; Watanabe, 2006). Tylenchid anhydrobiotic stages may consist of infective cysts, each cyst containing several hundred unhatched second-stage larvae (L2) within their individual eggshells (e.g. *Globodera*, *Meliodogyne*), or seed galls, each gall containing several thousand desiccated L2 (e.g. *A. tritici*), or 'eelworm wool', which comprises clumps of fourth-stage larvae (L4) of *D. dipsaci* (reviewed by Perry, 1999; see Perry and Moens, Chapter 1, this volume). Although plant-parasitic nematodes are generally inconvenient model systems for studies on the biochemistry and molecular biology of anhydrobiosis, *D. dipsaci*, which reproduces abundantly on infected plants, has been widely used for physiological, biochemical and ultrastructural studies (e.g. Perry, 1977; Barrett, 1982; Wharton and Lemmon, 1998).

Two broad categories of anhydrobiotic nematodes are recognized: fast- and slow-dehydration strategists (Womersley, 1987; see Perry and Moens, Chapter 1, this volume for discussion). Fast-dehydration-strategist nematodes inhabit environments exposed to frequent desiccation and they can survive immediate and prolonged exposure to rapid dehydration. These nematodes include *Panagrolaimus rigidus*, a nematode that inhabits extremely exposed environments on the surface of moss cushions (Ricci and Pagani, 1997), and the L4 of *D. dipsaci*, which can remain in an anhydrobiotic state for up to 23 years (Perry, 1977). Fast-dehydration strategists may be pre-adapted at a cellular level to survive desiccation, possibly by constitutively expressing protective molecules required for anhydrobiosis or by sequestering the mRNAs required for the synthesis of these protective molecules. Alternatively, they may have behavioural (coiling, clumping) or morphological adaptations which slow the rate of water loss (Perry, 1999). Slow-dehydration-strategist nematodes are unable to survive exposure to extreme desiccation unless they have first experienced a period of preconditioning to moderate reductions in relative humidity (RH). This preconditioning period enables the nematodes to induce the necessary biochemical changes required to survive in an anhydrobiotic state. Such nematodes are found in soil habitats that experience slow rates of water loss. The majority of anhydrobiotic nematodes are believed to fall into this latter group (Womersley, 1987).

The dauer larval stages of rhabditid nematodes are adapted for motile dispersal, longevity, reduced metabolic activity and food finding (see Grant and Viney, Chapter 5, this volume). While they may have some desiccation resistance (depending on the species), no examples of fully anhydrobiotic rhabditid dauer larvae have been described. For example, exposure of *C. elegans* dauer larvae to 97% RH for 24h resulted in 30% mortality of the treated worms (Gal *et al.*, 2004). Several research groups have investigated the desiccation tolerances of the dauer infective juvenile stages (IJ) of a variety of species and strains of the entomopathogenic nematodes (EPNs) *Heterorhabditis* and *Steinernema* (reviewed by Glazer, 2002; Grewal *et al.*, 2006). When preconditioned at high relative humidity, EPN IJs can resist moderate amounts of desiccation, but typically they are unable to survive exposure to <85% RH for more than a few days. Thus, EPN IJ are not anhydrobiotic nematodes *sensu stricto*. EPN can be readily cultured in large quantities in the laboratory and have been used as convenient models to study gene induction in response to mild dehydration stress (Gal *et al.*, 2003, 2005; Tyson *et al.*, 2007; Somvanshi *et al.*, 2008). In general, *Heterorhabditis* IJs tend to be more desiccation sensitive than those of *Steinernema* (O'Leary *et al.*, 2001; Liu *et al.*, 2002, Piggott *et al.*, 2002; Somvanshi *et al.*, 2008). However, *Heterorhabditis bacteriophora* has several advantages for desiccation studies: an RNAi protocol has been developed for this species (Ciche and Sternberg, 2007), its genome is currently being sequenced (Ciche, 2007) and a large *H. bacteriophora* EST data set has recently been published (Bai *et al.*, 2009). Although *C. elegans* is sensitive to desiccation and osmotic stress, the molecular resources available for this model nematode have been utilized very skilfully to identify signalling pathways (Solomon *et al.*, 2004; Lamitina *et al.*, 2006) and gene products required for survival during acute hypertonic stress (Choe and Strange, 2007a,b, 2008). Similarly Gal *et al.* (2004) have used RNAi to demonstrate a role for *Ce-lea-1* (a gene encoding a LEA protein) in the ability of *C. elegans* dauer larvae to withstand desiccation, and osmotic and heat stress. *Caenorhabditis elegans* could also provide a suitable desiccation-sensitive reporter system to investigate the effects of transgenic expression of candidate anhydrobiotic genes from other nematodes.

Previous work on the biochemistry and molecular genetics of anhydrobiosis in nematodes has been focused primarily on the fungal-feeding soil nematode *A. avenae*, which is a slow-dehydration strategist (Crowe and Madin, 1975). *Aphelenchus avenae* can be cultured in large quantities in flasks of autoclaved wheat seeds inoculated with the fungus *Rhizoctonia solani*. The large volume of nematodes recovered makes *A. avenae* ideal for biochemical studies, but it takes ~4 weeks to obtain a harvestable culture, microscopic observation of the nematodes *in situ* is difficult, being obscured by the fungal mycelium, and *A. avenae* is not amenable to RNAi by feeding (Reardon *et al.*, 2010) unless a suitable fungal vector system to deliver dsRNA can be developed. The genus *Panagrolaimus* contains both fast-dehydration and slow-dehydration anhydrobiote strategists (Wharton and Barclay, 1993; Ricci and Pagani, 1997; Shannon *et al.*, 2005). Panagrolaimids feed on bacteria and occupy a variety of niches, ranging from Antarctic, temperate and semi-arid soils to terrestrial mosses. The Antarctic species *Panagrolaimus davidi*, which has been used extensively by Wharton and colleagues (e.g. Wharton and Brown, 1991; Wharton *et al.*, 2003) to study the physiology of freezing tolerance, is also a slow-dehydration-strategist anhydrobiote (Wharton and Barclay, 1993). *Panagrolaimus* spp. have a short generation time (~10 days), and these nematodes can be grown in large quantities either on agar plates or in liquid culture, with *E. coli* as a food source, using protocols developed for *C. elegans*. Shannon *et al.* (2008) have also found that *Panagrolaimus* is amenable to gene knockout by RNAi when fed on *E. coli* expressing dsRNA for essential conserved genes. Phylogenetic analysis shows that the anhydrobiotic species and strains of *Panagrolaimus* characterized to date belong to a single clade (Shannon *et al.*, 2005), which will facilitate comparative transcriptome analyses of the molecular basis of dehydration tolerance within a single genus. Adhikari *et al.* (2009) have successfully cultured the Antarctic nematode *P. murrayi*, which is freezing and desiccation tolerant, and they have characterized an EST data set from this nematode (see Adhikari and Adams, Chapter 9, this volume). *Plectus murrayi* feeds on bacteria, and in the laboratory its life cycle takes ~6–8 weeks at 16°C. As a member of the Order Chromadoria, which is a basal grouping to clades III, IV and V (Blaxter *et al.*, 1998), *P. murrayi* will be a particularly interesting model to study the molecular evolution of freezing tolerance and anhydrobiosis in nematodes.

6.8 Conclusions and Future Directions

All organisms have some ability to respond to environmental fluctuations such as temperature changes or exposure to oxidative, osmotic or desiccation stress. This includes the capacity to detoxify ROS, to protect membranes, macromolecules and organelles, and to repair or remove damaged molecules. The potential to synthesize compatible solutes in response to desiccation and osmotic stress is also widespread in nature (Yancey *et al.*, 1982). Organisms living in environments subject to severe desiccation are under strong selection pressure to evolve mechanisms to survive during periods of seasonal or temporary water stress. The ability to undergo anhydrobiosis is the ultimate manifestation of such an adaptive response. This raises the question: is anhydrobiosis the result of an enhanced expression of the generalized stress response to dehydration which occurs in desiccation-sensitive organisms or do anhydrobiotes possess unique or novel molecules or specialist adaptations? Our analysis of the research literature suggests the former, since no unique adaptation has yet been found in anhydrobiotes. For example, trehalose is a particular example of a compatible solute, which is used by many organisms to counteract osmotic stress; LEA proteins are found in desiccation-sensitive nematodes like *C. elegans* and in non-tolerant plant seeds; and antioxidants are a feature of most stress responses in most organisms. Of course, there are many uncharacterized sequences currently being uncovered in genomics studies on anhydrobiotes, and the key adaptations essential for anhydrobiosis – and only found in anhydrobiotes – might be found among these. However, we might by now expect to have seen some correspondence between the uncharacterized sequences of different anhydrobiotes, but this is not the case.

All biological kingdoms contain some anhydrobiotic taxa; thus, anhydrobiosis is a phylogenetically ancient survival strategy. Oliver *et al.* (2000) hypothesize that a capacity for anhydrobiosis in vegetative tissues was primitively present in land plants (bryophytes) but was lost in the evolution of vascular plants. The successful radiation of vascular plants is generally considered to result from the evolution of anhydrobiotic seeds and pollen, and the evolution of efficient water transport mechanisms. However, a vegetative desiccation-tolerant phenotype has been independently re-acquired at least ten times by angiosperm resurrection plants (Oliver *et al.*, 2005). This re-evolution of desiccation tolerance appears to have been achieved by the expression in the resurrection plant vegetative tissues of genes involved in seed desiccation tolerance (Oliver *et al.*, 2000, 2005; Illing *et al.*, 2005). Each time a desiccation-tolerant phenotype re-evolved in a resurrection plant lineage, the response patterns and biochemical adaptations differed. Although our knowledge of the genetic response profiles of nematodes and other anhydrobiotic invertebrates is limited, it seems likely that, similar to resurrection plants, invertebrate taxa can evolve anhydrobiotic phenotypes by expressing varying combinations of functionally equivalent molecules. Possibly the major requirement for the successful evolution of an anhydrobiotic nematode is the development of sensitive control mechanisms which would: (i) coordinate the expression of a suite of effector molecules and processes required for cellular protection and repair; and (ii) ensure that these protective mechanisms function appropriately at different stages in the desiccation and rehydration processes (Fig. 6.2).

Gene induction studies in plants show that the signal transduction pathways responsible for vegetative desiccation tolerance and seed anhydrobiosis are highly complex (reviewed by Yamaguchi-Shinozaki and Shinozaki, 2006; Phillips *et al.*, 2007; Mazzucotelli *et al.*, 2008; Moore *et al.*, 2009). Such signal

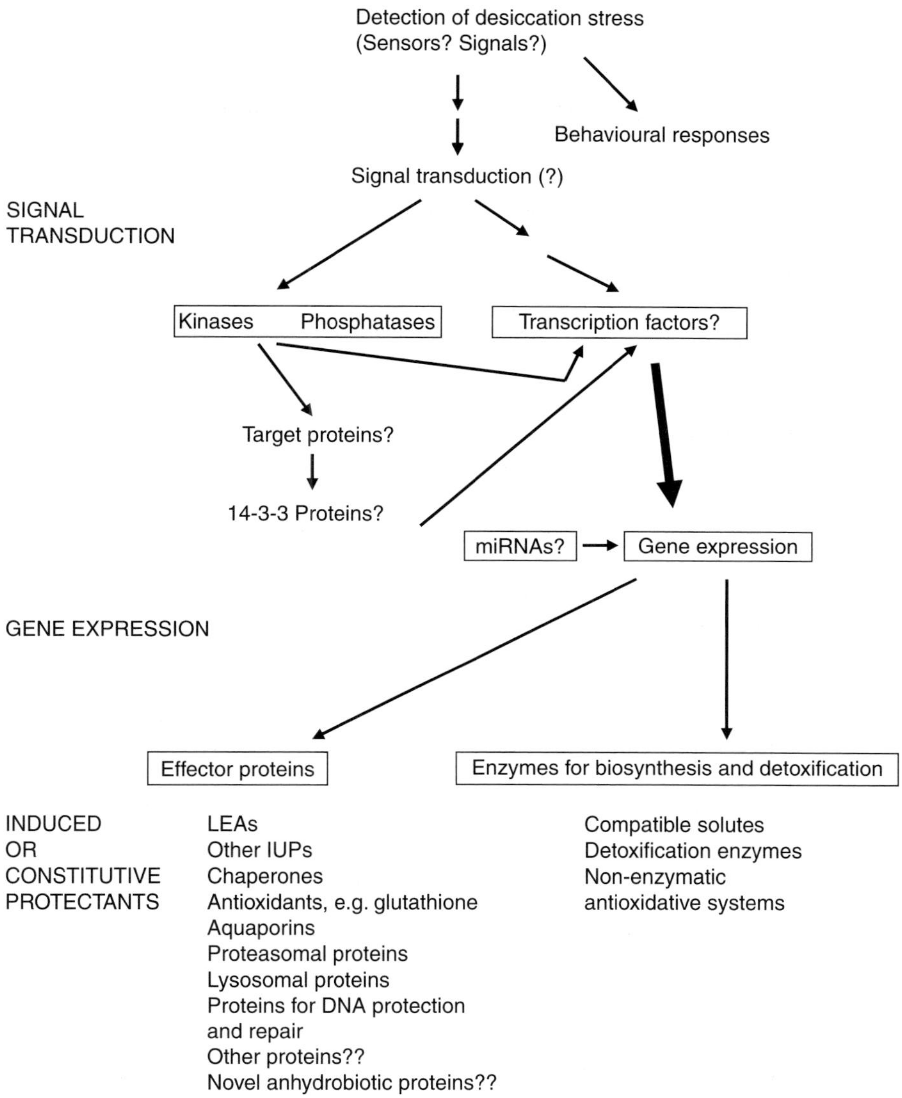

Fig. 6.2. Possible steps involved in the desiccation stress responses of anhydrobiotic nematodes (LEA, late embryogenesis abundant; IUP, intrinsically unfolded protein).

transduction pathways control the synthesis and functional activities of the effector proteins, enzymes and other molecules which provide anhydrobiotic protection. The expression of these effector molecules is tightly controlled and coordinated through complex layers of regulatory interactions, beginning with hormonal signalling pathways and including positive and negative interactions from regulatory systems such as protein kinase pathways, transcription factors, alternative splicing regulators, RNA binding proteins, miRNAs and protein sumoylation or ubiquitination.

Whole genome sequence data are currently available for the single-celled anhydrobiotes *D. radiodurans* and *S. cerevisiae*. In addition to *A. thaliana*, full genome sequence data are also available for seven other plant species with anhydrobiotic seeds (www.ncbi.nlm.nih.gov/genomes/PLANTS/PlantList.html), and a considerable amount of transcriptome and proteome data are also available for angiosperm resurrection plants. These data have provided significant insights into the molecular mechanisms involved in desiccation tolerance. However, genome, transcriptome and proteomic information for anhydrobiotic animals is still scant (or absent), and information on the regulatory pathways is lacking. This situation will change in the near future as anhydrobiotic taxa from all the main groups of animal anhydrobiotes – rotifers, tardigrades, brine shrimps and nematodes – are being investigated at the molecular level and several EST and genome sequencing projects are currently underway. New DNA sequencing technologies provide the opportunity to generate large amounts of genomic and transcriptomic data rapidly for non-model organisms. Access to high-throughput transcriptomic data also enables investigators to track gene expression profiles on a whole genome scale; when applied to an anhydrobiotic nematode, such an approach would identify its 'desiccome'. Desiccome data would then need to be aligned with proteome data and subjected to further *in vitro* and *in vivo* validation. Comparisons of transcriptome data from anhydrobiotic taxa from different phyla will also aid in identifying core anhydrobiotic processes and in distinguishing these from 'private' species-specific novelties or refinements.

6.9 Acknowledgments

Research in our laboratories is funded by the Leverhulme Trust (A.T. and A.M.B.), by Science Foundation Ireland (A.M.B.) and by the European Research Council (A.T.).

6.10 References

Adhikari, B.N., Wall, D.H. and Adams, B.J. (2009) Desiccation survival in an Antarctic nematode: molecular analysis using expressed sequenced tags. *BMC Genomics* 10, 69.

Alberts, B., Johnson, A., Lewis, J., Raff, M., Roberts, K. and Walter, P. (2008) *Molecular Biology of the Cell*, 5th edn. Garland Science, New York.

Amuti, K.S. and Pollard, C.J. (1977) Soluble carbohydrates of dry and developing seeds. *Phytochemistry* 16, 529–532.

Arakawa, T. and Timasheff, S.N. (1982) Stabilization of protein structure by sugars. *Biochemisty* 21, 6536–6544.

Aroian, R.V., Carta, L., Kaloshian, I. and Sternberg, P.W. (1993) A free-living *Panagrolaimus* sp. from Armenia can survive in anhydrobiosis for 8.7 years. *Journal of Nematology* 25, 500–502.

Bagniewska-Zadworna, A. (2008) The root microtubule cytoskeleton and cell cycle analysis through desiccation of *Brassica napus* seedlings. *Protoplasma* 233, 177–185.

Bahrndorff, S., Tunnacliffe, A., Wise, M.J., McGee, B., Holmstrup, M. and Loeschcke, V. (2009) Bioinformatics and protein expression analyses implicate LEA proteins in the drought response of Collembola. *Journal of Insect Physiology* 55, 210–217.

Bai, X.D., Adams, B.J., Ciche, T.A., Clifton, S., Gaugler, R., Hogenhout, S.A., Spieth, J., Sternberg, P.W., Wilson, R.K. and Grewal, P.S. (2009) Transcriptomic analysis of the entomopathogenic nematode *Heterorhabditis bacteriophora* TTO1. *BMC Genomics* 10, 205.

Barrett, J. (1982) Metabolic responses to anabiosis in the fourth stage juveniles of *Ditylenchus dipsaci* (Nematoda). *Proceedings of the Royal Society of London Series B – Biological Sciences* 216, 159–177.

Barrett, J. (1991) Anhydrobiotic nematodes. *Agricultural Zoology Reviews* 4, 161–176.

Barrett, J. and Butterworth, P.E. (1985) DNA stability in the anabiotic fourth-stage juveniles of *Ditylenchus dipsaci. Annals of Applied Biology* 106, 121–124.

Battaglia, M., Olvera-Carrillo, Y., Garciarrubio, A., Campos, F. and Covarrubias, A.A. (2008) The enigmatic LEA proteins and other hydrophilins. *Plant Physiology* 148, 6–24.

Battista, J.R., Park, M.J. and McLemore, A.E. (2001) Inactivation of two homologues of proteins presumed to be involved in the desiccation tolerance of plants sensitizes *Deinococcus radiodurans* R1 to desiccation. *Cryobiology* 43, 133–139.

Benaroudj, N., Lee, D.H. and Goldberg, A.L. (2001) Trehalose accumulation during cellular stress protects cells and cellular proteins from damage by oxygen radicals. *Journal of Biological Chemistry* 276, 24261–24267.

Berlett, B.S. and Stadtman, E.R. (1997) Protein oxidation in aging, disease, and oxidative stress. *Journal of Biological Chemistry* 272, 20313–20316.

Bies-Ethève, N., Gaubier-Comella, P., Debures, A., Lasserre, E., Jobet, E., Raynal, M., Cooke, R. and Delseny, M. (2008) Inventory, evolution and expression profiling diversity of the LEA (late embryogenesis abundant) protein gene family in *Arabidopsis thaliana. Plant Molecular Biology* 67, 107–124.

Billi, D. (2009) Subcellular integrities in *Chroococcidiopsis* sp. CCMEE 029 survivors after prolonged desiccation revealed by molecular probes and genome stability assays. *Extremophiles* 13, 49–57.

Blaxter, M.L., De Ley, P., Garey, J.R. *et al.* (1998) A molecular evolutionary framework for the phylum Nematoda. *Nature* 392, 71–75.

Bolen, D.W. (2004) Effects of naturally occurring osmolytes on protein stability and solubility: issues important in protein crystallization. *Methods* 34, 312–322.

Bolen, D.W. and Baskakov, I.V. (2001) The osmophobic effect: natural selection of a thermodynamic force in protein folding. *Journal of Molecular Biology* 310, 955–963.

Browne, J., Tunnacliffe, A. and Burnell, A. (2002) Plant desiccation gene found in a nematode. *Nature* 416, 38.

Browne, J.A., Dolan, K.M., Tyson, T., Goyal, K., Tunnacliffe, A. and Burnell, A.M. (2004) Dehydration-specific induction of hydrophilic protein genes in the anhydrobiotic nematode *Aphelenchus avenae. Eukaryotic Cell* 3, 966–975.

Bryant, G., Koster, K.L. and Wolfe, J. (2001) Membrane behaviour in seeds and other systems at low water content: the various effects of solutes. *Seed Science Research* 11, 17–25.

Buitink, J. and Leprince, O. (2004) Glass formation in plant anhydrobiotes: survival in the dry state. *Cryobiology* 48, 215–220.

Burg, M.B., Ferraris, J.D. and Dmitrieva, N.I. (2007) Cellular response to hyperosmotic stresses. *Physiological Reviews* 87, 1441–1474.

Cadenas, E. and Davies, K.J.A. (2000) Mitochondrial free radical generation, oxidative stress, and aging. *Free Radical Biology and Medicine* 29, 222–230.

Caffrey, M., Fonseca, V. and Leopold, A.C. (1988) Lipid–sugar interactions – relevance to anhydrous biology. *Plant Physiology* 86, 754–758.

Caprioli, M., Katholm, A.K., Melone, G., Ramløv, H., Ricci, C. and Santo, N. (2004) Trehalose in desiccated rotifers: a comparison between a bdelloid and a monogonont species. *Comparative Biochemistry and Physiology Part A* 139, 527–532.

Chakrabortee, S., Boschetti, C., Walton, L.J., Sarkar, S., Rubinsztein, D.C. and Tunnacliffe, A. (2007) Hydrophilic protein associated with desiccation tolerance exhibits broad protein stabilization function. *Proceedings of the National Academy of Sciences of the USA* 104, 18073–18078.

Chaplin, M. (2006) Do we underestimate the importance of water in cell biology? *Nature Reviews Molecular Cell Biology* 7, 861–866.

Chen, S.B., Glazer, I., Gollop, N., Cash, P., Argo, E., Innes, A., Stewart, E., Davidson, I. and Wilson, M.J. (2006) Proteomic analysis of the entomopathogenic nematode *Steinernema feltiae* IS-6IJs under evaporative and osmotic stresses. *Molecular and Biochemical Parasitology* 145, 195–204.

Choe, K.P. and Strange, K. (2007a) Evolutionarily conserved WNK and Ste20 kinases are essential for acute volume recovery and survival after hypertonic shrinkage in *Caenorhabditis elegans*. *American Journal of Physiology – Cell Physiology* 293, C915–C927.

Choe, K.P. and Strange, K. (2007b) Molecular and genetic characterization of osmosensing and signal transduction in the nematode *Caenorhabditis elegans*. *FEBS Journal* 274, 5782–5789.

Choe, K.P. and Strange, K. (2008) Genome-wide RNAi screen and *in vivo* protein aggregation reporters identify degradation of damaged proteins as an essential hypertonic stress response. *American Journal of Physiology and Cell Physiology* 295, C1488–C1498.

Ciche, T. (2007) The biology and genome of *Heterorhabditis bacteriophora*. In: The *C. elegans* Research Community (ed.) *WormBook*, doi/10.1895/wormbook.1.135.1, http://www.wormbook.org

Ciche, T.A. and Sternberg, P.W. (2007) Postembryonic RNAi in *Heterorhabditis bacteriophora*: a nematode insect parasite and host for insect pathogenic symbionts. *BMC Developmental Biology* 7, 101.

Clark, M.S., Thorne, M.A.S., Purac, J., Burns, G., Hillyard, G., Popovic, Z.D., Grubor-Lajsic, G. and Worland, M.R. (2009) Surviving the cold: molecular analyses of insect cryoprotective dehydration in the Arctic springtail *Megaphorura arctica* (Tullberg). *BMC Genomics*, 10, 328.

Clegg, J.S. (1967) Metabolic studies of cryptobiosis in encysted embryos of *Artemia salina*. *Comparative Biochemistry and Physiology* 20, 801–809.

Clegg, J.S. (1973) Do dried cryptobiotes have a metabolism? In: Crowe, J.H. and Clegg, J.S. (eds) *Anhydrobiosis*. Dowden, Hutchinson and Ross, Stroudsburg, Pennsylvania, pp. 141–146.

Clegg, J.S. (1979) Metabolism and the intracellular water environment: the vicinal–water network model. In: Drost-Hansen, W. and Clegg, J.S. (eds) *Cell Associated Water*. Academic Press, New York, pp. 363–413.

Clegg, J.S. (2001) Cryptobiosis – a peculiar state of biological organization. *Comparative Biochemistry and Physiology Part B* 128, 613–624.

Coucheney, F., Gal, L., Beney, L., Lherminier, J., Gervais, P. and Guzzo, J. (2005) A small HSP, Lo18, interacts with the cell membrane and modulates lipid physical state under heat shock conditions in a lactic acid bacterium. *Biochimica et Biophysica Acta – Biomembranes* 1720, 92–98.

Creighton, T.E. (1993) *Proteins: Structures and Molecular Properties*, 2nd edn. W.H. Freeman, New York.

Crichton, R.R., Wilmet, S., Legssyer, R. and Ward, R.J. (2002) Molecular and cellular mechanisms of iron homeostasis and toxicity in mammalian cells. *Journal of Inorganic Biochemistry* 91, 9–18.

Crowe, J.H. and Crowe, L.M. (1984) Effects of dehydration on membranes and membrane stabilization at low water activities. In: Chapman, D. (ed.) *Biological*

Membranes, Vol. 5. Academic Press, London, pp. 58–103.
Crowe, J.H. and Crowe, L.M. (1986) Stabilization of membranes in anhydrobiotic organisms. In: Leopold, A.C. (ed.) *Membranes, Metabolism and Dry Organisms*. Cornell University Press, New York, pp. 188–209.
Crowe, J.H. and Madin, K.A.C. (1975) Anhydrobiosis in nematodes – evaporative water-loss and survival. *Journal of Experimental Zoology* 193, 323–333.
Crowe, J.H., Crowe, L.M. and Chapman, D. (1984) Preservation of membranes in anhydrobiotic organisms – the role of trehalose. *Science* 223, 701–703.
Crowe, J.H., Crowe, L.M., Carpenter, J.F., Rudolph, A.S., Wistrom, C.A., Spargo, B.J. and Anchordoguy, T.J. (1988) Interactions of sugars with membranes. *Biochimica et Biophysica Acta* 947, 367–384.
Crowe, J.H., Crowe, L.M. and Hoekstra, F.A. (1989) Phase-transitions and permeability changes in dry membranes during rehydration. *Journal of Bioenergetics and Biomembranes* 21, 77–91.
Crowe, J.H., Hoekstra, F.A. and Crowe, L.M. (1992) Anhydrobiosis. *Annual Review of Physiology* 54, 579–599.
Crowe, J.H., Carpenter, J.F. and Crowe, L.M. (1998) The role of vitrification in anhydrobiosis. *Annual Review of Physiology* 60, 73–103.
Davies, M.J. (2005) The oxidative environment and protein damage. *Biochimica et Biophysica Acta – Proteins and Proteomics* 1703, 93–109.
Du, J. and Gebicki, J.M. (2004) Proteins are major initial cell targets of hydroxyl free radicals. *International Journal of Biochemistry & Cell Biology* 36, 2334–2343.
Ellis, R.J. and Hartl, F.-U. (2003) Protein folding and chaperones. In: Cooper, D.N. (ed.) *Nature Encyclopedia of the Human Genome*. Nature Publishing Group, London, pp. 806–810.
Escote, X., Zapater, M., Clotet, J. and Posas, F. (2004) Hog1 mediates cell-cycle arrest in G1 phase by the dual targeting of Sic1. *Nature Cell Biology* 6, 997–1002.
Förster, F., Liang, C., Shkumatov, A. *et al.* (2009) Tardigrade workbench: comparing stress-related proteins, sequence-similar and functional protein clusters as well as RNA elements in tardigrades. *BMC Genomics* 10, 469.
França, M.B., Panek, A.D. and Eleutherio, E.C.A. (2007) Oxidative stress and its effects during dehydration. *Comparative Biochemistry and Physiology A – Molecular and Integrative Physiology* 146, 621–631.
Fredrickson, J.K., Li, S.-M.W., Gaidamakova, E.K., Matrosova, V.Y., Zhai, M., Sulloway, H.M., Scholten, J.C., Brown, M.G., Balkwill, D.L. and Daly, M.J. (2008) Protein oxidation: key to bacterial desiccation resistance? *ISME Journal* 2, 393–403.
Gal, T.Z., Glazer, I. and Koltai, H. (2003) Differential gene expression during desiccation stress in the insect-killing nematode *Steinernema feltiae* IS-6. *Journal of Parasitology* 89, 761–766.
Gal, T.Z., Glazer, I. and Koltai, H. (2004) An LEA group 3 family member is involved in survival of *C. elegans* during exposure to stress. *FEBS Letters* 577, 21–26.
Gal, T.Z., Glazer, I., Sherman, A. and Koltai, H. (2005) Protein interaction of nucleosome assembly protein 1 and casein kinase 2 during desiccation response in the insect-killing nematode *Steinemema feltiae* IS-6. *Journal of Parasitology* 91, 691–693.
Ghezzi, P. (2005) Regulation of protein function by glutathionylation. *Free Radical Research* 39, 573–580.
Gladyshev, E. and Meselson, M. (2008) Extreme resistance of bdelloid rotifers to ionizing radiation. *Proceedings of the National Academy of Sciences of the USA* 105, 5139–5144.
Glazer, I. (2002) Survival biology. In: Gaugler, R. (ed.) *Entomopathogenic Nematology*. CAB International, Wallingford, UK, pp. 169–188.
Godwin, H. (1968) Evidence for longevity of seeds. *Nature* 220, 708–709.
Goyal, K., Tisi, L., Basran, A., Browne, J., Burnell, A., Zurdo, J. and Tunnacliffe, A. (2003) Transition from natively unfolded

to folded state induced by desiccation in an anhydrobiotic nematode protein. *Journal of Biological Chemistry* 278, 12977–12984.

Goyal, K., Walton, L.J. and Tunnacliffe, A. (2005a) LEA proteins prevent protein aggregation due to water stress. *Biochemical Journal* 388, 151–157.

Goyal, K., Browne, J.A., Burnell, A.M. and Tunnacliffe, A. (2005b) Dehydration-induced *tps* gene transcripts from an anhydrobiotic nematode contain novel spliced leaders and encode atypical GT-20 family proteins. *Biochimie* 87, 565–574.

Grelet, J., Benamar, A., Teyssier, E., Avelange-Macherel, M.-H., Grunwald, D. and Macherel, D. (2005) Identification in pea seed mitochondria of a late-embryogenesis abundant protein able to protect enzymes from drying. *Plant Physiology* 137, 157–167.

Grewal, P.S., Bornstein-Forst, S., Burnell, A.M., Glazer, I. and Jagdale, G.B. (2006) Physiological, genetic, and molecular mechanisms of chemoreception, thermobiosis, and anhydrobiosis in entomopathogenic nematodes. *Biological Control* 38, 54–65.

Groppa, M.D. and Benavides, M.P. (2008) Polyamines and abiotic stress: recent advances. *Amino Acids* 34, 35–45.

Guindon, S. and Gascuel, O. (2003) A simple, fast, and accurate algorithm to estimate large phylogenies by maximum likelihood. *Systematic Biology* 52, 696–704.

Gunstone, F.D. (1996) *Fatty Acid and Lipid Chemistry*. Blackie Academic and Professional, London.

Guzhova, I., Krallish, I., Khroustalyova, G., Margulis, B. and Rapoport, A. (2008) Dehydration of yeast: changes in the intracellular content of Hsp70 family proteins. *Process Biochemistry* 43, 1138–1141.

Haaning, S., Radutoiu, S., Hoffmann, S.V., Dittmer, J., Giehm, L., Otzen, D.E. and Stougaard, J. (2008) An unusual intrinsically disordered protein from the model legume *Lotus japonicus* stabilizes proteins *in vitro*. *Journal of Biological Chemistry* 283, 31142–31152.

Halliwell, B. and Gutteridge, J.M.C. (2007) *Free Radicals in Biology and Medicine*, 3rd edn. Oxford University Press, Oxford, UK.

Hara, M., Terashima, S., Fukaya, T. and Kuboi, T. (2003) Enhancement of cold tolerance and inhibition of lipid peroxidation by citrus dehydrin in transgenic tobacco. *Planta* 217, 290–298.

Hara, M., Fujinaga, M. and Kuboi, T. (2004) Radical scavenging activity and oxidative modification of citrus dehydrin. *Plant Physiology and Biochemistry* 42, 657–662.

Harris, D.R., Pollock, S.V., Wood, E.A. *et al.* (2009) Directed evolution of ionizing radiation resistance in *Escherichia coli*. *Journal of Bacteriology* 191, 5240–5252.

Haslbeck, M., Walke, S., Stromer, T., Ehrnsperger, M., White, H.E., Chen, S., Saibil, H.R. and Buchner, J. (1999) Hsp26: a temperature-regulated chaperone. *EMBO Journal* 18, 6744–6751.

Helm, R.F., Huang, Z., Edwards, D., Leeson, H., Peery, W. and Potts, M. (2000) Structural characterization of the released polysaccharide of desiccation-tolerant *Nostoc commune* DRH-1. *Journal of Bacteriology* 182, 974–982.

Hendriksen, N.B. (1983) Anhydrobiosis in nematodes: studies on *Plectus* sp. In: Lebrun, P., André, H.M., De Medts, A., Grégoire-Wibo, C. and Wauthy, G. (eds) *New Trends in Soil Biology*. Dieu-Brichart, Ottignies-Louvain-la-Neuve, France, pp. 387–394.

Hengherr, S., Heyer, A.G., Koehler, H.R. and Schill, R.O. (2008) Trehalose and anhydrobiosis in tardigrades – evidence for divergence in responses to dehydration. *FEBS Journal* 275, 281–288.

Herdeiro, R.S., Pereira, M.D., Panek, A.D. and Eleutherio, E.C.A. (2006) Trehalose protects *Saccharomyces cerevisiae* from lipid peroxidation during oxidative stress. *Biochimica et Biophysica Acta – General Subjects* 1760, 340–346.

Hoekstra, F.A. and Golovina, E.A. (2002) The role of amphiphiles. *Comparative Biochemistry and Physiology A – Molecular and Integrative Physiology* 131, 527–533.

Hoekstra, F.A., Golovina, E.A. and Buitink, J. (2001) Mechanisms of plant desiccation tolerance. *Trends in Plant Science* 6, 431–438.

Holterman, M., van der Wurff, A., van den Elsen, S., van Megen, H., Bongers, T., Holovachov, O., Bakker, J. and Helder, J. (2006) Phylum-wide analysis of SSU rDNA reveals deep phylogenetic relationships among nematodes and accelerated evolution toward crown clades. *Molecular Biology and Evolution* 23, 1792–1800.

Horvath, I., Multhoff, G., Sonnleitner, A. and Vigh, L. (2008) Membrane-associated stress proteins: more than simply chaperones. *Biochimica et Biophysica Acta – Biomembranes* 1778, 1653–1664.

Hundertmark, M. and Hincha, D.K. (2008) LEA (late embryogenesis abundant) proteins and their encoding genes in *Arabidopsis thaliana. BMC Genomics* 9, 118.

Illing, N., Denby, K.J., Collett, H., Shen, A. and Farrant, J.M. (2005) The signature of seeds in resurrection plants: a molecular and physiological comparison of desiccation tolerance in seeds and vegetative tissues. *Integrative and Comparative Biology* 45, 771–787.

Ingram, J. and Bartels, D. (1996) The molecular basis of dehydration tolerance in plants. *Annual Review of Plant Physiology and Plant Molecular Biology* 47, 377–403.

Iturriaga, G., Cushman, M.A.F. and Cushman, J.C. (2006) An EST catalogue from the resurrection plant *Selaginella lepidophylla* reveals abiotic stress-adaptive genes. *Plant Science* 170, 1173–1184.

Jäger, K. and Potts, M. (1988) Distinct fractions of genomic DNA from cyanobacterium *Nostoc commune* that differ in the degree of methylation. *Gene* 74, 197–201.

Jönsson, K.I. and Schill, R.O. (2007) Induction of Hsp70 by desiccation, ionising radiation and heat-shock in the eutardigrade *Richtersius coronifer. Comparative Biochemistry and Physiology Part B* 146, 456–460.

Jönsson, K.I., Harms-Ringdahl, M. and Torudd, J. (2005) Radiation tolerance in the eutardigrade *Richtersius coronifer. International Journal of Radiation Biology* 81, 649–656.

Keilin, D. (1959) The problem of anabiosis or latent life: history and current concept. *Proceedings of the Royal Society of London B* 150, 149–191.

Knecht, E., Aguado, C., Cárcel, J., Esteban, I., Esteve, J.M., Ghislat, G., Moruno, J.F., Vidal, J.M. and Sáez, R. (2009) Intracellular protein degradation in mammalian cells: recent developments. *Cellular and Molecular Life Sciences* 66, 2427–2443.

Koster, K.L., Webb, M.S., Bryant, G. and Lynch, D.V. (1994) Interactions between soluble sugars and POPC (1-palmitoyl-2-oleoylphosphatidylcholine) during dehydration – vitrification of sugars alters the phase-behavior of the phospholipid. *Biochimica et Biophysica Acta – Biomembranes* 1193, 143–150.

Koster, K.L., Lei, Y.P., Anderson, M., Martin, S. and Bryant, G. (2000) Effects of vitrified and nonvitrified sugars on phosphatidylcholine fluid-to-gel phase transitions. *Biophysical Journal* 78, 1932–1946.

Kovacs, D., Kalmar, E., Torok, Z. and Tompa, P. (2008) Chaperone activity of ERD10 and ERD14, two disordered stress-related plant proteins. *Plant Physiology* 147, 381–390.

Kranner, I. and Birtic, S. (2005) A modulating role for antioxidants in desiccation tolerance. *Integrative and Comparative Biology* 45, 734–740.

Kranner, I., Beckett, R.P., Wornik, S., Zorn, M. and Pfeifhofer, H.W. (2002) Revival of a resurrection plant correlates with its antioxidant status. *Plant Journal* 31, 13–24.

Kranner, I., Birtic, S., Anderson, K.M. and Pritchard, H.W. (2006) Glutathione half-cell reduction potential: a universal stress marker and modulator of programmed cell death? *Free Radical Biology and Medicine* 40, 2155–2165.

Lamitina, S.T. and Strange, K. (2005) Transcriptional targets of DAF-16 insulin signaling pathway protect *C. elegans* from extreme hypertonic stress. *American Journal of Physiology – Cell Physiology* 288, C467–C474.

Lamitina, T., Huang, C.G. and Strange, K. (2006) Genome-wide RNAi screening identifies protein damage as a regulator of osmoprotective gene expression. *Proceedings of the National Academy of Sciences of the USA* 103, 12173–12178.

Lapinski, J. and Tunnacliffe, A. (2003) Anhydrobiosis without trehalose in bdelloid rotifers. *FEBS Letters* 553, 387–390.

Lenne, T., Garvey, C.J., Koster, K.L. and Bryant, G. (2009) Effects of sugars on lipid bilayers during dehydration – SAXS/WAXS measurements and quantitative model. *Journal of Physical Chemistry B* 113, 2486–2491.

Liang, P., Amons, R., Clegg, J.S. and MacRae, T.H. (1997) Molecular characterization of a small heat shock/α-crystallin protein in encysted *Artemia* embryos. *Journal of Biological Chemistry* 272, 19051–19058.

Liu, Q.Z., Piggott, S.J., Solomon, A. and Glazer, I. (2002) Physiological and biochemical changes in nematodes of the genus *Heterorhabditis* following desiccation. *Phytoparasitica* 30, 253–261.

Lopez-Martinez, G., Benoit, J.B., Rinehart, J.P., Elnitsky, M.A., Lee, R.E. and Denlinger, D.L. (2009) Dehydration, rehydration, and overhydration alter patterns of gene expression in the Antarctic midge, *Belgica antarctica*. *Journal of Comparative Physiology B: Biochemical, Systemic, and Environmental Physiology* 179, 481–491.

Los, D.A. and Murata, N. (1998) Structure and expression of fatty acid desaturases. *Biochimica et Biophysica Acta – Lipids and Lipid Metabolism* 1394, 3–15.

Luo, Y., Li, W.M., and Wang, W. (2008) Trehalose: protector of antioxidant enzymes or reactive oxygen species scavenger under heat stress? *Environmental and Experimental Botany* 63, 378–384.

Macrae, M., Kramer, D.L. and Coffino, P. (1998) Developmental effect of polyamine depletion in *Caenorhabditis elegans*. *Biochemical Journal* 333, 309–315.

Mattimore, V. and Battista, J.R. (1996) Radioresistance of *Deinococcus radiodurans*: functions necessary to survive ionizing radiation are also necessary to survive prolonged desiccation. *Journal of Bacteriology* 178, 633–637.

Mazzucotelli, E., Mastrangelo, A.A., Crosatti, C., Guerra, D., Stanca, A.M. and Cattivelli, L. (2008) Abiotic stress response in plants: when post-transcriptional and post-translational regulations control transcription. *Plant Science* 174, 420–431.

Moore, J.P., Le, N.T., Brandt, W.F., Driouich, A. and Farrant, J.M. (2009) Towards a systems-based understanding of plant desiccation tolerance. *Trends in Plant Science* 14, 110–117.

Mukhopadhyay, R., Kumar, S. and Hoh, J.H. (2004) Molecular mechanisms for organizing the neuronal cytoskeleton. *Bioessays* 26, 1017–1025.

Nakayama, K., Okawa, K., Kakizaki, T. and Inaba, T. (2008) Evaluation of the protective activities of a late embryogenesis abundant (LEA) related protein, Cor15am, during various stresses *in vitro*. *Bioscience Biotechnology and Biochemistry* 72, 1642–1645.

Nauser, T., Koppenol, W.H. and Gebicki, J.M. (2005) The kinetics of oxidation of GSH by protein radicals. *Biochemical Journal* 392, 693–701.

Neumann, S., Reuner, A., Brümmer, F. and Schill, R.O. (2009) DNA damage in storage cells of anhydrobiotic tardigrades. *Comparative Biochemistry and Physiology, Part A* 153, 425–429.

Niki, E. (2009) Lipid peroxidation: physiological levels and dual biological effects. *Free Radical Biology and Medicine* 47, 469–484.

Norton, D.C. (1978) *Ecology of Plant-parasitic Nematodes*. John Wiley & Sons, New York.

O'Leary, S.A., Power, A.P., Stack, C.M. and Burnell, A.M. (2001) Behavioural and physiological responses of infective juveniles of the entomopathogenic nematode *Heterorhabditis* to desiccation. *Biocontrol* 46, 345–362.

Oliver, M.J., Tuba, Z. and Mishler, B.D. (2000) The evolution of vegetative desiccation tolerance in land plants. *Plant Ecology* 151, 85–100.

Oliver, M.J., Velten, J. and Mishler, B.D. (2005) Desiccation tolerance in bryophytes: a reflection of the primitive strategy for plant survival in dehydrating habitats? *Integrative and Comparative Biology* 45, 788–799.

Oliver, M.J., Hudgeons, J., Dowd, S.E. and Payton, P.R. (2009) A combined subtractive suppression hybridization and expression profiling strategy to identify novel desiccation response transcripts from *Tortula ruralis* gametophytes. *Physiologia Plantarum* 136, 437–460.

O'Mahony, P.J. and Oliver, M.J. (1999) The involvement of ubiquitin in vegetative desiccation tolerance. *Plant Molecular Biology* 41, 657–667.

Parsegian, A., Rau, D. and Zimmerber, J. (1986) Structural transitions induced by osmotic stress. In: Leopold, A.C. (ed.) *Membranes, Metabolism and Dry Organisms*. Cornell University Press, New York, pp. 306–317.

Patten, E.E., Willems, A.R. and Tyers, M. (1998) Combinatorial control in ubiquitin-dependent proteolysis: don't Skp the F-box hypothesis. *Trends in Genetics* 14, 236–243.

Pereira, E.D., Panek, A.D. and Eleutherio, E.C.A. (2003) Protection against oxidation during dehydration of yeast. *Cell Stress and Chaperones* 8, 120–124.

Perry, R.N. (1977) Desiccation survival of larval and adult stages of plant parasitic nematodes, *Ditylenchus dipsaci* and *Ditylenchus myceliophagus*. *Parasitology* 74, 139–148.

Perry, R.N. (1999) Desiccation survival of parasitic nematodes. *Parasitology* 119, S19–S30.

Phillips, J.R., Dalmay, T. and Bartels, D. (2007) The role of small RNAs in abiotic stress. *FEBS Letters* 581, 3592–3597.

Piggott, S.J., Liu, Q.Z., Glazer, I. and Wright, D.J. (2002) Does osmoregulatory behaviour in entomopathogenic nematodes predispose desiccation tolerance? *Nematology* 4, 483–487.

Potts, M. (1994) Desiccation tolerance of prokaryotes. *Microbiological Reviews* 58, 755–805.

Potts, M. (1999) Mechanisms of desiccation tolerance in cyanobacteria. *European Journal of Phycology* 34, 319–328.

Pouchkina-Stantcheva, N.N., McGee, B.M., Boschetti, C. *et al.* (2007) Functional divergence of former alleles in an ancient asexual invertebrate. *Science* 318, 268–271.

Pressel, S., Ligrone, R. and Duckett, J.G. (2006) Effects of de- and rehydration on food-conducting cells in the moss *Polytrichum formosum*: a cytological study. *Annals of Botany* 98, 67–76.

Rand, R.P. and Parsegian, V.A. (1989) Hydration forces between phospholipid bilayers. *Biochimica et Biophysica Acta* 988, 351–376.

Ratnakumar, S. and Tunnacliffe, A. (2006) Intracellular trehalose is neither necessary nor sufficient for desiccation tolerance in yeast. *FEMS Yeast Research* 6, 902–913.

Reardon, W., Chakrabortee, S., Campos Pereira, T. *et al.* (2010) Expression profiling and cross-species RNA interference (RNAi) of desiccation-induced transcripts in the anhydrobiotic nematode *Aphelenchus avenae*. *BMC Molecular Biology* 11, 6.

Reyes, J.L., Rodrigo, M.-J., Colmenero-Flores, J.M., Gil, J.-V., Garay-Arroyo, A., Campos, F., Salamini, F., Bartels, D. and Covarrubias, A.A. (2005) Hydrophilins from distant organisms can protect enzymatic activities from water limitation effects *in vitro*. *Plant Cell and Environment* 28, 709–718.

Reyes, J.L., Campos, F., Wei, H., Arora, R., Yang, Y., Karlson, D.T. and Covarrubias, A.A. (2008) Functional dissection of hydrophilins during *in vitro* freeze protection. *Plant Cell and Environment* 31, 1781–1790.

Ricci, C. and Pagani, M. (1997) Desiccation of *Panagrolaimus rigidus* (Nematoda): survival, reproduction and the influence on the internal clock. *Hydrobiologia* 347, 1–13.

Robinson, A.F., Orr, C.C. and Heintz, C.E. (1984) Some factors affecting survival of desiccation by infective juveniles of *Orrina phyllobia. Journal of Nematology* 16, 86–91.

Rorat, T. (2006) Plant dehydrins – tissue location, structure and function. *Cell and Molecular Biology Letters* 11, 536–556.

Rouhier, N., Koh, C.S., Gelhaye, E., Corbier, C., Favier, F., Didiejean, C. and Jacquot, J.P. (2008) Redox based anti-oxidant systems in plants: biochemical and structural analyses. *Biochimica et Biophysica Acta – General Subjects* 1780, 1249–1260.

Sallon, S., Solowey, E., Cohen, Y., Korchinsky, R., Egli, M., Woodhatch, I., Simchoni, O. and Kislev, M. (2008) Germination, genetics, and growth of an ancient date seed. *Science* 320, 1464.

Sanchez-Ballesta, M.T., Rodrigo, M.J., Lafuente, M.T., Granell, A. and Zacarias, L. (2004) Dehydrin from *Citrus*, which confers *in vitro* dehydration and freezing protection activity, is constitutive and highly expressed in the flavedo of fruit but responsive to cold and water stress in leaves. *Journal of Agricultural and Food Chemistry* 52, 1950–1957.

Sanders, S.W. and Maxcy, R.B. (1979) Isolation of radiation-resistant bacteria without exposure to irradiation. *Applied and Environmental Microbiology* 38, 436–439.

Sarkar, S., Davies, J.E., Huang, Z., Tunnacliffe, A. and Rubinsztein, D.C. (2007) Trehalose, a novel mTor-independent autophagy enhancer, accelerates the clearance of mutant huntingtin and α-synuclein. *Journal of Biological Chemistry* 282, 5641–5652.

Sato, M. and Bremner, I. (1993) Oxygen free-radicals and metallothionein. *Free Radical Biology and Medicine* 14, 325–337.

Schafer, F.Q. and Buettner, G.R. (2001) Redox environment of the cell as viewed through the redox state of the glutathione disulfide/glutathione couple. *Free Radical Biology and Medicine* 30, 1191–1212.

Schill, R.O., Steinbrück, G.H.B. and Köhler, H.-R. (2004) Stress gene (*hsp70*) sequences and quantitative expression in *Milnesium tardigradum* (Tardigrada) during active and cryptobiotic stages. *Journal of Experimental Biology* 207, 1607–1613.

Shannon, A.J., Browne, J.A., Boyd, J., Fitzpatrick, D.A. and Burnell, A.M. (2005) The anhydrobiotic potential and molecular phylogenetics of species and strains of *Panagrolaimus* (Nematoda, Panagrolaimidae). *Journal of Experimental Biology* 208, 2433–2445.

Shannon, A.J., Tyson, T., Dix, I., Boyd, J. and Burnell, A.M. (2008) Systemic RNAi mediated gene silencing in the anhydrobiotic nematode *Panagrolaimus superbus. BMC Molecular Biology* 9, 58.

Sharon, M.A., Kozarova, A., Clegg, J.S., Vacratsis, P.O. and Warner, A.H. (2009) Characterization of a group 1 late embryogenesis abundant protein in encysted embryos of the brine shrimp *Artemia franciscana. Biochemistry and Cell Biology* 87, 415–430.

Shen-Miller, J., Mudgett, M.B., Schopf, W., Clarke, S. and Berger, R. (1995) Exceptional seed longevity and robust growth: ancient sacred lotus from China. *American Journal of Botany* 82, 1367–1380.

Shih, M., Hoekstra, F. and Hsing, Y. (2008) Late embryogenesis abundant proteins. *Advances in Botanical Research* 48, 211–255.

Shirkey, B., McMaster, N.J., Smith, S.C., Wright, D.J., Rodriguez, H., Jaruga, P., Birincioglu, M., Helm, R.F. and Potts, M. (2003) Genomic DNA of *Nostoc commune* (Cyanobacteria) becomes covalently modified during long-term (decades) desiccation but is protected from oxidative damage and degradation. *Nucleic Acids Research* 31, 2995–3005.

Solomon, A., Salomon, R., Paperna, I. and Glazer, I. (2000) Desiccation stress of entomopathogenic nematodes induces the accumulation of a novel heat-stable protein. *Parasitology* 121, 409–416.

Solomon, A., Bandhakavi, S., Jabbar, S., Shah, R., Beitel, G.J. and Morimoto, R.I. (2004) *Caenorhabditis elegans* OSR-1 regulates behavioral and physiological responses to hyperosmotic environments. *Genetics* 167, 161–170.

Somero, G.N. (1986) Protons, osmolytes, and fitness of internal milieu for protein function. *American Journal of Physiology* 251, R197–R213.

Somvanshi, V.S., Koltai, H. and Glazer, I. (2008) Expression of different desiccation-tolerance related genes in various species of entomopathogenic nematodes. *Molecular and Biochemical Parasitology* 158, 65–71.

Sun, W.Q. and Leopold, A.C. (1997) Cytoplasmic vitrification and survival of anhydrobiotic organisms. *Comparative Biochemistry and Physiology* 117B, 327–333.

Tanford, C. (1978) Hydrophobic effect and organization of living matter. *Science* 200, 1012–1018.

Tanford, C. (1980) *The Hydrophobic Effect: Formation of Micelles and Biological Membranes*, 2nd edn. Wiley, New York.

Teramoto, N., Sachinvala, N.D. and Shibata, M. (2008) Trehalose and trehalose-based polymers for environmentally benign, biocompatible and bioactive materials. *Molecules* 13, 1773–1816.

Timasheff, S.N. (1992) Water as ligand: preferential binding and exclusion of denaturants in protein unfolding. *Biochemistry* 31, 9857–9864.

Tolleter, D., Jaquinod, M., Mangavel, C. *et al.* (2007) Structure and function of a mitochondrial late embryogenesis abundant protein are revealed by desiccation. *Plant Cell* 19, 1580–1589.

Tompa, P. (2002) Intrinsically unstructured proteins. *Trends in Biochemical Sciences* 27, 527–533.

Tsvetkova, N.M., Horvath, I., Torok, Z., Wolkers, W.F., Balogi, Z., Shigapova, N., Crowe, L.M., Tablin, F., Vierling, E., Crowe, J.H. and Vigh, L. (2002) Small heat-shock proteins regulate membrane lipid polymorphism. *Proceedings of the National Academy of Sciences of the USA* 99, 13504–13509.

Tunnacliffe, A. and Lapinski, J. (2003) Resurrecting Van Leeuwenhoek's rotifers: a reappraisal of the role of disaccharides in anhydrobiosis. *Philosophical Transactions of the Royal Society of London Series B* 358, 1755–1771.

Tunnacliffe, A. and Wise, M.J. (2007) The continuing conundrum of the LEA proteins. *Naturwissenschaften* 94, 791–812.

Tunnacliffe, A., Hincha, D.K., Leprince, O. and Macherel, D. (2010) LEA proteins: versatility of form and function. In: Lubzens, E., Cerda, J. and Clark, M.S. (eds) *Dormancy and Resistance in Harsh Environments*. Topics in Current Genetics. Springer, Berlin, pp. 91–108.

Tyson, T., Reardon, W., Browne, J.A. and Burnell, A.M. (2007) Gene induction by desiccation stress in the entomopathogenic nematode *Steinernema carpocapsae* reveals parallels with drought tolerance mechanisms in plants. *International Journal of Parasitology* 37, 763–776.

Vicré, M., Farrant, J.M. and Driouich, A. (2004) Insights into the cellular mechanisms of desiccation tolerance among angiosperm resurrection plant species. *Plant Cell and Environment* 27, 1329–1340.

Wallin, E. and von Heijne, G. (1998) Genome-wide analysis of integral membrane proteins from eubacterial, archaean, and eukaryotic organisms. *Protein Science* 7, 1029–1038.

Wang, C., Li, J.J. and Yuan, M. (2007) Salt tolerance requires cortical microtubule reorganization in *Arabidopsis*. *Plant and Cell Physiology* 48, 1534–1547.

Wang, J. and Kim, S.K. (2003) Global analysis of dauer gene expression in *Caenorhabditis elegans*. *Development* 130, 1621–1634.

Watanabe, M. (2006) Anhydrobiosis in invertebrates. *Applied Entomology and Zoology* 41, 15–31.

Welch, W.J. and Brown, C.R. (1996) Influence of molecular and chemical chaperones on protein folding. *Cell Stress and Chaperones* 1, 109–115.

Wharton, D.A. and Barclay, S. (1993) Anhydrobiosis in the free-living Antarctic nematode *Panagrolaimus davidi* (Nematoda, Rhabditida). *Fundamental and Applied Nematology* 16, 17–22.

Wharton, D.A. and Brown, I.M. (1991) Cold-tolerance mechanisms of the Antarctic nematode *Panagrolaimus davidi. Journal of Experimental Biology* 155, 629–641.

Wharton, D.A. and Lemmon, J. (1998) Ultrastructural changes during desiccation of the anhydrobiotic nematode *Ditylenchus dipsaci. Tissue and Cell* 30, 312–323.

Wharton, D.A., Goodall, G. and Marshall, C.J. (2003) Freezing survival and cryoprotective dehydration as cold tolerance mechanisms in the Antarctic nematode *Panagrolaimus davidi. Journal of Experimental Biology* 206, 215–221.

Wise, M.J. and Tunnacliffe, A. (2004) POPP the question: what do LEA proteins do? *Trends in Plant Science* 9, 13–17.

Wolfe, J. and Bryant, G. (1999) Freezing, drying, and/or vitrification of membrane–solute–water systems. *Cryobiology* 39, 103–129.

Womersley, C. (1987) A reevaluation of strategies employed by nematode anhydrobiotes in relation to their natural environment. In: Veech, J.A and Dickson, D.W. (eds) *Vistas on Nematology*. Society of Nematologists Inc., Hyattsville, Maryland, pp. 165–173.

Worland, M.R. (1996) The relationship between water content and cold tolerance in the arctic collembolan *Onychiurus arcticus* (Collembola: Onychiuridae). *European Journal of Entomology* 93, 341–348.

Xue, T.T., Li, X.Z., Zhu, W., Wu, C.G., Yang, G.G. and Zheng, C.C. (2009) Cotton metallothionein GhMT3a, a reactive oxygen species scavenger, increased tolerance against abiotic stress in transgenic tobacco and yeast. *Journal of Experimental Botany* 60, 339–349.

Yaakov, G., Duch, A., Garcia-Rubio, M., Clotet, J., Jimenez, J., Aguilera, A. and Posas, F. (2009) The stress-activated protein kinase Hog1 mediates S phase delay in response to osmostress. *Molecular Biology of the Cell* 20, 3572–3582.

Yamaguchi-Shinozaki, K. and Shinozaki, K. (2006) Transcriptional regulatory networks in cellular responses and tolerance to dehydration and cold stresses. *Annual Review of Plant Biology* 57, 781–803.

Yancey, P.H., Clark, M.E., Hand, S.C., Bowlus, R.D. and Somero, G.N. (1982) Living with water-stress – evolution of osmolyte systems. *Science* 217, 1214–1222.

7 Longevity and Stress Tolerance of Entomopathogenic Nematodes

Parwinder S. Grewal,[1] Xiaodong Bai[1]
and Ganpati B. Jagdale[2]

[1]*Department of Entomology, The Ohio State University, Wooster, Ohio, USA;*
[2]*Department of Plant Pathology, University of Georgia, Athens, Georgia, USA*

7.1 Introduction

Entompathogenic nematodes (EPNs) are lethal endoparasites with potential for biological control of insect pests (Grewal *et al.*, 2005) and as emerging model systems for studies in biological sciences. The two families of EPNs, Heterorhabditidae (Strongyloidea) and Steinernematidae (Strongyloidoidea) (*sensu* De Ley and Blaxter, 2002), are mutualistically associated with the entomopathogenic bacteria (EPB) *Photorhabdus* and *Xenorhabdus*, respectively, in the Enterobacteriaceae (Boemare, 2002). The infective stage of the life cycle of EPNs has traditionally been termed infective juveniles (IJs), and this term will be retained in this chapter. The IJs transmit the bacterial symbionts to the insect hosts. Following entry through the cuticle or natural body openings, the IJs release the symbiotic bacteria into the haemocoel, where the bacteria grow and kill the host within 24–48 h (Goodrich-Blair and Clarke, 2007). Nematodes feed on the symbiotic bacteria, complete one to three generations in the host cadaver and, as food resources are depleted, new IJs are produced, which disperse

 Molecular and Physiological Basis of Nematode Survival (eds R.N. Perry and D.A. Wharton)

in search of a new host (Poinar, 1990). The symbiotic bacteria interact with EPNs in at least two states (Forst and Clarke, 2002): the phoretic state, in which the bacteria persist in the gut of the non-feeding IJs (Boemare *et al.*, 1996), and the vegetative state, in which the bacteria produce an arsenal of virulence factors, ensuring rapid insect mortality (ffrench-Constant and Waterfield, 2006; Goodrich-Blair and Clarke, 2007; An *et al.*, 2009). Bioconversion of the insect cadaver by bacterial exoenzymes allows the bacteria to multiply and serve as food for the nematodes. During this state, the bacteria also produce secondary metabolites to prevent invasion of the insect cadaver by competing soil microbes (Webster *et al.*, 2002), enabling the nematodes and bacteria to re-associate in a protected niche.

The IJ is the only free-living stage and serves several important functions in the life cycle of the EPN–EPB symbiotic complex. The IJs retain and nourish the EPB, survive under environmental conditions detrimental to other life stages, disperse, find, select and invade hosts, and finally transmit the EPB into the insect haemocoel. The natural habitat for EPNs is the soil, which is a difficult environment for the IJs to persist in and locate insect hosts. Nevertheless, EPN IJs have been isolated from soils throughout the world, in ecosystems ranging from subarctic to arid, and temperate to tropical climates (Poinar, 1990; Hominick, 2002). Therefore, during the course of evolution, EPN IJs must have adopted unique survival mechanisms to resist environmental stresses.

EPN IJs are similar in ontogeny to the 'dauer' (enduring or non-ageing) larvae of the free-living, bacteria-feeding nematode *Caenorhabditis elegans*, which has been investigated intensively as a model system. In this species, dauers are produced in response to food limitation as the dauers can live without food for many months. Dauers are induced only in the first- and early second-stage larvae, by a constitutively produced population pheromone and food limitation (Riddle, 1988; Grant and Viney, Chapter 5, this volume). The dauers differ from the adults in many ways, including an arrested growth. Dauers contain intestinal granules that are thought to store food (Riddle, 1988). The dauers are encased in a dauer-specific cuticle, which is relatively resistant to dehydration. They have reduced metabolic rates (O'Riordan and Burnell, 1989, 1990), elevated levels of superoxide dismutase and are relatively resistant to oxidative stress (Larsen, 1993; Vanfleteren, 1993). They also possess elevated levels of several heat shock proteins (Hsps) (Dalley and Golomb, 1992). Nematodes that exit from the dauer state resume growth and have subsequent lifespans that are similar to those that have not arrested at the dauer stage (Klass and Hirsh, 1976). Although several mutants that show extended longevity in the adults have been isolated in *C. elegans*, the relationship between the adult and dauer larva longevity has not yet been explored in these mutants. However, the findings that several dauer-constitutive mutations can increase the lifespan of fertile adults raises the possibility that dauers do not live longer simply because their growth is arrested but rather because they express an active longevity programme. In fact, activity of a gene, *daf-16*, which plays a significant role in the ageing of adult nematodes, is required for the maintenance of the dauer state of *C. elegans* (Kenyon, 1997).

In this chapter, we summarize research on the longevity and stress tolerance of EPN IJs and discuss physiological, genetic and molecular mechanisms controlling IJ longevity and tolerance to major environmental stresses. An enhanced understanding of the physiological mechanisms and underlying genetic factors that control IJ longevity and stress toletrance can lead to the identification of species and strains of EPNs with superior survival traits, leading to the development of products with a longer shelf life and field persistence potential for use in biological pest control.

7.2 Longevity of Infective Juveniles

Like the dauers of *C. elegans*, the EPN IJs are capable of surviving for weeks to months without feeding. Studies show that the innate IJ longevity differs substantially in different EPN species and strains (Grewal, 2000a; Grewal *et al.*, 2002). Grewal *et al.* (2002) found that IJ longevity at 25°C in tap water varied from 4.5 to 11 weeks at the LT_{50} (time to 50% IJ mortality) and 6 to 16 weeks at the LT_{90} (time to 90% mortality) levels among 15 strains of *Heterorhabditis bacteriophora*, with commercial strains Lewiston-IBCS and HP88-MRD having the shortest IJ longevity (Table 7.1). Compared with *H. bacteriophora*, species of *Steinernema* tend to have greater IJ longevity in water. Grewal (2000b) reported that at 25°C, 50% of *Steinernema feltiae* IJs survived for 30 weeks (~7 months), while 10% survived up to 36 weeks (~9 months). *Steinernema riobrave* and *Steinernema carpocapsae* had intermediate longevities, with 10% of the IJs surviving for 26 and 22 weeks, respectively (Table 7.2).

Table 7.1. Longevity of infective juveniles (IJs) of different *Heterorhabditis bacteriophora* strains at 25°C in tap water. Data are estimated time (in weeks) to 90% (LT_{90}) and 50% (LT_{50}) mortality of IJs with standard errors (S.E.) and 95% confidence intervals (C.I.). Lewiston-IBCS and HP88-MRD populations were derived from commercial products. (Adapted from Grewal *et al.*, 2002.)

Strain	LT_{90}	S.E.	95% C.I.	LT_{50}	S.E.	95% C.I.
Lewiston-IBCS	6.4	0.07	6.28–6.57	4.5	0.04	4.39–4.53
HP88-MRD	6.8	0.10	6.60–6.98	4.3	0.05	4.19–4.37
HP88	8.3	0.13	8.03–8.55	5.0	0.06	4.88–5.11
Riwaka	10.2	0.16	9.93–10.56	7.5	0.07	7.33–7.63
Oswego	10.3	0.14	10.00–10.57	7.0	0.07	6.89–7.14
GPS3	10.7	0.15	10.38–10.97	7.6	0.07	7.46–7.73
NC1	10.9	0.17	10.62–11.31	7.6	0.08	7.50–7.79
OH25	12.2	0.23	11.80–12.69	9.6	0.11	9.40–9.84
GPS5	11.3	0.17	11.01–11.68	8.1	0.08	7.92–8.22
GPS2	11.4	0.19	11.10–11.83	8.1	0.08	7.92–8.25
KMD10	12.2	0.24	11.81–12.76	9.4	0.12	9.15–9.61
Acows	12.4	0.22	12.04–12.90	9.3	0.01	9.15–9.55
GPS11	12.7	0.24	12-23–13.18	9.4	0.11	9.17–9.62
GPS1	14.6	0.38	13.92–15.42	10.2	0.18	9.87–10.58
KMD19	16.1	0.65	14.98–17.56	11.1	0.33	10.54–11.87

Table 7.2. Longevity (in weeks) of non-desiccated and desiccated infective juveniles (IJs) of different *Steinernema* species at 25 and 5°C. Nematodes were desiccated in water-dispersible granular formulation, whilst non-desiccated nematodes were held in tap water. (Data are adapted from Grewal, 2000.)

Nematode species	Strain	Temperature (°C)	Desiccated	No. of weeks at which 50% of IJs survived	No. of weeks at which 10% of IJs survived
S. feltiae	SN	25	No	27–30	34–36
S. feltiae	SN	5	No	42–44	49–51
S. feltiae	SN	25	Yes	13–15	19–21
S. feltiae	SN	5	Yes	34–36	40–42
S. riobrave	Texas	25	No	19–20	23–26
S. riobrave	Texas	5	No	19–21	23–25
S. riobrave	Texas	25	Yes	20–21	26–29
S. riobrave	Texas	5	Yes	17–19	20–22
S. carpocapsae	All	25	No	16–18	19–22
S. carpocapsae	All	5	No	50–52	>52
S. carpocapsae	All	25	Yes	25–28	33–35
S. carpocapsae	All	5	Yes	42–44	52–53

The extended survival of IJs without food at temperatures that are conducive to normal growth and reproduction of EPNs raises interesting questions about the potential mechanisms and genetic factors regulating metabolism and survival of the IJs. For example, which environmental factors influence IJ longevity? What are the molecular mechanisms/factors modulating IJ longevity and survival at optimum and extreme environmental conditions? Can IJ longevity and stress tolerance be genetically manipulated? These and other related questions concerning longevity and stress tolerance in IJs of EPNs are addressed below.

7.3 Factors Affecting Longevity of Infective Juveniles

7.3.1 Stored energy reserves

As EPN IJs do not feed, one of the major factors determining their longevity is the amount of stored energy reserves. Lipid constitutes about 60% of the dry weight of IJs (Selvan *et al.*, 1993) and is considered as one of the major energy reserves. The amount of lipid in the IJs varies with the nematode species (Grewal and Georgis, 1998). In commercial production systems, lipid content of IJs can also be influenced by factors such as the amount and type of media components and antifoam used, temperature and dissolved oxygen during fermentation (Grewal, 2002). The rate of lipid utilization can also differ between EPN species (Grewal, 2000b) and among individuals within a species (Patel *et al.*, 1997), and is affected by temperature, oxygen and nematode activity during storage (Grewal and Georgis, 1998). Patel and Wright (1997) reported that EPN IJs also have appreciable amounts of glycogen. They reported that glycogen levels varied from 8% dry weight in *S. riobrave* to ~18% in *Steinernema glaseri*. IJs of both *S. carpocapsae* and *S. riobrave* survived for 120–135 days and utilized ~90% of their glycogen reserves at an almost constant rate during a 112-day storage period. The IJs of *S. feltiae* and *S. glaseri* lived for much longer (>450 days) but their glycogen content decreased by 27 and 40%, respectively, during a 250-day storage period. In contrast with the other species, the rate of lipid decline preceded that of glycogen in *S. carpocapsae*.

7.3.2 Temperature

Temperature is a major environmental factor influencing life processes of all organisms. Temperature adaptation of EPN species appears to be a continuum, with species representing both the cold and warm extremes (Grewal *et al.*, 2004, 2006). Grewal *et al.* (1994) determined the thermal infection (host penetration and host death), establishment (development of the IJs to adults in the host following infection) and reproduction niche breadths of eight species and concluded that *S. feltiae* represented the coldest extreme and *S. riobrave* represented the warmest extreme. EPN thermal niche breadths were progressively narrower for movement (4–40°C), host penetration (5–39°C), host death (8–39°C) and reproduction (10–35°C) (Grewal *et al.*, 2004, 2006). Grewal *et al.* (2004) concluded that EPNs must adopt alternative survival strategies to withstand temperature extremes, as their reproduction occurs at a very narrow temperature range (10–35°C), with some species, such as *S. carpocapsae*, reproducing only between 20 and 30°C. Studies show that EPN IJs can withstand temperatures as low as −80°C to as high as 40°C. Thus, IJs serve an important survival function by withstanding temperature extremes that are lethal to all other life stages of EPNs.

Temperature has a strong influence on IJ longevity and this influence varies with EPN species (Grewal, 2000a). For example, 50% of *S. feltiae* IJs survived in tap water for 42–44 and 27–30 weeks at 5 and 25°C, respectively (Table 7.2). Similarly, 50% of *S. carpocapsae* IJs survived in tap water for 16–18 weeks at 25°C and for 50–52 weeks at 5°C. By contrast, the survival of *S. riobrave* IJs was not much different between 25 and 5°C (Table 7.2), perhaps due to its lack of cold adaptation (Grewal, 2002). Heat tolerance also varies substantially among species and strains of EPNs. The survival of IJs exposed to 40°C in tap water for 2 h varied between 13% and 90% among 15 *H. bacteriophora* strains (Grewal *et al.*, 2002) and between 37% and 82% in 15 *S. carpocapsae* strains (Somasekhar *et al.*, 2002) (see Table 7.3).

Temperature tolerance of EPNs can be modulated by preconditioning of the IJs at sublethal temperatures (Jagdale and Grewal, 2003). Warm (35°C for 1 day) and cold (5°C for 2 days) acclimation enhanced heat (40°C for 8 h) and freezing (−20°C for 4 h) tolerance of *S. carpocapsae*. By contrast, warm and cold acclimation enhanced heat but not freezing tolerance of *S. feltiae* and freezing but not heat tolerance of *S. riobrave* (Jagdale and Grewal, 2003).

Table 7.3. Stress tolerance of infective juveniles (IJs) of different strains of *Steinernema carpocapsae* and *Heterorhabditis bacteriophora*. Data are mean (±S.E.) per cent survival of IJs after treatment with various stressors. For heat tolerance, the IJs were exposed to 40°C for 2 h. For desiccation tolerance, the IJs were exposed at 25°C to 25% glycerol for 72 h (*H. bacteriophora*) and 40% glycerol for 96 h (*S. carpocapsae*). For hypoxia tolerance, the IJs were held at 25°C under anoxic conditions for 96 h (*H. bacteriophora*) and 10 days (*S. carpocapsae*). Strains All, Lewiston-IBCS and HP-88-MRD were obtained from commercial sources. (Data for *H. bacteriophora* strains are from Grewal *et al.*, 2002 and for *S. carpocapsae* strains are from Somasekhar *et al.*, 2002.)

Strain	Heat	Desiccation	Hypoxia
H. bacteriophora			
Lewiston-IBCS	38 (±8.2)	90 (±7.8)	42 (±1.5)
HP88-MRD	35 (±11.3)	74 (±0.6)	58 (±12.0)
HP88	13 (±6.3)	71 (±0.5)	40 (±4.2)
Riwaka	52 (±4.6)	69 (±5.0)	26 (±4.0)
Oswego	73 (±8.5)	85 (±0.2)	51 (±0.4)
GPS3	65 (±4.0)	78 (±2.2)	15 (±4.8)
NC1	58 (±4.0)	83 (±3.6)	11 (±3.5)
OH25	91 (±4.5)	20 (±12.8)	62 (±3.2)
GPS5	70 (±0.5)	72 (±4.2)	51 (±6.3)
GPS2	75 (±2.4)	63 (±1.8)	25 (±4.5)
KMD10	82 (±4.2)	85 (±4.5)	85 (±8.2)
Acows	83 (±4.0)	82 (±6.8)	88 (±9.8)
GPS11	87 (±4.2)	77 (±3.8)	79 (±2.3)
GPS1	90 (±5.2)	85 (±0.2)	72 (±2.2)
KMD19	83 (±4.3)	25 (±14.3)	90 (±8.3)
S. carpocapsae			
All	60 (±0.2)	42 (±0.5)	52 (±0.5)
KMD2	58 (±0.2)	24 (±0.5)	38 (±0.4)
KMD3	54 (±0.2)	38 (±0.5)	72 (±0.2)
KMD4	60 (±0.1)	22 (±0.4)	51 (±0.2)
KMD5	65 (±3.0)	43 (±1.5)	47 (±0.5)
KMD7	77 (±0.5)	60 (±0.5)	20 (±0.2)
KMD11	64 (±2.0)	42 (±0.4)	50 (±0.2)
KMD14	73 (±0.8)	30 (±0.5)	52 (±0.1)
KMD18	40 (±0.2)	24 (±0.4)	21 (±0.1)
KMD38	37 (±1.0)	25 (±0.5)	52 (±2.0)
KMD52	75 (±1.0)	42 (±0.5)	35 (±0.1)
KMD26	56 (±0.2)	25 (±0.5)	56 (±0.4)
KMD28	78 (±0.5)	61 (±0.5)	66 (±0.5)
KMD30	74 (±0.5)	58 (±0.5)	58 (±0.5)
KMD33	82 (±0.7)	61 (±0.5)	65 (±0.5)

7.3.3 Desiccation

Desiccation can have a strong influence on EPN IJ longevity. Compared with nematode species which can undergo complete anyhydrobiosis (see Perry and Moens, Chapter 1, and Burnell and Tunnacliffe, Chapter 6, this volume), EPNs can withstand only limited desiccation (Simons and Poinar, 1973;

Womersley, 1990). Simons and Poinar (1973) were the first to demonstrate that *S. carpocapsae*, when air-dried slowly at 97% relative humidity (RH) at room temperature, can withstand subsequent exposure to much lower RH. Using glycerol as an osmolyte, it has been demonstrated that desiccation tolerance varies substantially among species and strains of EPNs (see Table 7.3). For example, *H. bacteriophora* IJs could only withstand exposure to 25% glycerol for 72h at 25°C, and the survival varied between 25% and 90% among 15 *H. bacteriophora* strains (Grewal *et al.*, 2002). By contrast, *S. carpocapsae* IJs were able to withstand exposure to 40% glycerol for 96h, under which conditions survival varied between 22% and 61% in 15 *S. carpocapsae* strains (Somasekhar *et al.*, 2002). Similarly, Yan *et al.* (2010) reported large differences among species of *Steinernema* and *Heterorhabditis* in osmotic desiccation in 25% glycerol at 15°C.

Induction of anhydrobiosis has been considered as an approach to extend storage stability (longevity) of EPNs (Georgis, 1990; Grewal, 2000a,b, 2002). Although direct contact desiccation (air-drying) was not successful on a commercial scale, slow desiccation at an initial water activity of 0.970 in water-dispersible granules resulted in extension of IJ longevity in some EPN species (Grewal, 2000a,b). For example, desiccation in water-dispersible granules at 25°C extended IJ longevity in *S. carpocapsae* and *S. riobrave* but shortened it in *S. feltiae* (Table 7.2). In *S. carpocapsae*, 50% of the IJs survived in water-dispersible granules for 25–28 weeks, compared with only 16–18 weeks in tap water at 25°C. However, the longevity of desiccated IJs in water-dispersible granules was reduced compared with IJs that were kept in tap water (non-desiccated) in all three species examined at 5°C (see Table 7.2). In fact, direct desiccation of *S. riobrave* IJs was lethal at 5°C, and IJs of this species could only withstand cold storage when preconditioned at 5°C prior to desiccation (Grewal and Jagdale, 2002).

Grewal and Jagdale (2002) reported that cold preacclimation at 5°C for 2 days enhanced osmotic desiccation survival of *S. feltiae* in 25% glycerol at both 5 and 25°C and of *S. carpocapsae* and *S. riobrave* only at 5°C. Non-acclimated *S. carpocapsae* and *S. riobrave* were extremely sensitive to desiccation directly in 25% glycerol at 5°C, resulting in over 98% mortality within 6 days, but *S. feltiae* was more sensitive to desiccation at 25°C than at 5°C. Cold preacclimation increased survival of all three species in the water-dispersible granular formulation at both 5 and 25°C. The survival of *S. riobrave* at 5°C in the water-dispersible granular formulation was positively correlated with the length of preacclimation period at 5°C ($R^2 = 0.99$) and with the amount of trehalose accumulated (also see below) during cold preacclimation ($R^2 = 0.81$). These results support the hypothesis that cold preacclimation enhances desiccation survival of EPNs at cold temperatures and the increased survival correlates well with the increased trehalose accumulation. Interestingly, Jagdale and Grewal (2007) found that both cold (5°C for 2 days) and warm (35°C for 1 day) preacclimation of IJs in tap water can influence osmotic desiccation ability of EPNs in both cold (5°C) and warm (35°C) conditions. Both cold and warm preacclimation enhanced cold (5°C) osmotic desiccation survival of *S. carpocapsae*, *S. feltiae* and *S. riobrave*; however, only warm preacclimation enhanced osmotic desiccation at 35°C in *S. feltiae* and *S. riobrave* and only cold preacclimation in *S. carpocapsae*.

These findings suggest that the effect of cold and warm preacclimation on desiccation survival differs with EPN species.

7.3.4 Hypoxia

Since nematodes are aerobic organisms, hypoxic conditions can reduce nematode survival and longevity. Burman and Pye (1980) reported that *S. carpocapsae* can withstand oxygen tensions of as low as 0.5% saturation at 20°C. In sandy soil, survival of *S. carpocapsae* and *S. glaseri* decreased as oxygen concentration decreased from 20% to 1% (Kung *et al.*, 1990). There are large differences in the ability of EPN species and strains to withstand hypoxic conditions (Table 7.3). While *H. bacteriophora* can withstand anoxic conditions in water at 25°C for only 4 days (Grewal *et al.*, 2002), *S. carpocapsae* can tolerate such conditions for up to 10 days (Somasekhar *et al.*, 2002). Also there appears to be more variation in hypoxia tolerance among strains of *H. bacteriophora* (Grewal *et al.*, 2002) than those of *S. carpocapsae* (Somasekhar *et al.*, 2002). Qiu and Bedding (1999a) reported that *S. carpocapsae* IJs incubated in M9 buffer at 23°C under absolute anaerobic conditions were fully inactivated in 16 h but could be revived when returned to aerobic conditions if exposure to anaerobic conditions was not more than 7 days. The survival time under anaerobic conditions was significantly affected by temperature, with 90% survival times being 20, 7 and 5 days at 5, 23 and 28°C, respectively.

7.4 Physiological Mechanisms of Longevity and Stress Tolerance

7.4.1 Physiology of longevity

Selvan *et al.* (1993) reported that the water content of EPN IJs increases over time during storage in water. The IJs of *S. carpocapsae*, *S. glaseri* and *H. bacteriophora* had 14–16% more water than the freshly emerged IJs. Selvan *et al.* (1993) also found that the percentage of unsaturated fatty acids increased, whereas saturated fatty acids decreased with an increase in storage time.

7.4.2 Physiology of temperature tolerance

Studies suggest that EPNs may use homeoviscous adaptation to maintain membrane fluidity by altering the proportions of saturated and unsaturated fatty acids under temperature stress conditions (Grewal *et al.*, 2006). For example, the increased proportion of unsaturated fatty acids in the phospholipids of *S. carpocapsae* was correlated with increased membrane fluidity when they were reared at 18°C compared with their normal rearing temperature of 25°C (Fodor *et al.*, 1994). Increased unsaturation indices of total and phospholipids were observed in *S. feltiae* and *S. carpocapsae* as their culture or storage temperature decreased from 25 to 5°C (Jagdale and Gordon, 1997a) and in *Heterorhabditis megidis* when stored at 5°C for 5 weeks (Fitters *et al.*, 1997). It is also likely that EPNs modify the kinetic properties of some metabolic enzymes to adapt to temperature changes (see Grewal *et al.*, 2006). Jagdale and Gordon (1997b) reported enhanced specific activities of both glucose-6-phosphate dehydrogenase and hexokinase, with their lowest K_m (Michaelis–Menten constant) values in *S. feltiae, S. carpocapsae* and *S. riobrave* when recycled at cooler than at warmer temperatures. Such shifts in specific enzyme activities may be necessary to meet the increased energy metabolism demand under temperature stress conditions, as noted by Hochachka and Somero (1984).

Temperature tolerance in EPN IJs also appears to be facilitated by trehalose accumulation (see Grewal *et al.*, 2006). IJs of many EPN species have been shown to accumulate trehalose when exposed to sublethal cold temperatures (Ogura and Nakashima, 1997; Qiu and Bedding, 1999b; Grewal and Jagdale, 2002; Jagdale and Grewal, 2003). Qiu and Bedding (1999b) reported that IJs of *S. carpocapsae* synthesized trehalose but not glycerol at low temperatures. In IJs incubated aerobically in tap water at temperatures ranging from 2 to 14°C, their trehalose levels increased from 1.9% dry weight to equilibrium levels ranging from 3.4% at 14°C to 6.0% at 5°C. When ageing IJs, which had lower energy reserves than fresh ones, were exposed in the same way to 5°C for 7 days, their trehalose levels were lower than those of fresh IJs but the survival rates of the IJs did not drop substantially. Changes in lipid, glycogen and protein levels of IJs during cold induction and subsequent recovery indicated that trehalose was not synthesized from glycogen but from lipids and/or proteins. As the cold-preacclimated nematodes also survive better even at cooler temperatures (Ogura and Nakashima, 1997; Grewal and Jagdale, 2002; Jagdale and Grewal, 2003), a role for trehalose in cold hardiness has been postulated. Jagdale and Grewal (2003) found that cold acclimation induced trehalose accumulation and increased freezing tolerance in most species of EPN but not in all. They also found that, although cold acclimation significantly increased freezing tolerance of *S. riobrave*, it was not correlated with increased trehalose levels, suggesting that the increased freezing tolerance may be due to the accumulation of other compounds such as polyols acting as colligative cryoprotectants, as observed in some insect species (Lee and Denlinger, 1991; Storey and Storey, 1992).

A relationship between trehalose accumulation and the acquisition of heat tolerance has also been discovered (Jagdale and Grewal, 2003). *Steinernema feltiae, S. carpocapsae* and *S. riobrave* accumulated trehalose when acclimated at either 5 or 35°C, but the amount of trehalose accumulation differed by species and temperature (Jagdale and Grewal, 2003). Enhanced heat tolerance (at 40°C for 8 h) was positively correlated with the increased trehalose levels in warm- and cold-acclimated *S. carpocapsae* and *S. feltiae* but not in *S. riobrave*. Similarly, the enhanced freezing tolerance (–20°C for 4 h) was positively correlated with the increased trehalose levels in warm- and cold-acclimated *S. carpocapsae* and warm-acclimated *S. riobrave* but not in *S. feltiae*. These data indicate that trehalose accumulation is not only a cold-associated phenomenon but is a more general response of EPNs to thermal stress.

Jagdale *et al.* (2005) explored the relationship between heat (35°C) or cold (1 and 10°C) shock (short exposure) and trehalose metabolism in IJs of *H. bacteriophora* and reported that the trehalose accumulation increased in both heat- and cold-shocked nematodes within 3h of exposure compared with nematodes maintained at 25°C (culture temperature). In both heat- and cold-shocked nematodes, the activity of trehalose-6-phosphate synthase (T6PS), an enzyme involved in the synthesis of trehalose, was significantly increased and the activity of trehalase, an enzyme involved in the degradation of trehalose, was significantly decreased during the first 3h of exposure. Generally, the trehalose levels and T6PS and trehalase activity returned to their original levels when nematodes were transferred back to 25°C. These results demonstrate that the trehalose concentrations in *H. bacteriophora* are influenced by both heat and cold shocks and are regulated by the action of two trehalose-metabolizing enzymes, T6PS and trehalase.

Although trehalose may also act as an energy reserve in many organisms (Behm, 1997), it is mainly considered to be a protectant against environmental stresses. For example, in yeast, high concentrations of trehalose have been linked with dehydration, freezing, osmotic pressure and ethanol shocks (Singer and Lindquist, 1998). Genetic evidence also suggests that heat-shock-induced accumulation of trehalose and its biosynthetic enzymes may play thermoprotective roles in yeast (De Virgilio *et al.*, 1993, 1994; Hottiger *et al.*, 1994; Majara *et al.*, 1996). The genes *tps1* and *tps2*, which encode the T6PS complex, are now believed to be Hsps. Studies show that the deletion of either gene causes an inability to accumulate trehalose during heat shock, which, in turn, significantly reduces thermotolerance in yeast (Bell *et al.*, 1992; De Virgilio *et al.*, 1993; Hottiger *et al.*, 1994). Since *H. bacteriophora* accumulates trehalose and increases T6PS activity after heat shock (Jagdale *et al.*, 2005), it is possible that this enzyme in nematodes, as in yeast, may also be acting as a Hsp and performing a role in thermotolerance. More details of thermotolerance and Hsps in nematodes are given by Devaney, Chapter 10, this volume.

From an ecological standpoint, both increases and decreases in ambient temperature may provide cues for the organism to prepare for desiccation. An elevation in environmental temperature may lead to increased evaporation, and a decrease in temperature may lead to freezing, both of which may potentially lead to desiccation. Many invertebrates, such as collembolla (Worland *et al.*, 1998), insects (Lee and Denlinger, 1991) and nematodes (Ogura and Nakashima, 1997; Grewal and Jagdale, 2002; Jagdale and Grewal, 2003), have been shown to accumulate trehalose when exposed to the cold, perhaps to survive anticipated subsequent freezing. Grewal and Jagdale (2002) evaluated the effect of cold (5°C) and warm (35°C) acclimation on desiccation tolerance of *S. carpocapsae*, *S. feltiae* and *S. riobrave* in 25% glycerol. Both cold and warm acclimation enhanced desiccation tolerance of *S. feltiae* at both 5 and 35°C, and of *S. carpocapsae* and *S. riobrave* at only 5°C. However, desiccation tolerance of *S. carpocapsae* and *S. riobrave* at 35°C was increased by either cold or warm acclimation in the two species, respectively. This increased desiccation tolerance was positively correlated with the acclimation-induced trehalose accumulation. Jagdale *et al.* (2005) found that the levels of both trehalose and its biosynthetic enzyme T6PS increased when nematodes were cold-shocked slightly above freezing (1°C). This suggests that the accumulation of both trehalose and T6PS is not a transient response but may relate to protection against cold-/freezing-stress-induced desiccation. It has also been shown that trehalose and glycerol protect against harmful effects of desiccation in insects (Storey and Storey, 1991) and nematodes (Womersley, 1990; Crowe *et al.*, 1992; Solomon *et al.*, 2000). The increase in trehalose concentration in *H. bacteriophora* subjected to heat and cold shock may be indicative of a nematode's preparation for desiccation. Therefore, both heat- and cold-shock-induced accumulations of trehalose may also be involved in the survival of EPNs during desiccation, induced by freezing or excessive evaporation due to heat.

7.4.3 Physiology of desiccation tolerance

Although the physiological mechanisms involved in the induction of anhydrobiosis are not fully understood, a relationship between accumulation of polyols and sugars and their role in protection of biological membranes and intracellular proteins during dehydration has been documented in many anhydrobiotic nematodes (Womersley, 1987; 1990; Barrett, 1991; Crowe and Crowe, 1992; Behm, 1997). For example, Crowe and Madin (1975) showed a strong correlation between accumulation of glycerol and trehalose during dehydration and survival of a mycophagous nematode, *Aphelenchus avenae*, in dry air. However, Higa and Womersley (1993) suggest that the production of glycerol in this nematode is the result of anoxic conditions in large aggregates rather than a response to desiccation. A correlation between glycerol or trehalose accumulation and increased desiccation tolerance has also been observed in *H. megidis*, *Heterorhabditis indica* and *S. carpocapsae* (O'Leary *et al.*, 2001). Preconditioning of *S. feltiae*, *S. carpocapsae* and *H. bacteriophora* at 97% RH for 3 days enhanced their survival at 85% and 75% RH (Popiel *et al.*, 1987; Womersley, 1990; Solomon *et al.*, 1999), which has been correlated with the synthesis of trehalose, glycerol or water-stress-related proteins (Solomon *et al.*, 1999, 2000; O'Leary *et al.*, 2001; Grewal *et al.*, 2006). According to O'Leary *et al.* (2001), *S. carpocapsae* accumulated trehalose and *Heterorhabditis* spp. (*H. megidis* and *H. indica*) accumulated glycerol when preacclimated at 98% RH, and these compounds were responsible for increased desiccation tolerance of these nematodes at 57% RH. Again, these results need to be interpreted with caution as all the research on EPN desiccation has been done in large clumps, suggesting that glycerol may also be a response to anoxia or a general stress response in these species. Nevertheless, the lack of induction of trehalose or a similar disaccharide upon desiccation in *H. megidis* and *H. indica* is consistent with the view that heterorhabditid nematodes are at best limited anhydrobiotes and can tolerate only a moderate amount of slow desiccation by accumulating glycerol. Although glycerol has many desirable properties as a 'compatible solute', it is one of the least effective carbohydrates at preserving membrane stability at low water activities (Crowe *et al.*, 1984).

Grewal and Jagdale (2002) reported that cold preacclimation (5°C) enhanced osmotic desiccation tolerance of *S. carpocapsae*, *S. feltiae* and *S. riobrave* at 5°C, which was positively correlated with the increased trehalose levels during cold preacclimation. Jagdale and Grewal (2007) explored the relationship between cold (5°C) and warm (35°C) acclimation-induced trehalose accumulation and osmotic desiccation tolerance of *S. feltiae*, *S. carpocapsae* and *S. riobrave* at 5 and 35°C. They found that the desiccation tolerance of all three species at 5°C was enhanced by both the cold and warm acclimation. Desiccation tolerance of *S. feltiae* and *S. riobrave* at 35°C was enhanced by only warm acclimation and of *S. carpocapsae* by only cold acclimation. Trehalose content of both warm- and cold-acclimated *S. feltiae*, cold-acclimated *S. carpocapsae* and warm-acclimated *S. riobrave* was positively correlated with their desiccation tolerance at 5 and 35°C (Jagdale and Grewal, 2007).

Synthesis and accumulation of specific proteins during the desiccation process have been characterized among bacteria, fungi, yeast and plant seeds (Dure, 1993; Close, 1996). The late embryogenic abundant (LEA) proteins are a diverse group of water-stress-related proteins that are expressed in maturing seeds and in water-deficit-stressed vegetative tissues of higher plants (Chandler *et al.*, 1988; Close, 1996; Burnell and Tunnacliffe, Chapter 6, this volume). Solomon *et al.* (2000) showed that *S. feltiae* acclimated at 5°C for 3–4 weeks accumulated large amounts of stress-related proteins (Desc47) along with trehalose when exposed to 97% RH (predesiccation) for 3 days and suggested that the acquisition of desiccation tolerance in *S. feltiae* was due to increased accumulation of both Desc47 protein and trehalose. Similarly, Serwe-Rodriguez *et al.* (2004) demonstrated the induction of several novel proteins, including a 37 kDa protein in *S. carpocapsae* in desiccated host cadavers. Gal *et al.* (2003) reported over 90 differentially expressed sequence tags (ESTs) in *S. feltiae* IS-6 strain in response to desiccation, and Chen *et al.* (2005, 2006) identified an array of water-stress-related proteins in *S. feltiae* IS-6 strain in response to desiccation.

The capability of anhydrobiotic organisms to tolerate desiccation is generally associated with the accumulation of carbohydrates, including trehalose (Sun and Leopold, 1997) and water-stress related proteins (Solomon *et al.*, 2000; Gal *et al.*, 2003; Chen *et al.*, 2005, 2006). Trehalose protects membranes and proteins from desiccation damage by replacing structural water associated with the phospholipid bilayer, maintaining membrane fluidity and keeping the bilayer in the liquid crystalline state and by forming glass (vitrification) to stabilize the cell content (Crowe *et al.*, 1996; Browne *et al.*, 2002). During desiccation, trehalose also protects proteins by replacing 'bound water' and reducing the reaction of dried glucose with amino acid side chains of proteins (known as the 'browning' or Maillard reaction) (Behm, 1997). Although the mechanisms of water-stress-related proteins in protecting organisms during desiccation stress are unclear, it is posssible that these proteins are acting as Hsps, which are known to maintain homeostasis during environmental stress in both eukaryotic and prokaryotic cells (Nagao *et al.*, 1990). Grewal and Jagdale (2002) reported a positive link between cold-storage-induced trehalose accumulation and enhanced desiccation tolerance of *S. riobrave* in a water-dispersible granular formulation at 5°C. Jagdale and Grewal (2007) found that even warm-storage-induced trehalose accumulation was positively correlated with the enhanced desiccation tolerance of *S. feltiae*, *S. carpocapsae* and *S. riobrave* at 5°C. It has been reported that the K_m values for certain enzymes from cold-hardy insect species tend to be less temperature sensitive than from the warm-adapted species (Storey and Storey, 1991). Thus, the inability of cold-stored *S. riobrave* to withstand desiccation at 35°C may partly be due to the poor biochemical adaptation of this warm-adapted species to cold.

7.4.4 Physiology of hypoxia tolerance

According to Qiu and Bedding (1999b), the levels of glycogen and trehalose in *S. carpocapsae* IJs declined rapidly under anaerobic conditions. However, the lactate level increased correspondingly, but the lipid levels remained unchanged. When anaerobically incubated IJs were returned to aerobic conditions, both glycogen and trehalose levels increased sharply, while the lipid and lactate levels decreased correspondingly. This suggests that, like most other animals, EPNs depend on carbohydrate reserves to provide energy under anaerobic conditions.

7.5 Genetic Selection for Temperature and Desiccation Tolerance

Although thermal niche breadths of EPN species appear to be conserved, variations in heat tolerance within the thermal niche breadth in the natural populations of *H. bacteriophora* (Grewal *et al.*, 2002) and *S. carpocapsae* (Somasekhar *et al.*, 2002) have been demonstrated. Glazer *et al.* (1991) found that heritability for heat tolerance for *H. bacteriophora* inbred lines was high (h^2 = 0.98). Both temperature tolerance (i.e. activity within the thermal niche breadth) and temperature activity ranges can be manipulated through genetic selection at constant temperatures in the laboratory. Grewal *et al.* (1996a) reported that cold tolerance of *S. feltiae* significantly improved during selection at 15°C, resulting in increased host mortality at 8 and 10°C, decreased LT_{50} values (time taken to kill 50% of the host larvae) at 10°C and increased nematode establishment in infected hosts across the entire thermal niche breadth. However, they found no change in the thermal infection or establishment niche breadths of *S. feltiae* during this cold selection. In another study on *H. bacteriophora* and *Steinernema arenarium* (= *Steinernema anomalae*), Grewal *et al.* (1996b) observed that both temperature tolerance and thermal activity ranges were malleable. They found that the thermal infection niche breadth of *H. bacteriophora* was extended from 10–32°C to 8–35°C during selection at 15°C and to 8–37°C during selection at 30°C. Extension in the thermal establishment niche breadth was observed only when the nematodes were selected at a cold temperature. Improvement in the overall establishment success of *H. bacteriophora* throughout the thermal niche breadth was observed following warm selection. Trade-offs in establishment success only occurred at warmer temperatures following selection at the cold temperature. Interestingly, both the extension in thermal niche breadth and overall improvement in *H. bacteriophora* reproduction success were obtained through both cold and warm selection. Jagdale and Gordon (1998) also observed improvements and trade-offs in survival of EPNs at temperature extremes following genetic selection. They found that long-term culturing of *S. feltiae*, *S. carpocapsae* and *S. riobrave* at cooler temperatures (10–20°C) increased the lower lethal temperature but decreased the upper lethal temperature limits.

7.6 Molecular Mechanisms of Desiccation Tolerance

Current knowledge about the gene function and molecular mechanisms involved in desiccation tolerance of EPNs is far less than that of physiological aspects. Gal *et al.* (2001) identified novel genes of *S. feltiae* IS-6 that exhibit changes in transcript levels upon dehydration. These included glycogen synthase (*Sf-gsy-1*), the rate-limiting enzyme in the synthesis of glycogen, which is likely to play a role in desiccation survival. In this study, Gal *et al.* (2001) established changes in the steady-state level of *Sf-gsy-1* transcripts upon dehydration. Our results suggest a shift from glycogen to trehalose synthesis upon dehydration, which is regulated, at least in part, by suppression of glycogen synthase transcription.

Using cDNA subtractive hybridization, Gal *et al.* (2003) identified ESTs differentially expressed in the semi-arid nematode *S. feltiae* IS-6 during exposure to desiccation stress. *Steinernema feltiae* IS-6 ESTs, differentially expressed during dehydration stress, were selected by subtractive hybridization. Two hybridizations were performed with RNA from nematodes dehydrated for 8 and 24 h versus controls. Ninety-two unique ESTs were identified. Some were homologous to known stress-related genes, including four entries of the water-stress-related protein late embryogenic abundant (*Sf-LEA-1*), the stress-responsive enzyme aldehyde dehydrogenase (*Sf-ALDH*), Hsp40, a zinc-binding protein required for disease resistance signalling in barley, casein kinase involved in the response to specific stresses, glycerol kinase involved in transfer of energy to the mitochondria, *ubq2* involved in the heat shock response of eukaryotes, glutathione peroxidase involved in exposure to oxidative stress in yeast, sodium bile acid transporter involved in ion channelling also during stress, and ten entries of cytochrome P450 involved in drought-stressed seedlings of rice. Other ESTs were homologous to *C. elegans* hypothetical proteins, with no known function, and 24 ESTs were novel, with no homology to known sequences in GenBank.

Determination of the expression profile of *S. feltiae* differentially expressed ESTs during desiccation stress demonstrated that all were upregulated during dehydration. Expression of all increased following 8 h of desiccation. After 24 h of desiccation, expression of some was further increased, whilst expression of others decreased. The stress-related *Sf-ALDH* (GenBank accession no. AF522285), *hsp40* (accession no. AF522286), zinc-binding protein (accession no. BQ563202), *ubq-2* (accession no. BQ563199) and glycerol kinase (accession no. BQ563201) were upregulated following 8h of desiccation and downregulated following 24h of desiccation, whereas expression of *Sf-LEA-1* (accession no. AF522287), cytochrome P450 (accession no. BQ563204) and glutathione peroxidase (accession no. BQ563205) was increased following 8h and further increased following 24h of desiccation. Of the stress-related genes, the increment of *Sf-LEA-1* and glutathione peroxidase was the greatest (313- and 136-fold upregulation in desiccated nematodes compared with non-desiccated controls). The highest increment was observed for EST accession no. BQ563209, encoding a homologue to hypothetical *C. elegans* protein (accession no. NM_067400) following 8 and 24h of desiccation (114- and 794-fold upregulation in desiccated nematodes compared with non-desiccated controls), and for a nucleosome binding protein (accession no. BQ579831) following 8 and 24h of desiccation (40- and 685-fold upregulation in desiccated nematodes compared with non-desiccated controls). The novel ESTs accession no. BQ563217 and BQ563218 were highly expressed following 24h, whereas the novel ESTs accession nos BQ563219 and BQ563220 were upregulated after 8h of desiccation. Others showed only low levels of elevated expression upon desiccation.

Strikingly, following 2h of *C. elegans* dehydration (and 85% of mortality), Gal *et al.* (2003) could detect only a very slight change in the steady-state level of *C. elegans* transcripts. Expression of the stress-related *C. elegans* proteins LEA-like and ALDH was only slightly elevated during dehydration (0.39- and 0.25-fold upregulation in desiccated nematodes compared with non-desiccated controls). Similarly, expression of *C. elegans* hypothetical proteins (accession nos NM_067400; T19316; NM_068209; T32747; T29492; T21199; T26447; T16543), whose homologues were upregulated during dehydration in *S. feltiae*, did not change markedly following dehydration of *C. elegans*. Some of these ESTs were known as stress-related, whilst others showed homology to hypothetical *C. elegans* proteins, thus assigning them a role in the stress response. Some were novel, suggesting their involvement in specific traits of *S. feltiae*. All analysed ESTs were upregulated during 8 and 24h of *S. feltiae* IS-6 dehydration. The response of *C. elegans* to dehydration was phenotypically different from that of *S. feltiae* IS-6. Significantly, genes whose homologues were upregulated in *S. feltiae* IS-6 did not show any increment in their expression level in *C. elegans* during dehydration, suggesting differences in the molecular and physiological mechanisms of response by *C. elegans* to desiccation stress, compared with the semi-arid *S. feltiae* IS-6. Our work unveiled some of the components of the genetic networks activated during desiccation, including different categories of transcripts that show different regulation in the environmentally tolerant dauer stage of nematodes.

Somvanshi *et al.* (2008) investigated expression of four genes, aldehyde dehydrogenase, nucleosome assembly protein 1, glutathione peroxidase and Hsp40, during desiccation stress in EPN species with differing stress tolerance ability. After 24h of desiccation, they found an inverse relationship between expression of the studied genes and phenotypic desiccation tolerance capability of the nematodes. *Heterorhabditis bacteriophora* TTO1 was most susceptible to desiccation but showed the highest expression of all the studied genes under desiccation; *S. carpocapsae* and *S. riobrave* showed the lowest expression of these genes but were the most desiccation tolerant.

7.7 Identification of Longevity and Stress Tolerance Genes

Over 1200 ESTs from *H. bacteriophora* GPS11 strain (Sandhu *et al.*, 2006; Bai *et al.*, 2007) and over 30,000 ESTs from *H. bacteriophora* TTO1 strain (Bai *et al.*, 2009), which have become available recently, reveal a number of interesting genes potentially affecting IJ longevity and stress tolerance in EPNs. Furthermore, over 13,000 protein-coding sequences have been predicted from the in-progress genome sequence of *H. bacteriophora* TTO1 strain (Bai *et al.*, unpublished data). Below we summarize some of the interesting findings related to the potential longevity and stress tolerance genes found in these developing genetic resources for EPNs.

7.7.1 Longevity genes

As expected due to the evolutionary conservation of longevity genes, comparative genomic approaches reveal that the majority of the longevity genes identified in *C. elegans* are also present in *H. bacteriophora* (Bai *et al.*, unpublished data). In *C. elegans*, adult longevity is affected by many genes, including those affecting the mitochondrial respiration rate, germline signalling pathway, insulin/IGF-1 signalling pathway, dietary restriction, TOR signalling, JNK signalling, SKN-1-dependent oxidative stress response, and others (Lee *et al.*, 2003; Baumeister *et al.*, 2006). The DAF-2 insulin/IGF signalling pathway is an important pathway regulating *C. elegans* longevity, dauer formation and stress response (Baumeister *et al.*, 2006; Gami and Wolkow, 2006). Almost all the genes in this important pathway have been found to be present in *H. bacteriophora* (Fig. 7.1). In this signalling pathway, the peptide hormones of insulin and IGF-1 are recognized by DAF-2, a predicted receptor tyrosine kinase. DAF-2 activation leads to the sequential activation of downstream phosphoinositide 3-kinase, 3-phosphoinositide-dependent kinase 1, and the serine/threonine kinases Akt and Sgk (Murphy *et al.*, 2003; Tullet *et al.*, 2008). The active DAF-2 signalling pathway produces the inactive form of phosphorylated DAF-16, a forkhead transcription factor. The mutation of DAF-2 or other components deactivates the signalling pathway and results in an active form of DAF-16, which serves in the nucleus as a transcriptional regulator of many downstream genes controlling longevity and stress responses (Murphy *et al.*, 2003; Baumeister *et al.*, 2006; Gami and Wolkow, 2006; Golden and Melov, 2007). DAF-16 is also the target of other signalling pathways (Barsyte *et al.*, 2001; Lin *et al.*, 2001; Garigan *et al.*, 2002; Larsen and Clarke, 2002).

7.7.2 Stress tolerance genes

The genes involved in stress responses are also conserved from archaebacteria to mammals (Prahlad and Morimoto, 2009). We have found many stress response proteins in *H. bacteriophora* EST sequences and the in-progress genome (Bai *et al.*, unpublished data). The *H. bacteriophora* genome harbours

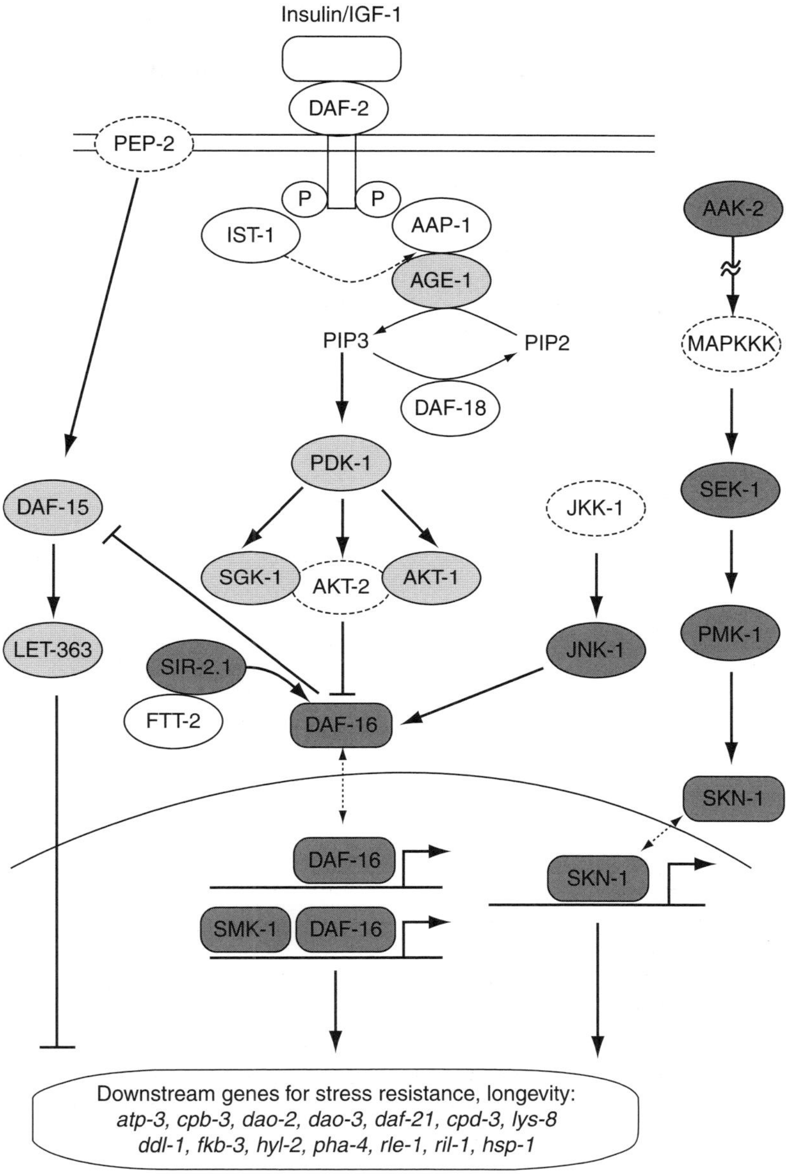

Fig. 7.1. Components of the insulin/IGF signalling pathway identified in *Heterorhabditis bacteriophora* (adapted from Baumeister *et al.*, 2006). Genes and proteins in solid-lined circles have been identified in *H. bacteriophora* TTO1. Proteins in light grey and dark grey circles are negative and positive regulators of *C. elegans* longevity, respectively.

a number of repair protein-encoding genes, whose products are used to correct the damage generated by exposure to endogenous stresses (e.g. reactive oxygen species) or exogenous stresses (e.g. ultraviolet light). Based on the damage being targeted, the repair mechanisms can be classified into three groups: base excision repair, nucleotide excision repair and DNA mismatch repair. The base excision repair is the process of correcting damage to a single base caused by oxidation, alkylation, hydrolysis or deamination (Baute and Depicker, 2008). The repair process involves the removal of the damage caused by a DNA glycosylase, the removal of phosphodiester bond by an apurinic/apyrimidinic (AP) endonuclease, resynthesis of the missing base by a DNA polymerase and the final gap-sealing by a DNA ligase (Baute and Depicker, 2008). In the *H. bacteriophora* in-progress genome, we have identified all the components of the base excision repair pathway. The *ung-1* uracil DNA glycosylase has the specificity to remove from DNA molecules the uracil generated from cytosine deamination, thereby initiating the base excision repair pathway. The *exo-3* AP endonuclease family 1 gene is one of the four types of AP endonucleases that have been classified according to their sites of incision. Also present in the *H. bacteriophora* in-progress genome are four types of ATP-dependent DNA ligases (I, II, III and IV) and five types of DNA polymerases (alpha, delta, epsilon, kappa and eta).

The nucleotide excision repair pathway is used for the correction of a wide range of chemically and structurally distinct DNA lesions in eukaryotic genomes (Hess *et al.*, 1998). Multiple genes are involved in the pathway, and the mutations of these genes are associated with genetic instability and disease of the organisms (Cleaver *et al.*, 2009). Depending on where the repair occurs, the nucleotide excision repair pathway could be divided into global genomic repair and transcription-coupled repair. The transcription-coupled repair process involves the arrest of the transcription process by ubiquitylation of the RNA polymerase II and damage detection by DNA damage-binding protein XPC, followed by the cleavage of the DNA around the damaged site and resynthesis of the cleavage site and ligation. In the *H. bacteriophora* in-progress genome, we have identified a transcription-coupled repair protein CSB, a nucleotide excision repair protein XPC, five subunits of the nucleotide excision repair factor TFIIH (RAD3, SSL2, TFB1, TFB2 and RING finger containing E3 ubiquitin ligase), multiple DNA polymerases and DNA ligases for DNA resynthesis and ligation, and a specialized DNA ligase XRCC1 for X-ray repair cross-complementing protein 1. DNA mismatch repair deals with the DNA mismatches generated during normal DNA metabolism or aberrant DNA processing reactions, including DNA replication, recombination and repair (Tullet *et al.*, 2008). We have identified DNA mismatch repair proteins MLH1, MSH2, MSH5, MSH6 and PMS2.

We have identified a number of Hsps in the *H. bacteriophora* genome, including Hsp1, Hsp3, Hsp6, Hsp12.2, Hsp16.2, Hsp17, Hsp25, Hsp43, Hsp60, Hsp70, and other Hsps such as STI-1, STC-1, DAF-21, DNJ-12 and DNJ-13. Although named as prokaryotic Hsps, DNJ-12 and DNJ-13 proteins are commonly found in eukaryotic organisms, from nematodes to humans. In *C. elegans*, DNJ-12 is associated with the processes of mitotic spindle organization and embryonic development and DNJ-13 with hermaphrodite genitalia development and reproduction.

In the *H. bacteriophora* in-progress genome, a number of transcription factors/regulators involved in stress responses have also been found.

HSF1 is a heat shock transcription factor functioning as a transcriptional regulator of stress-induced gene expression and is involved in nematode larval development, innate immunity and regulation of adult lifespan (Baugh and Sternberg, 2006; Singh and Aballay, 2006). SKN-1, skinhead transcription factor, functions in the p38 mitogen-activated protein kinase pathway to regulate the oxidative stress response and in parallel to DAF-16 in the DAF-2-mediated insulin/IGF-1-like signalling pathway to regulate adult lifespan (Tullet *et al.*, 2008). SKN-1 is similar to the basic region of bZIP transcriptional factors. XBP-1 is a bZIP transcription factor that is required for the unfolded protein response that counteracts cellular stress induced by accumulation of unfolded proteins in the endoplasmic reticulum (Calfon *et al.*, 2002). Other transcriptional factors/regulators identified in *H. bacteriophora* include GFL-1, CUL-4, LAG-1, DIN-1 and DJR-1.1.

In the *H. bacteriophora* in-progress genome, we have also identified two 2-cys peroxiredoxins, encoded by *prdx-2* and *prdx-3* genes. Peroxiredoxins are peroxidase enzymes that reduce hydrogen peroxide and contribute to the oxidative stress response of multicellular organisms (Olahova *et al.*, 2008). Other oxidative stress response proteins include, among others, CTL-2 catalase, EGL-9 dioxygenase, PXN-2 peroxidase, TRX-2 thioredoxin, TRXR-2 thioredoxin reductase and GST-1 glutathione *S*-transferase.

7.8 Conclusions and Future Directions

Great progress has been made in our understanding of the longevity and stress tolerance of EPNs during the past four decades. Much research during this period has focused on physiological mechanisms of stress tolerance in the IJs. Studies on molecular mechanisms controlling IJ longevity and stress tolerance have only begun recently. The genomic studies on EPNs are now opening new doors for hypothesis-driven functional genomics research and for potential genetic manipulations of EPNs for improved IJ longevity and stress tolerance for their use in biological pest control.

The innate capacity of EPN IJs to live at optimum growth temperature varies with nematode species and strains. The IJs of some EPN species can only live for up to 6 weeks, while those of other species can live for up to 9 months in water at 25°C. This IJ longevity is influenced by both internal and external factors, including the amount of stored energy reserves, rate of metabolism, activity, culture method, temperature, oxygen availability, desiccation and ultraviolet radiation.

Interesting discoveries have been made with respect to the association of trehalose with temperature and desiccation tolerance of EPN IJs. It has been discovered that trehalose accumulation is not only a cold-associated phenomenon. Trehalose also accumulates rapidly when the EPN IJs are heat-shocked. It appears that trehalose accumulation is part of the general strategy of EPN IJs to deal with multiple stresses, including cold, freezing, heat, desiccation and ultraviolet radiation. It has also been discovered that partially desiccated IJs have enhanced cold, heat and ultraviolet radiation tolerance.

With the recent availability of ESTs for EPNs, in-progress genome sequence data, demonstration of gene knockout with RNA interference and gene expression profiling during desiccation, the field of EPN research is moving into a new era of genomics and functional genomics. With the availability of these tools, the stage is now set for rapid progress in our understanding of the genetic and molecular factors controlling longevity and stress tolerance in EPN IJs, with new and interesting discoveries likely.

7.9 References

An, R., Sreevatsan, S. and Grewal, P.S. (2009) Comparative in-vivo gene expression of the closely related bacteria *Photorhabdus temperata* and *Xenorhabdus koppenhoeferi* upon infection of the same insect host, *Rhizotrogus majalis*. *BMC Genomics* 10, 433.

Bai, X., Grewal, P.S., Hogenhout, S.A., Adams, B.J., Ciche, T.A., Gaugler, R. and Sternberg. P.W. (2007) Expressed sequence tag analysis of gene representation in insect parasitic nematode *Heterorhabditis bacteriophora*. *Journal of Parasitology* 93, 1343–1349.

Bai, X., Adams, B.J., Ciche, T.A., Clifton, S. *et al.* (2009) Transcriptomic analysis of the entomopathogenic nematode, *Heterorhabditis bacteriophora* TTO1. *BMC Genomics* 10, 205.

Barrett, J. (1991) Anhydrobiotic nematodes. *Agricultural Zoology Reviews* 4, 161–176.

Barsyte, D., Lovejoy, D.A. and Lithgow, G.J. (2001) Longevity and heavy metal resistance in *daf-2* and *age-1* long-lived mutants of *Caenorhabditis elegans*. *FASEB Journal* 15, 627–634.

Baugh, L.R. and Sternberg, P.W. (2006) DAF-16/FOXO regulates transcription of *cki-1*/Cip/Kip and repression of *lin-4* during *C. elegans* L1 arrest. *Current Biology* 16, 780–785.

Baumeister, R., Schaffitzel, E. and Hertweck, M. (2006) Endocrine signaling in *Caenorhabditis elegans* controls stress response and longevity. *Journal of Endocrinology* 190, 191–202.

Baute, J. and Depicker, A. (2008) Base excision repair and its role in maintaining genome stability. *Critical Reviews in Biochemistry and Molecular Biology* 43, 239–276.

Behm, C.A. (1997) The role of trehalose in the physiology of nematodes. *International Journal for Parasitology* 27, 215–229.

Bell, W., Klaassen, P., Ohnacker, M., Bollern, T. and Herweijer, M. (1992) Characterization of the 56-kDa subunit of yeast trehalose-6-phosphate synthase and cloning of its gene reveal its identity with the product of CIF1, a regulator of carbon catabolite inactivation. *European Journal of Biochemistry* 209, 951–959.

Boemare, N. (2002) Biology, taxonomy and systematics of *Photorhabdus* and *Xenorhabdus*. In: Gaugler, R. (ed.). *Entomopathogenic Nematology*. CAB International, Wallingford, UK, pp. 35–56.

Boemare, N., Laumond, C. and Mauleon, H. (1996) The entomopathogenic nematode–bacterium complex: biology, life cycle and vertebrate safety. *Biocontrol Science and Technology* 6, 333–345.

Browne, J., Tunnacliffe, A. and Burnell, A. (2002) Anhydrobiosis – plant desiccation gene found in a nematode. *Nature* 416, 38–38.

Burman, M. and Pye, A.E. (1980) *Neoaplectana carpocapsae*: respiration of infective juveniles. *Nematologica* 26, 214–218.

Calfon, M., Zeng, H., Urano, F., Till, J.H., Hubbard, S.R., Harding, H.P., Clark, S.G. and Ron, D. (2002) IRE1 couples endoplasmic reticulum load to secretory capacity by processing the XBP-1 mRNA. *Nature* 415, 92–96.

Chandler, P.M., Walker-Simmons, M., King, R.W., Crouch, M. and Close, T.J. (1988) Expression of ABA-inducible genes in water stressed cereal seedlings. *Journal of Cell Biology Supplement* 12C, 143.

Chen, S., Gollop, N. and Glazer, I. (2005) Cross-stress tolerance and expression of stress-related proteins in osmotically desiccated entomopathogenic *Steinernema feltiae* IS-6. *Parasitology* 131, 695–703.

Chen, S., Glazer, I., Gollop, N., Cash, P., Argo, E., Innes, A., Stewart, E., Davidson, I. and Wilson, M.J. (2006) Proteomic analysis of the entomopathogenic nematode *Steinernema feltiae* IS-6 IJs under evaporative and osmotic stresses. *Molecular and Biochemical Parasitology* 145, 195–204.

Cleaver, J.E., Lam, E.T. and Revet, I. (2009) Disorders of nucleotide excision repair: the genetic and molecular basis of heterogeneity. *Nature Review Genetics* 10, 756–768.

Close, T.J. (1996) Dehydrins: emergence of a biochemical role of a family of plant dehydration proteins. *Physiological Plantarum* 97, 795–803.

Crowe, J.H. and Crowe, L.M. (1992) Membrane integrity in anhydrobiotic organisms: towards a mechanism for stabilizing dry cells. In: Somero, G.N., Osmond, C.B. and Bolis, C.L. (eds) *Water and Life*. Springer-Verlag, Berlin, Heidelberg, Germany, pp. 87–103.

Crowe, J.H. and Madin, K.A.C. (1975) Anhydrobiosis in nematodes: evaporative water loss and survival. *Journal of Experimental Zoology* 193, 323–334.

Crowe, J.H., Crowe, L.M. and Chapman, D. (1984) Infrared spectroscopic studies on interactions of water and carbohydrates with a biological membrane. *Archives of Biochemistry and Biophysics* 232, 400–407.

Crowe, J.H., Hoekstra, F.A. and Crowe, L.M. (1992) Anhydrobiosis. *Annual Reviews in Physiology* 54, 579–599.

Crowe, J.H., Hoekstra, F.A., Nguyen, K.H.N. and Crowe, L.M. (1996) Is vitrification involved in depression of the phase-transition temperature in dry phospholipids. *Biochimica et Biophysica Acta – Biomembranes* 1280, 187–196.

Dalley, B.K. and Golomb, M. (1992) Gene expression in *Caenorhabditis elegans* 'dauer' larva: developmental regulation of Hsp90 and other genes. *Developmental Biology* 151, 80–90.

De Ley, P. and Blaxter, M. (2002) Systematic position and phylogeny. In: Lee, D.L. (ed.) *The Biology of Nematodes*. Taylor & Francis, London, pp. 1–30.

De Virgilio, C., Bürckert, N., Bell, W., Boller, T. and Wiemken, A. (1993) Distribution of *TPS2*, the gene encoding the 100-kDa sub-unit of the trehalose-6-phosphate synthase/phosphatase complex in *Saccharomyces cerevisiae*, causes accumulation of trehalose-6-phosphate and loss of trehalose-6-phosphatase activity. *European Journal of Biochemistry* 212, 315–323.

De Virgilio, C., Hottiger, T., Dominguez, J., Boller, T. and Wiemken, A. (1994) The role of trehalose synthesis for the acquisition of thermotolerance in yeast. I. Genetic evidence that trehalose is a thermoprotectant. *European Journal of Biochemistry* 219, 179–186.

Dure, L. (1993) Plant response to cellular dehydration during environmental stress: structural motifs in LEA proteins. In: Close, T.J. and Bray, E.A. (eds) *Current Topics in Plant Physiology, Vol. 10*. American Society of Plant Physiologists, Rockville, Maryland, pp. 91–103.

ffrench-Constant, R. and Waterfield, N. (2006) An ABC guide to the bacterial toxin complexes. *Advances in Applied Microbiology* 58, 169–183.

Fitters, P.F.L., Meijer, E.M.J., Wright, D.J. and Griffin, C.T. (1997) Estimation of lipid reserves in unstained living and dead nematodes by image analysis. *Journal of Nematology* 29, 160–167.

Fodor, A., Dey, I., Farkas, T. and Chitwood, D.J. (1994) Effects of temperature and dietary lipids on phospholipid fatty acids and membrane fluidity in *Steinernema carpocapsae*. *Journal of Nematology* 26, 278–285.

Forst, S. and Clarke, D. (2002) Bacteria–nematode symbioses. In: Gaugler, R. (ed.) *Entomopathogenic Nematology*. CAB International, Wallingford, UK, pp. 57–77.

Gal, T.Z., Solomon, A., Glazer, I. and Koltai, H. (2001) Reduction in the levels of glycogen and glycogen synthase transcripts during desiccation in the insect-killing, desiccation tolerant nematode *Steinernema feltiae* IS-6. *Journal of Parasitology* 87, 725–732.

Gal, T.Z., Glazer, I. and Koltai, H. (2003) Differential gene expression during desiccation stress in *Steinernema feltiae* IS-6 and *Caenorhabditis elegans* nematodes. *Journal of Parasitology* 89, 761–766.

Gami, M.S. and Wolkow, C.A. (2006) Studies of *Caenorhabditis elegans* DAF-2/insulin signaling reveal targets for pharmacological manipulation of lifespan. *Aging Cell* 5, 31–37.

Garigan, D., Hsu, A.L., Fraser, A.G., Kamath, R.S., Ahringer, J. and Kenyon, C. (2002) Genetic analysis of tissue aging in *Caenorhabditis elegans*: a role for heat-shock factor and bacterial proliferation. *Genetics* 161, 1101–1112.

Georgis, R. (1990) Formulation and application technology. In: Gaugler, R. and Kaya, H.K. (eds) *Entomopathogenic Nematodes in Biological Control*. CRC Press, Boca Raton, Florida, pp. 173–191.

Glazer, I., Gaugler, R. and Segal, D. (1991) Genetics of the nematode *Heterorhabditis bacteriophora* strain HP88: the diversity of beneficial traits. *Journal of Nematology* 23, 324–333.

Golden, T.R. and Melov, S. (2007) Gene expression changes associated with aging in *C. elegans*. In: The *C. elegans* Research Community (ed.) *WormBook*, doi/10.1895/wormbook.1.127.1, http://www.wormbook.org.

Goodrich-Blair, H. and Clarke, D.J. (2007) Mutualism and pathogenesis in *Xenorhabdus* and *Photorhabdus*: two roads to the same destination. *Molecular Microbiology* 64, 260–268.

Grewal, P.S. (2000a) Enhanced ambient storage stability of an entomopathogenic nematode through anhydrobiosis. *Pest Management Science* 56, 401–406.

Grewal, P.S. (2000b) Anhydrobiotic potential and long-term storage stability of entomopathogenic nematodes (Rhabditida: Steinernematidae). *International Journal for Parasitology* 30, 995–1000.

Grewal, P.S. (2002) Formulation and application technology. In: Gaugler, R. (ed.) *Entomopathogenic Nematology*. CAB International, Wallingford, UK, pp. 265–287.

Grewal, P.S. and Georgis, R. (1998) Entomopathogenic nematodes. In: Hall, F.R. and Menn, J. (eds) *Methods in Biotechnology: Biopesticides: Use and Delivery*. Humana Press Inc., Totowa, New Jersey, pp. 271–299.

Grewal, P.S. and Jagdale, G.B. (2002) Enhanced trehalose accumulation and desiccation survival of entomopathogenic nematodes through cold preacclimation. *Biocontrol Science and Technology* 12, 533–545.

Grewal, P.S., Selvan, S. and Gaugler, R. (1994) Thermal adaptation of entomopathogenic nematodes: niche breadth for infection, establishment, and reproduction. *Journal of Thermal Biology* 19, 245–253.

Grewal, P.S., Gaugler, R. and Wang, Y. (1996a) Enhanced cold tolerance of the entomopathogenic nematode *Steinernema feltiae* through genetic selection. *Annals of Applied Biology* 129, 335–341.

Grewal, P.S., Gaugler, R. and Shupe, C. (1996b) Rapid evolution of thermal sensitivity of entomopathogenic nematodes in response to selection at temperature extremes. *Journal of Invertebrate Pathology* 68, 65–73.

Grewal, P.S., Wang, X. and Taylor, R.A.J. (2002) Dauer juvenile longevity and stress tolerance in natural populations of entomopathogenic nematodes: is there a relationship? *International Journal for Parasitology* 32, 717–725.

Grewal, P.S., Power, K.T., Grewal, S.K., Suggars, A. and Haupricht, S. (2004) Enhanced predictability in biological control of white grubs with new strains of entomopathogenic nematodes. *Biological Control* 30, 73–82.

Grewal, P.S., Ehlers, R.-U. and Shapiro-Ilan, D.I. (2005) *Nematodes as Biocontrol Agents*. CAB International, Wallingford, UK.

Grewal, P.S., Bornstein-Forst, S., Burnell, A.M., Glazer, I. and Jagdale, G.B. (2006) Physiological, genetic, and molecular mechanisms of chemoreception, thermobiosis and anhydrobiosis in entomopathogenic nematodes. *Biological Control* 38, 54–65.

Hess, N.J., Buchko, G.W., Conradson, S.D., Espinosa, F.J., Ni, S., Thrall, B.D. and Kennedy, M.A. (1998) Human nucleotide excision repair protein XPA: extended X-ray absorption fine-structure evidence for a metal-binding domain. *Protein Science* 7, 1970–1975.

Higa, L.M. and Womersley, C. (1993) New insights into the anhydrobiotic phenomenon: the effects of trehalose content and differential rates of water loss on the survival of *Aphelenchus avenae*. *Journal of Experimental Zoology* 267, 120–129.

Hochachka, P.W. and Somero, G.N. (1984) *Biochemical Adaptation*. Princeton University Press, Princeton, New Jersey.

Hominick, W.M. (2002) Biogeography. In: Gaugler, R. (ed.) *Entomopathogenic Nematology*. CAB International, Wallingford, UK, pp. 115–143.

Hottiger, T., De Virgilio, C., Hall, M.N., Boller, T. and Wiemken, A. (1994) The role of trehalose synthesis for the acquisition of thermotolerance in yeast II. Physiological concentrations of trehalose increase the thermal stability of proteins *in vitro*. *European Journal of Biochemistry* 219, 187–193.

Jagdale, G.B. and Gordon, R. (1997a) Effect of temperature on the composition of fatty acids in total lipids and phospholipids of entomopathogenic nematodes. *Journal of Thermal Biology* 22, 245–251.

Jagdale, G.B. and Gordon, R. (1997b) Effect of temperature on the activities of glucose-6-phosphate dehydrogenase and hexokinase in entomopathogenic nematodes (Nematoda: Steinernematidae). *Comparative Biochemistry and Physiology* 118A, 1151–1156.

Jagdale, G.B. and Gordon, R. (1998) Effect of propagation temperatures on temperature tolerances of entomopathogenic nematodes. *Fundamentals and Applied Nematology* 21, 177–183.

Jagdale, G.B. and Grewal, P.S. (2003) Acclimation of entomopathogenic nematodes to novel temperature: trehalose accumulation and the acquisition of thermotolerance. *International Journal of Parasitology* 33, 145–152.

Jagdale, G.B. and Grewal, P.S. (2007) Storage temperature influences desiccation and ultra violet radiation tolerance of entomopathogenic nematodes. *Journal of Thermal Biology* 32, 20–27.

Jagdale, G.B., Grewal, P.S. and Salminen, S. (2005) Both heat-shock and cold-shock influence trehalose metabolism in an entomopathogenic nematode. *Journal of Parasitology* 91, 151–157.

Kenyon, C. (1997) Environmental factors and gene activities that influence life span. In: Riddle, D.L., Blumenthal, T., Meyer, B.J. and Priess, J.R. (eds) C. elegans *II*. Cold Spring Harbor Laboratory Press, Cold Spring Harbor, New York, pp. 791–813.

Klass, M. and Hirsh, D. (1976) Non-ageing developmental variant of *Caenorhabditis elegans*. *Nature* 260, 523–525.

Kung, S.P., Gaugler, R. and Kaya, H.K. (1990) Influance of soil pH and oxygen on entomopathogenic nematode persistence. *Journal of Nematology* 22, 440–445.

Larsen, P.L. (1993) Aging and resistance to oxidative damage in *Caenorhabditis elegans*. *Proceedings of the National Academy of Sciences of the USA* 90, 8905–8909.

Larsen, P.L. and Clarke, C.F. (2002) Extension of life-span in *Caenorhabditis elegans* by a diet lacking coenzyme Q. *Science* 295, 120–123.

Lee, R.E. and Denlinger, D.L. (1991) *Insects at Low Temperature*. Chapman & Hall, New York.

Lee, S.S., Lee, R.Y.N., Fraser, A.G., Kamath, R.S., Ahringer, J. and Ruvkun, G. (2003) A systematic RNAi screen identifies a critical role for mitochondria in *C. elegans* longevity. *Nature Genetics* 33, 40–48.

Lin, K., Hsin, H., Libina, N. and Kenyon, C. (2001) Regulation of the *Caenorhabditis elegans* longevity protein DAF-16 by insulin/IGF-1 and germline signaling. *Nature Genetics* 28, 139–145.

Majara, M., O'Connor-Cox, E.S.C. and Axcell, B.C. (1996) Trehalose – a stress protectant and stress indicator compound for yeast exposed to adverse conditions. *Journal of American Society of Brewing Chemistry* 54, 221–227.

Murphy, C.T., McCarroll, S.A., Bargmann, C.I., Fraser, A., Kamath, R.S., Ahringer, J., Li, H. and Kenyon, C. (2003) Genes that act downstream of DAF-16 to influence the lifespan of *Caenorhabditis elegans*. *Nature* 424, 277–283.

Nagao, R.T., Kimpel, J.A. and Key, J.L. (1990) Molecular and cellular biology of the heat-shock response. In: Scandalios, J.G. (ed.) *Advances in Genetics, Genomic Responses to Environmental Stress*. Academic Press, New York, pp. 235–274.

Ogura, N. and Nakashima, T. (1997) Cold tolerance and preconditioning of infective juveniles of *Steinernema kushidai* (Nematoda: Steinernematidae). *Nematologica* 43, 107–115.

Olahova, M., Taylor, S.R., Khazaipoul, S., Wang, J., Morgan, B.A., Matsumoto, K., Blackwell, T.K. and Veal, E.A. (2008) A redox-sensitive peroxiredoxin that is important for longevity has tissue- and stress-specific roles in stress resistance. *Proceedings of the National Academy of Sciences of the USA* 105, 19839–19844.

O'Leary, S.A., Power, A.P., Stack, C.M. and Burnell, A.M. (2001) Behavioral and physiological responses of infective juveniles of the entomopathogenic nematode *Heterorhabditis* to desiccation. *Biocontrol* 46, 345–362.

O'Riordan, V. and Burnell, A.M. (1989) Intermediary metabolism in the 'dauer' larva of the nematode *C. elegans*. I. Glycolysis, gluconeogenesis, oxidative phosphorylation, and tricarboxylic acid cycle. *Comparative Biochemistry and Physiology* 92B, 233–238.

O'Riordan, V. and Burnell, A.M. (1990) Intermediary metabolism in the 'dauer' larva of the nematode *C. elegans*. II. The glycoxylate cycle and fatty acid oxidation. *Comparative Biochemistry and Physiology* 95B, 125–130.

Patel, M.N. and Wright, D.J. (1997) Glycogen: its importance in the infectivity of aged juveniles of *Steinernema carpocapsae*. *Parasitology* 114, 591–596.

Patel, M.N., Perry, R.N. and Wright, D.J. (1997) Desiccation survival and water contents of entomopathogenic nematodes, *Steinernema* spp. (Rhabditida: Steinernematidae). *International Journal for Parasitology* 27, 61–70.

Poinar, G.O. Jr (1990) Taxonomy and biology of Steinernematidae and Heterorhabditidae. In: Gaugler, R. and Kaya, H.K. (eds) *Entomopathogenic Nematodes in Biological Control*. CRC Press, Boca Raton, Florida, pp. 23–61.

Popiel, I., Holtemann, K.D., Glaser, I. and Womersley, C. (1987) Commercial storage and shipment of entomogenous nematodes. U.S. Patent Application PCT/ o587/02043.

Prahlad, V. and Morimoto, R.I. (2009) Integrating the stress response: lessons for neurodegenerative diseases from *C. elegans*. *Trends in Cell Biology* 19, 52–61.

Qiu, L. and Bedding, R. (1999a) The relationship between energy metabolism and survival of the infective juveniles of *Steinernema carpocapsae* under unstressed-aerobic and anaerobic conditions. In: Glazer, I., Richardson, P., Boemare, N. and Coudert, F. (eds) *Survival Strategies of Entomopathogenic Nematodes*. EUR 18855 EN Report, pp.149–156.

Qiu, L.H. and Bedding, R. (1999b) Low temperature induced cryoprotectant synthesis by the infective juveniles of *Steinernema carpocapsae*: biological significance and mechanisms involved. *Cryo-Letters* 20, 393–404.

Riddle, D.L. (1988) The 'dauer' larva. In: Wood, W.B. (ed.) *The Nematode* Caenorhabditis elegans. Cold Spring Harbor Laboratory, Cold Spring Harbor, New York, pp. 393–412.

Sandhu, S.K., Jagdale, G.B., Hogenhout, S. and Grewal, P.S. (2006) Comparative analysis of the expressed genome of the entomopathogenic nematode, *Heterorhabditis bacteriophora*. *Molecular and Biochemical Parasitology* 145, 239–244.

Selvan, S., Gaugler, R. and Grewal, P.S. (1993) Water content and fatty acid composition of infective juvenile entomopathogenic nematodes during storage. *Journal of Parasitology* 79, 510–516.

Serwe-Rodriguez, J., Appleman, B. and Bornstein-Forst, S. (2004) Effects of host desiccation on development, survival, and infectivity of entomopathogenic nematode *Steinernema carpocapsae*. *Journal of Invertebrate Pathology* 85, 175–181.

Simons, W.R. and Poinar, G.O. Jr (1973) The ability of *Neoaplectana carpocapsae* (Steinernematidae: Nematodea) to survive extended periods of desiccation. *Journal of Invertebrate Pathology* 22, 228–230.

Singer, M.A. and Lindquist, S. (1998) Thermotolerance in *Saccharomyces cerevisiae*: the yin and yang of trehalose. *Trends in Biotechnology* 6, 460–468.

Singh, V. and Aballay, A. (2006) Heat-shock transcription factor (HSF)-1 pathway required for *Caenorhabditis elegans* immunity. *Proceedings of the National Academy of Sciences of the USA* 103, 13092–13097.

Solomon, A., Paperna, I. and Glazer, I. (1999) Desiccation survival of the entomopathogenic nematode *Steinernema feltiae*: induction of anhydrobiosis. In: Glazer, I., Richardson, P., Boemare, N. and Coudert, F. (eds) *Survival Strategies of Entomopathogenic Nematodes*. EUR 18855 EN Report, pp. 83–98.

Solomon, A., Solomon, R., Paperna, I. and Glazer, I. (2000) Desiccation stress of entomopathogenic nematodes induces the accumulation of a novel heat-stable-protein. *Parasitology* 121, 409–416.

Somasekhar, N., Grewal, P.S. and Klein, M.G. (2002) Genetic variability in stress tolerance and fitness among natural populations of *Steinernema carpocapsae*. *Biological Control* 23, 303–310.

Somvanshi, V.S., Koltai, H. and Glazer, I. (2008) Expression of different desiccation-tolerance related genes in various species of entomopathogenic nematodes. *Molecular and Biochemical Parasitology* 158, 65–71.

Storey, K.B. and Storey, J.M. (1991) Biochemistry of cryoprotectants. In: Lee, R.E. and Denlinger, D.L. (eds) *Insects at Low Temperature*. Chapman & Hall, New York, pp. 64–93.

Storey, K.B. and Storey, J.M. (1992) Biochemical adaptations for winter survival in insects. *Advances in Low Temperature Biology* 1, 101–140.

Sun, W.Q. and Leopold, A.C. (1997) Cytoplasmic vitrification and survival of anhydrobiotic organisms. *Comparative Biochemistry and Physiology* 117A, 327– 333.

Tullet, J.M., Hertweck, M., An, J.H., Baker, J., Hwang, J.Y., Liu, S., Oliveira, R.P., Baumeister, R. and Blackwell, T.K. (2008) Direct inhibition of the longevity-promoting factor SKN-1 by insulin-like signaling in *C. elegans*. *Cell* 132, 1025–1038.

Vanfleteren, J.R. (1993) Oxidative stress and aging in *Caenorhabditis elegans*. *Biochemical Journal* 292, 605–608.

Webster, J., Chen, G., Hu, K. and Li, J. (2002) Bacterial metabolites. In: Gaugler, R. (ed.) *Entomopathogenic Nematology*. CAB International, Wallingford, UK, pp. 79–98.

Womersley, C.Z. (1987) A reevaluation of strategies employed by nematode anhydrobiotes in relation to their natural environment. In: Veech, J.A. and Dickson, D.W. (eds) *Vista on Nematology: a Commemoration of the 25th Anniversary of the Society of Nematology*. Society of Nematology, Hyattsville, Maryland, pp.165–173.

Womersley, C.Z. (1990) Dehydration survival and anhydrobiotic potential. In: Gaugler, R. and Kaya, H.K. (eds) *Entomopathogenic Nematodes in Biological Control*. CRC Press, Boca Raton, Florida, pp. 117–137.

Worland, M.R., Grubor-Lajsic, G. and Montiel, P.O. (1998) Partial desiccation induced by sub-zero temperatures as a component of the survival strategy of the Arctic collembolan *Onychiurus arcticus* (Tullberg). *Journal of Insect Physiology* 44, 211–219.

Yan, X., Liu, X., Han, R., Chen, S., De Clercq, P. and Moens, M. (2010) Osmotic induction of anhydrobiosis in entomopathogenic nematodes of the genera *Heterorhabditis* and *Steinernema*. *Biological Control* 53, 325–330.

8 Cold Tolerance

DAVID A. WHARTON

Department of Zoology, University of Otago, Dunedin, New Zealand

8.1 Introduction

Free-living nematodes, the free-living stages of parasitic nematodes and, in some cases, their parasitic stages (Wharton, 1999) are exposed to low temperatures in many parts of the world. Nematodes are ectotherms and are at the same temperature as their surroundings, the temperature of which may vary on a seasonal and a daily basis.

Low temperatures bring several problems for the survival of organisms. Some problems are associated with low temperature per se, which may cause changes in the viscosity, phase and organization of membranes, with a corresponding loss of function. There may also be changes in the structure and function of proteins, and a general reduction in metabolic activity as the temperature falls (Grout and Morris, 1987; Ramløv, 2000). Once the temperature falls below the melting point of its body fluids, the animal is at risk of freezing. This involves a change in phase from a liquid to a solid (ice), which may result in mechanical damage to cells. Ice excludes most solutes and hence they become concentrated in the remaining unfrozen portion of the solution. This freeze concentration effect, and the resulting osmotic stress, is a major cause of cell damage during freezing (Mazur, 1984). Intracellular freezing is usually considered lethal (Mazur, 1984) but some examples of the survival of intracellular freezing have been discovered (Wharton and Ferns, 1995; Salinas-Flores *et al.*, 2008).

Habitats where nematodes are likely to be frozen for varying periods of time include: terrestrial sites in the Antarctic (Wharton, 2003) and the Arctic (Coulson and Birkemoe, 2000), alpine sites (Hoschitz and Kaufmann, 2004), sea ice (Gradinger, 2001), cryoconite holes that form by the surface melting of glaciers (Christner *et al.*, 2003; Hodson *et al.*, 2008) and in temperate habitats that experience sub-zero temperatures over winter (MacGuidwin and Forge, 1991; Dimander *et al.*, 1999) or at other times of the year. Although deep-sea nematodes are not exposed to the risk of freezing, they are exposed to relatively low temperatures of about 2°C.

In this chapter I will describe the strategies that nematodes use to survive low temperatures, the mechanisms that may be involved and how these mechanisms contribute to the strategies of cold tolerance. My focus will be on the survival at temperatures below 0°C, where the animal is at risk of freezing.

8.2 Cold Tolerance Strategies

8.2.1 How many strategies?

There are a number of strategies by which nematodes may survive sub-zero temperatures (Fig. 8.1), depending on the rate of cooling, the moisture content of the substrate and whether the nematode can prevent inoculative freezing (the freezing of an organism as a result of ice from its surroundings travelling across its surface; Wharton, 2002). In common with other animals, the two main cold tolerance strategies are freeze avoidance and freezing tolerance. Freeze-avoiding animals reduce their risk of freezing at sub-zero temperatures, often by removing or masking ice nucleators (substances that cause ice nucleation, the initial process which results in the formation of an ice crystal), and survive in a supercooled state (where their body fluids are still liquid at temperatures below their melting point) but die if their body fluids freeze. Freezing-tolerant animals survive the freezing of at least part of their body fluids.

Nematodes are essentially aquatic organisms and, unless they are desiccated, face the risk of inoculative freezing by the ice in their surroundings seeding, via body orifices, the freezing of their body fluids. Some nematodes have a physical barrier, such as an eggshell or a sheath, that protects the nematode against inoculative freezing and allows its body fluids to supercool to low temperatures, even though the animal is encased in ice (strategy 2, Fig. 8.1). *Globodera rostochiensis* and *Globodera pallida* are examples of this strategy, where the infective larva is enclosed within an eggshell and a cyst wall. The eggshell prevents inoculative freezing and allows the larva to supercool to temperatures as low as −38 °C, even though the eggs are surrounded by ice (Wharton *et al.*, 1993; Wharton and Ramløv, 1995; Devine, 2010). In *Trichostronglus colubriformis* and *Heterorhabditis zealandica*, a sheath prevents inoculative freezing, at least under some circumstances, and allows

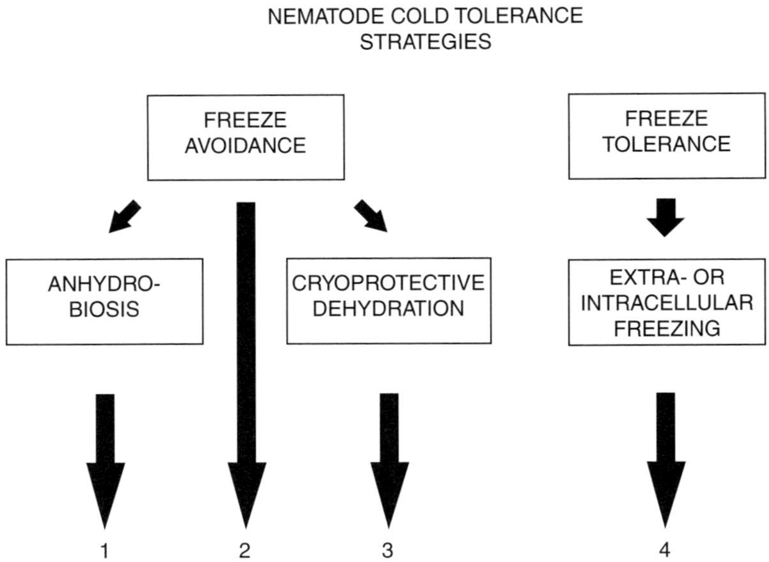

Fig. 8.1. Cold tolerance mechanisms in nematodes. Cold-tolerant nematodes are either freeze-avoiding (their body fluids remain liquid at temperatures below their melting point) or freezing tolerant (they survive the freezing of at least part of their body fluids: strategy 4). Classical freeze avoidance (strategy 2) involves keeping the body fluids liquid at temperatures well below their melting point. Since nematodes are often exposed to low temperatures whilst surrounded by water, this strategy is usually restricted to species and stages which possess a barrrier (such as an eggshell or sheath) that can prevent inoculative freezing from ice in their surroundings. If the nematode has been exposed to desiccation before sub-zero temperatures, there may be no freezable water in their bodies and they survive in a state of anhydrobiosis (strategy 1). If freezing of the nematode's surroundings occurs at a high sub-zero temperature, there may be insufficient force for inoculative freezing to occur; the nematodes will desiccate rather than freeze (due to the difference in vapour pressure between water and ice at the same temperature) and survive by cryoprotective dehydration (strategy 3).

the nematode to supercool in the presence of external ice (Wharton and Allan, 1989; Wharton and Surrey, 1994).

If the nematode is exposed to desiccation before exposure to low temperatures and it can survive anhydrobiotically (surviving a cessation of metabolism due to water loss; Wharton, 2002), it will survive without freezing since all freezable water will have been lost (strategy 1, Fig. 8.1). Anhydrobiotic nematodes lose all their water (or at least as much as can be measured). The proportion of osmotically inactive (unfreezable) water in animals is 10–25% (Wharton and Worland, 2001). A nematode that can survive the loss of osmotically active (freezable) but not osmotically inactive (unfreezable) water, and hence is not anhydrobiotic but merely desiccation tolerant, could nevertheless survive low temperatures in a similar fashion. Such a situation has yet to be described in nematodes but it does occur in earthworm cocoons, which will survive the loss of osmotically active but not osmotically inactive water (Holmstrup and Westh, 1995).

Where the nematode is surrounded by a large volume of water and/or the presence of ice nucleators (e.g. in soil) results in freezing at a relatively high sub-zero temperature, the freezing of its surroundings occurs slowly. This may not produce sufficient force for inoculative freezing to occur and the body contents of the nematode remain liquid, even though it is surrounded by ice. The vapour pressure of ice is lower than that of supercooled water at the same temperature (Fig. 8.2). The nematode thus loses water to its surroundings and dehydrates rather than freezes, a cold tolerance strategy known as 'cryoprotective dehydration' (Wharton *et al.*, 2003: strategy 3, Figs 8.1 and 8.2), earlier known as 'the protective dehydration mechanism of cold hardiness' (Holmstrup and Westh, 1994). Cryoprotective dehydration has only been demonstrated in one species of nematode, the Antarctic nematode *Panagrolaimus davidi* (Wharton *et al.*, 2003). However, a shrunken appearance in other species after freezing and thawing (Scholander *et al.*, 1953; Forge and MacGuidwin, 1992b) suggests that it may occur more widely.

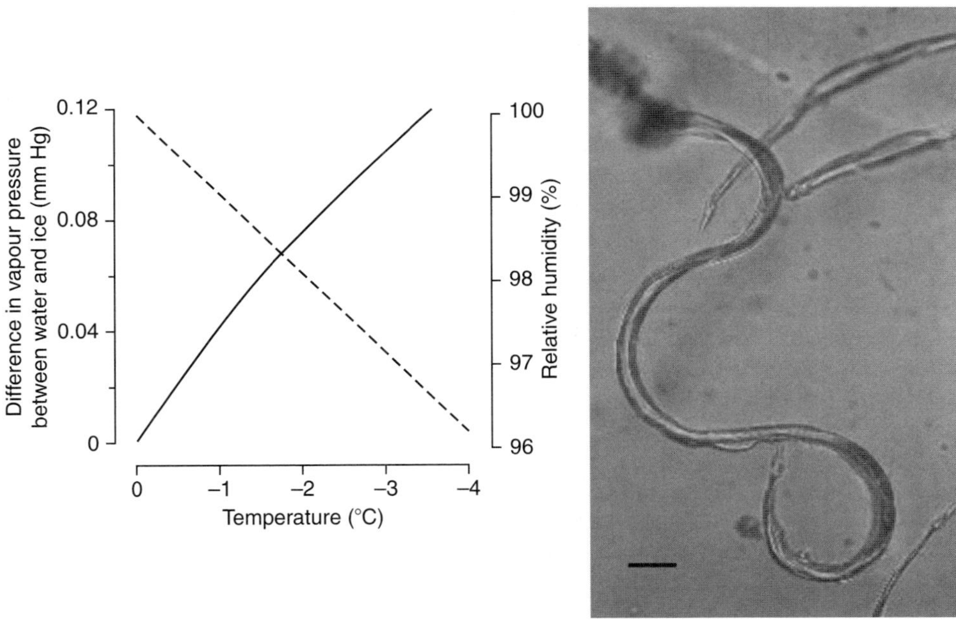

Fig. 8.2. Cryoprotective dehydration: the difference in vapour pressure between water and ice at the same temperature increases as the temperature decreases (graph, solid line), producing a desiccating force equivalent to the relative humidities shown (graph, dotted line). At –1°C the desiccating force is equivalent to a relative humidity of 99%. The photo shows specimens of *Panagrolaimus davidi* in which the water surrounding the nematode was frozen at –1°C for 30 min and then the sample cooled to –10°C at 0.5°C/min. The nematodes remain unfrozen but dehydrate as water is lost to the surrounding ice (scale bar = 50 μm). (The graph is redrawn from Wharton, 2003, using data from Weast, 1989.)

Where the nematode cannot prevent freezing and yet survives, it is freezing tolerant (strategy 4, Fig. 8.1). Intracellular freezing is usually considered to be fatal to organisms (Mazur, 1984). However, *P. davidi* has been shown to survive extensive intracellular ice formation (Wharton and Ferns, 1995; Wharton *et al.*, 2003; 2005b; Fig. 8.3). This remains the only organism known to have this ability, although survival of intracellular freezing has been demonstrated in isolated tissues and cells of some other organisms (Lee *et al.*, 1993; Acker and McGann, 2002; Salinas-Flores *et al.*, 2008; Sinclair and Renault, 2010). Survival whilst exposed to freezing in contact with water (Convey and Worland, 2000) and cryomicroscope observations (Wharton and Block, 1993) indicate that survival of intracellular freezing may be widespread amongst Antarctic nematodes, and perhaps amongst nematodes in general. *Panagrolaimus davidi* can also survive extracellular ice formation, where intracellular ice is absent (Wharton *et al.*, 2005b). The relative importance of extracellular and intracellular ice in this and other species is unclear.

8.2.2 What is the dominant strategy of nematode cold tolerance?

Cold tolerance abilities clearly vary between different nematode species. In a study of six nematode species the temperature at which 50% of the sample were killed (S_{50}) was more than 40°C lower in the best survivor than in the worst survivor, under the same experimental conditions (Smith *et al.*, 2008). Some species of *Panagrolaimus* can survive exposure to very low temperatures, whilst others cannot (Smith *et al.*, 2008).

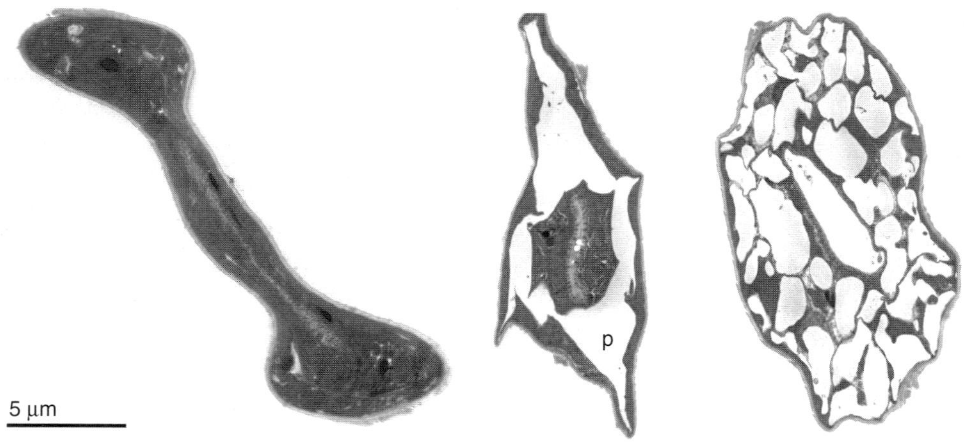

Fig. 8.3. Frozen samples of *Panagrolaimus davidi* prepared for transmission electron microscopy using a freeze substitution technique, which preserves the location of ice as white spaces: left – a nematode undergoing cryoprotective dehydration (no ice crystals, shrunken appearance), centre – extracellular freezing (p – pseudocoel), right – intracellular freezing. (Reprinted from Wharton *et al.* (2005b) with permission from Elsevier.)

There are relatively few species of nematode for which there is sufficient information to state which cold tolerance strategy they use. *Globodera rostochiensis* and *G. pallida* second-stage larvae (L2) within the egg use a freeze-avoiding strategy (Wharton *et al.*, 1993; Wharton and Ramløv, 1995; Devine, 2010). Infective third-stage larvae (L3) of *T. colubriformis* can be either freeze-avoiding or freezing tolerant (Wharton and Allan, 1989). Several species of Antarctic nematodes appear to be either freeze-avoiding or freezing tolerant when they are free of surface water (Pickup, 1990a,b,c), but when in contact with water they are freezing tolerant (Wharton and Block, 1993; Convey and Worland, 2000).

The most-studied nematode from a cold tolerance perspective is *P. davidi*. This species was originally isolated from Ross Island, East Antarctica. The strain in culture in my laboratory is parthenogenetic and appears to be related to other parthenogenetic *Panagrolaimus* strains and most closely to two strains isolated from California (Lewis *et al.*, 2009). In *P. davidi* from the field, however, males are present and there are differences in the size and proportions of females from cultured *P. davidi* and those from field samples (Wharton, 1998). A genetic analysis has recently revealed that field and cultured *P. davidi* are in fact separate clades (Raymond, Marshall and Wharton, unpublished results).

Panagrolaimus davidi is anhydrobiotic and is an external dehydration strategist (Wharton and Barclay, 1993; see Perry and Moens, Chapter 1, this volume for an explanation of the phrase 'external dehydration strategist'). Whilst desiccated at 99% relative humidity it will survive exposure to −20 or −80°C for at least 30 days, with no significant difference in the decline in survival with time between desiccated nematodes exposed to −80°C and controls exposed to desiccation alone (at 15°C; Wharton and Brown, 1991). Fully desiccated (anhydrobiotic) *P. davidi* would thus be expected to survive low temperatures by strategy 1 (Fig. 8.1). The eggshell of *P. davidi* can resist inoculative freezing and allow the enclosed embryo or first-stage larva to supercool in the presence of ice (Wharton, 1994; strategy 2, Fig. 8.1). If exposed to freezing in a situation where the propagation of ice in its surroundings is relatively rapid, inoculative freezing occurs and *P. davidi* is freezing tolerant (Wharton and Ferns, 1995; Wharton *et al.*, 2003). It survives both intracellular and extracellular ice formation (Wharton *et al.*, 2005b; strategy 4, Figs 8.1 and 8.3). If freezing occurs at a high sub-zero temperature, ice formation is slow, there is no inoculative freezing and it survives by cryoprotective dehydration (Wharton *et al.*, 2003, 2005b; strategy 3, Figs 8.1 and 8.3).

Panagrolaimus davidi thus has several strategies that enable it to survive low temperatures in response to the properties of its microenvironment and the sequence of changes in the thermal and hydric conditions to which it is exposed. Which strategy is dominant is hard to say, given the difficulty of assessing exactly the microenvironmental conditions to which the nematode is exposed. In moist soil the presence of large numbers of ice nucleators and a large volume of water would seem to favour freezing at a high sub-zero temperature and slow freezing rates, restricting the possibility of inoculative freezing and hence favouring cryoprotective dehydration. As the soil dries, the nematode may find itself in smaller volumes of water, with fewer ice nucleators, freezing at lower temperatures, producing faster freezing rates, inoculative freezing and favouring freezing tolerance (Wharton, 2003). If the soil dries out further, the nematode must survive anhydrobiotically.

Cryoprotective dehydration has also been described in several other soil invertebrates, including: earthworm cocoons (Holmstrup and Westh, 1994), Collembola (Holmstrup and Sømme, 1998; Worland *et al.*, 1998), enchytraeids (Pedersen and Holmstrup, 2003) and an Antarctic midge larva (Elnitsky *et al.*, 2008). It could be the dominant cold tolerance strategy in soil animals. However, some of these animals are also capable of freezing tolerance (Pedersen and Holmstrup, 2003; Lee *et al.*, 2006), suggesting a similar flexibility in cold tolerance strategy to that shown by *P. davidi*.

8.2.3 Ice nucleation

The state and control of ice nucleation is critical for which cold tolerance strategy a nematode employs. In freeze-avoiding strategies (strategy 1–3, Fig. 8.1), ice nucleation does not occur, or is prevented, and the water in the nematode's body is either lost or in a supercooled state. In freezing tolerance (strategy 4, Fig. 8.1), ice nucleation occurs but the nematode survives the resulting freezing event. Sources of ice nucleation may be endogenous (within the body of the nematode) or exogenous (from outside the body, such as via inoculative freezing from surrounding ice).

The nematode cuticle appears to act as a barrier to inoculative freezing. In frozen specimens of *P. davidi* processed for transmission electron microscopy using a freeze substitution technique (which preserves the location of ice crystals as spaces), ice crystals do not form in the cuticle (Wharton *et al.*, 2005b; Raymond and Wharton, unpublished results; Fig. 8.3). Video analysis of freezing events on a microscope cold stage indicate that nematode freezing is initiated at body openings and not randomly throughout the body, as might be expected if freezing occurred via the cuticle (Wharton and Ferns, 1995). The excretory pore is the commonest site for inoculative freezing, presumably being the weakest point through which ice can penetrate. At a high sub-zero temperature (−1°C) the force for inoculative freezing is not sufficient for it to penetrate body openings like the excretory pore and the nematode is able to remain unfrozen, even though it is surrounded by ice, and survive by cryoprotective dehydration (Wharton *et al.*, 2003, 2005b). However, *P. davidi* has only a weak ability to resist inoculative freezing, which occurs at temperatures lower than −1°C, and the nematode survives by freezing tolerance.

There have been few studies on the ability of other species of nematode to resist inoculative freezing. *Ditylenchus dipsaci* and *Panagrellus redivivus* have little ability to resist inoculative freezing, which occurs as soon as the surrounding water freezes (Wharton, unpublished results). The infective juveniles of *Steinernema feltiae*, *Steinernema anomalae* and *Heterorhabditis bacteriophora* are also susceptible to inoculative freezing (Brown and Gaugler, 1996). In a study of eight Antarctic nematode species (Wharton and Block, 1993), most could not resist inoculative freezing when the surrounding medium froze spontaneously (at temperatures from −2 to −10°C).

A *Ditylenchus* sp. (later identified as *Ditylenchus parcevivens*; Andrassy, 1998) was the only species that showed some ability to resist inoculative freezing. *Pseudoterranova decipiens* L3 freeze by inoculative freezing if cooled in contact with a solution containing an ice nucleator (Stormo *et al*., 2009). The capsule of host origin produced around *Anisakis* sp. L3 does not prevent ice nucleation of the nematode (Wharton and Aalders, 2002).

8.3 Cold Tolerance Mechanisms

8.3.1 Phenotypic plasticity

Phenotypic plasticity refers to 'the change in the expressed phenotype of a genotype as a function of the environment' (Scheiner, 1993). In a cold tolerance context this would involve acclimation (a physiological adjustment to a change in a physical factor induced in the laboratory) and acclimatization (such an adjustment to an environmental factor in nature) in response to low temperatures, rapid cold-hardening (an adjustment resulting from brief exposure to a low temperature) and cold-induced gene expression, which result in changes that produce a cold-hardy phenotype (Hawes and Bale, 2007). Later I will look at what these changes involve, or may involve, in nematodes, but first, is there any evidence that nematodes respond to low temperatures in a way that increases their cold hardiness?

Various species of Antarctic nematodes have lower supercooling points (the temperature at which supercooled body fluids freeze), in the absence of surface water, in winter than in summer (Pickup, 1990a,b,c). The eggs and infective larvae of *Nematodirus battus* (Ash and Atkinson, 1986) and *P. redivivus* adults free of surface water (Mabbett and Wharton, 1986) have lower supercooling points after low-temperature acclimation. Apart from the eggs of *N. battus*, whose eggshell protects against inoculative freezing, these species are usually exposed to sub-zero temperatures in the presence of water and hence do not have the opportunity to supercool unless they can avoid inoculative freezing. Lowered supercooling points in these studies, however, may indicate underlying physiological changes that constitute an acclimatization or acclimation response.

The freezing survival of *Meloidogyne hapla* larvae in soil or a polyethylene glycol medium is enhanced by low-temperature acclimation (Forge and MacGuidwin, 1990, 1992a,b). The S_{50} in water was significantly lower after acclimation in two (*P. redivivus* and *P. davidi*) of six species tested (Smith *et al*., 2008). The cold tolerance of *S. feltiae*, *S. anomalae* and *H. bacteriophora* was enhanced by acclimation (Brown and Gaugler, 1996).

Overall these studies suggest that some nematodes respond to low-temperature acclimation and to seasonal changes in temperature (acclimatization) in a manner that enhances their cold tolerance. Acclimation to low temperatures is, however, not always beneficial for freezing survival. Populations of *P. davidi* do not grow at temperatures below 6.8 °C (Brown *et al*., 2004) and, if not fed, their freezing tolerance declines due to starvation effects during long-term acclimation (Raymond and Wharton, unpublished results). Low-temperature acclimation has a detrimental affect on the cold tolerance of *H. zealandica* (Surrey, 1996).

Acclimatization refers to a long-term, usually seasonal, change that produces a cold-tolerant phenotype. Some organisms can produce a short-term response to low temperature on a timescale of minutes or hours, a phenomenon referred to as 'rapid cold-hardening' (Kelty and Lee, 1999; Worland and Convey, 2001). There is some evidence that a similar phenomenon occurs in some nematodes. The ability of L2 of *M. hapla* to survive freezing in 5% polyethylene glycol at −4°C for 24 h was enhanced within 12 h of exposure to 4°C (Forge and MacGuidwin, 1990, 1992a). This suggests rapid cold-hardening, although the response is not as rapid as that observed in some insects (Hawes *et al.*, 2007). Cold shock (1°C) elevates trehalose production by *H. bacteriophora* infective larvae after 3 h exposure (Jagdale *et al.*, 2005). This could be part of a rapid cold-hardening response. There is no rapid cold-hardening response in *P. redivivus* (Hayashi and Wharton, unpublished results).

The changes induced by cold exposure and similar environmental stresses that result in acclimation/acclimatization or rapid cold-hardening are presumably controlled by changes in gene expression. In nematodes the focus has been on desiccation-induced changes in gene expression (Goyal *et al.*, 2005a; Tyson *et al.*, 2007; Adhikari *et al.*, 2009; Reardon *et al.*, 2010), but attention is now turning to those triggered by cold and/or freezing (see Burnell and Tunnacliffe, Grewal *et al.* and Adkikari and Adams, Chapters 6, 7 and 9, respectively, this volume). Given that different cooling rates, nucleation temperatures and degrees of water stress can produce different survival strategies, it is critical to define the environmental conditions to which the nematodes have been exposed in these studies and to ensure that they relate to those experienced in their natural environment.

In organisms that are exposed to sudden and unpredictable changes in environmental conditions, stress responses may need to be expressed constitutively rather than induced by exposure to stress. For example, southern hemisphere freezing-tolerant insects maintain a moderate amount of freezing tolerance throughout the year, since their environment is influenced by proximity to the Southern Ocean and by El Niño southern oscillation events, which exposes them to both mild winter periods and summer cold snaps. This precludes extensive seasonal cold-hardening (Sinclair *et al.*, 2003).

Even where there is stress-induced gene expression, constitutively expressed genes could make important contributions to stress responses. In larvae of the Antarctic midge *Belgica antarctica*, heat shock proteins are constitutively expressed (Rinehart *et al.*, 2006) and will contribute to the overall stress response. In *P. davidi*, nematodes that had not been acclimated showed a considerable amount of freezing tolerance (S_{50} = −25.4°C), compared with other species tested (S_{50} = −0.7 to −3.8°C; Smith *et al.*, 2008). Acclimation does produce an increase in freezing tolerance (S_{50} = −43.6°C) but, nevertheless, a considerable proportion of the freezing tolerance is expressed constitutively.

8.3.2 Changes in phospholipid saturation

The ability of ectothermic animals to survive exposure to chilling, that is cold but not freezing temperatures, has been linked primarily to a change in membrane lipid composition (Cossins and Bowler, 1987; Hayward *et al.*, 2007). Unsaturated fatty acids are more fluid than are saturated fatty acids, and hence an increase in the proportion of unsaturated fatty acids enables the organism's membranes to resist solidification at lower temperatures (Margesin *et al.*, 2007). The membranes of chilling-sensitive nematodes (*Meloidogyne javanica, Caenorhabditis elegans, Aphelenchus avenae*) undergo phase transitions during cooling, whilst those of chilling-resistant nematodes (*Anguina tritici, M. hapla*) do not (Lyons *et al.*, 1975). The proportion of unsaturated fatty acids is increased at lower temperatures in *S. feltiae, Steinernema carpocapsae* and *Steinernema riobrave* (Jagdale and Gordon, 1997) and in *C. elegans* (Tanaka *et al.*, 1996). The synthesis of unsaturated fatty acids in the cysts of *G. rostochiensis* (Gibson *et al.*, 1995) and in the storage organs of mermithid nematodes (Gordon *et al.*, 1979) may be linked to their cold tolerance.

Chilling-resistant *C. elegans* can be produced by culturing at 10°C. These will survive exposure to 0°C, whilst those cultured at 25°C do not survive (Murray *et al.*, 2007). Chilling-resistant worms have an increased proportion of unsaturated fatty acids compared with chilling-sensitive worms. The proportions of saturated and unsaturated fatty acids are affected by the activity of Δ9-acyl desaturases. Genes for three of these enzymes are found in *C. elegans* (*fat-5, fat-6, fat-7*). Cold acclimation induces *fat-7* expression, whilst inhibiting *fat-7* expression (by RNA interference (RNAi)) reduces chilling resistance. The change in chilling tolerance during acclimation, however, is not fully explained by changes in lipid saturation, suggesting that other mechanisms are more dominant in producing chilling resistance (Murray *et al.*, 2007).

8.3.3 Heat shock proteins

Heat shock proteins (Hsps) are produced by nematodes in response to a variety of abiotic and biotic stressors (see Burnell and Tunnacliffe, Chapter 6, and Devaney, Chapter 10, this volume). They are induced by cold exposure in some organisms (Gross, 2004; Zhang and Guy, 2006). A temperature shift from 37 to 4°C increases the production of Hsp70 in the parasitic first-stage larvae of *Trichinella spiralis, Trichinella nativa* and *Trichinella nelsoni*, but levels of Hsp60 and Hsp90 decline or do not change (Martinez *et al.*, 2001). A 50 kDa protein that cross-reacts with Hsp90 antibody is induced by cold and osmotic stress in *T. spiralis* (Martinez *et al.*, 2002). A small Hsp (18.9 kDa) is induced by both heat and cold in *T. spiralis*, and a recombinant version of this protein has chaperone activity, inhibiting the heat-induced aggregation of citrate synthase (Wu *et al.*, 2007). Arctic species of *Trichinella*, in particular, have strong freezing tolerance abilities, surviving within the frozen carcasses of their hosts (Davidson *et al.*, 2008). It seems likely that cold-induced Hsps play a role in this ability.

In *Meloidogyne artiellia*, Hsp90 is constitutively expressed in all life stages but at higher levels in young egg masses and fourth-stage larvae (L4). Exposure to low temperature (5°C) increased Hsp90 expression in egg masses but not in L2 (De Luca *et al.*, 2009).

8.3.4 Organic osmolytes

Organic osmolytes (osmotically active substances) are low molecular weight organic compounds that are produced by organisms in response to water stress. A limited range of these solutes are associated with cold tolerance responses (Yancey, 2005). These include sugars (e.g. glucose, trehalose), polyols (e.g. glycerol, sorbitol) and amino acids (e.g. proline). In freeze-avoiding animals organic osmolytes act as colligative antifreezes, lowering the freezing and supercooling points of the animal's body fluids in direct proportion to their concentration, whilst in freezing-tolerant animals they act as cryoprotectants. Cryoprotectants decrease the proportion of ice formed, reducing the cellular dehydration that results from freeze concentration effects, and can have a direct protective effect on the structure of membranes and proteins during freezing and/or desiccation stress (Storey, 1997).

Trehalose appears to be the main cryoprotectant/antifreeze used by nematodes. However, trehalose plays other roles in nematode biology, including in the uptake, storage and utilization of carbohydrates, embryonic development, hatching processes and desiccation survival or anhydrobiosis (Behm, 1997). Its mere presence does not necessarily indicate a cryoprotective function. An increase in trehalose concentration during low-temperature acclimation has been shown in *P. davidi* (Wharton *et al.*, 2000a), *N. battus* eggs (Ash and Atkinson, 1983), *Steinernema kushidai* (Ogura and Nakashima, 1997), *S. carpocapsae* (Qiu and Bedding, 1999), *S. feltiae*, *S. riobrave* (Grewal and Jagdale, 2002; Jagdale and Grewal, 2003) and *Heterodera glycines* (Yen *et al.*, 1996), and as a cold shock response in *H. bacteriophora* (Jagdale *et al.*, 2005). Increased trehalose concentration is associated with increased freezing survival or cold tolerance in *P. davidi*, *S. kushidai* and *S. carpocapsae* but not in *S. feltiae* and *S. riobrave* (Ogura and Nakashima, 1997; Wharton *et al.*, 2000a; Jagdale and Grewal, 2003). Trehalose is at high concentration in the eggs of *G. rostochiensis*, where it may assist their supercooling capacity (Wharton and Ramløv, 1995). *Pseudoterranova decipiens* L3 produce trehalose, the levels of which are elevated after both cold and heat (37°C) acclimation (Stormo *et al.*, 2009).

The concentration of trehalose is mediated by trehalose-6-phosphate synthase (gene: *tps*), trehalose-6-phosphate phosphatase (producing trehalose) and trehalase (gene: *tre* – degrading trehalose). The presence of these enzymes has been demonstrated in a variety of nematodes (Behm, 1997). Five *tre* and two *tps* genes have been identified in *C. elegans*. RNAi experiments silencing these genes produced no obviously abnormal phenotypes, including no change in the ability to recover from a freezing stress, despite the silencing of *tps* genes producing a reduction in trehalose levels to 7% of normal (Pellerone *et al.*, 2003). The freezing stress employed in these experiments (exposure to −80°C in a freezing mixture which included glycerol), however, is not one to which the nematodes are exposed in nature. Cold shock increases the activity of trehalose-6-phosphate synthase and decreases the activity of trehalase in *H. bacteriophora*, resulting in an increase in trehalose concentration (Jagdale *et al.*, 2005). Trehalose concentrations also respond to heat, desiccation and other stresses, and it may be part of a global stress tolerance mechanism, as it is in other organisms (Grewal *et al.*, 2006).

There is less evidence for the role of other organic solutes in nematode cold tolerance. Nematodes synthesize a variety of polyols, including glycerol, inositol, ribitol and sorbitol (Barrett, 1981; Womersley, 1981). Glycerol is a common antifreeze or cryoprotectant in other organisms (Margesin *et al.*, 2007), but the concentrations of glycerol commonly found in nematodes are generally considered too low to have a cryoprotective effect (Wharton *et al.*, 1984). Glycerol concentrations in nematodes have been reported not to increase upon low-temperature acclimation (Qiu and Bedding, 1999; Wharton *et al.*, 2000a). Whilst there is no evidence yet of a role for glycerol in nematode cold tolerance, it does appear to be important in the response to osmotic stress (Lamitina *et al.*, 2004, see Wharton and Perry, Chapter 11, this volume).

8.3.5 Ice-active proteins

Many cold-tolerant organisms that are exposed to sub-zero temperatures produce ice-active proteins that control the formation or stability of ice in their bodies (Wharton *et al.*, 2005a; Margesin *et al.*, 2007). There are different types of ice-active proteins: ice-nucleating proteins trigger ice formation; antifreeze proteins inhibit ice formation by binding to ice or other ice nucleators, producing a thermal hysteresis (a difference between the melting and freezing point of a liquid in the presence of an ice crystal); whilst recrystallization-inhibiting proteins control the stability of ice after it has formed (Wharton *et al.*, 2005a).

Some freezing-tolerant animals produce ice-nucleating proteins, which control the site of ice formation in their bodies and ensure that freezing occurs at a relatively high sub-zero temperature (Lee, 1991; Duman, 2001). Nematodes are usually exposed to freezing in the presence of external water. Thus, they are likely to freeze by inoculative freezing from ice forming in their surroundings and have little need for ice-nucleating agents. There is no ice-nucleating activity in extracts prepared from *P. davidi* (Wharton and Worland, 1998).

Antifreeze activity, as indicated by a thermal hysteresis (Duman, 2001), has yet to be demonstrated in nematodes, although it could play a role in those that survive by cryoprotective dehydration. A homologue of type II antifreeze protein from fish (*Clupea harengus*, the Atlantic herring) has been identified by sequencing a library of transcripts from the Antarctic nematode *Plectus murrayi* after desiccation (Adhikari *et al.*, 2009; see Adhikari and Adams, Chapter 9, this volume). This gene is downregulated in response to desiccation but upregulated in response to freezing. However, in initial results, extracts of *P. murrayi* show little or no thermal hysteresis, although they do show recrystallization inhibition activity (Wharton and Raymond, unpublished results). Extracts of *P. davidi* inhibit organic ice nucleators (Wharton and Worland, 1998), a property of some antifreeze proteins (Duman, 2002), but have little thermal hysteresis activity (Wharton *et al.*, 2005a).

The freezing-tolerant nematode *P. davidi* shows strong recrystallization inhibition activity (Ramløv *et al.*, 1996) but no thermal hysteresis (Wharton *et al.*, 2005a). Recrystallization involves the growth of slightly larger ice crystals at the expense of smaller ones, resulting in the formation of fewer but larger crystals. This process could be quite harmful to a frozen organism (Knight *et al.*, 1995), and freezing-tolerant organisms may produce recrystallization-inhibiting proteins to minimize this damage. In *P. davidi* recrystallization inhibition activity is concentration and pH dependent, whilst heating produces a small loss of activity. Disulfide bonds do not appear to be critical to the activity and neither does glycosylation (Wharton *et al.*, 2005a). Unfortunately, our attempts to purify the (presumed) recrystallization-inhibiting protein responsible have proved unsuccessful since the activity is easily lost during chromatography and it does not bind to ice (Marshall, Wharton, Goodall and Clarke, unpublished results) in the ice-affinity purification technique (Kuiper *et al.*, 2003), unlike antifreeze proteins.

Recrystallization inhibition is not a general property of nematodes and is absent from extracts of *Anisakis* L3 (Wharton and Aalders, 2002). In a survey of recrystallization inhibition and freezing tolerance in various nematodes, the levels of recrystallization inhibition varied between species. Recrystallization inhibition did not correlate significantly with freezing survival, but the species that showed the greatest freezing tolerance (*P. davidi*) was also that which had the strongest recrystallization inhibition activity. In three species, the recrystallization inhibition activity increased upon low-temperature acclimation (Smith *et al.*, 2008). Overall, the evidence is that recrystallization-inhibiting proteins play a role in nematode cold tolerance but the proteins involved are yet to be isolated.

8.3.6 Other mechanisms of cold tolerance

Psychrophilic microorganisms (those that grow best at low temperatures) have cold-adapted enzymes that function at low temperatures (Margesin *et al.*, 2007). There is evidence that nematodes can become cold-adapted (Ehlers *et al.*, 2005) and that some nematodes modify the kinetic properties of some metabolic enzymes in response to low temperature and/or produce more cold-adapted isoenzymes (Grewal *et al.*, 2006). The preservation of ribosome function is an important component of cold adaptation in prokaryotes (Hayward *et al.*, 2007), but there is, as yet, no information on this in nematodes.

Both freezing and low temperatures are likely to affect both the ionic and water balance between different body compartments. In insects, chilling injury and pre-freeze mortality are correlated with changes in osmotic and ionic concentrations and in water distribution (Kostal *et al.*, 2004). It is likely that similar homeostatic mechanisms are important in the response of nematodes to low temperatures and freezing. Nematodes show a period of inactivity prior to recovery from desiccation, which is thought to represent a period of repair of damage and the restoration of normal physiological ionic conditions (Wharton *et al.*, 2000b). There does seem to be a similar lag phase after the recovery of nematodes from freezing stress (Wharton, unpublished observations), which may indicate that similar repair and restoration processes are occurring.

Aquaporins (water channels) are present in all animals, allowing water, and in some cases small solutes such as glycerol, to pass through cell membranes (Campbell *et al.*, 2008). There are eight aquaporin genes found in *C. elegans*, although, paradoxically, they do not appear to be essential for osmoregulation (Huang *et al.*, 2007). Aquaporins may be involved in the over-winter survival of the Arctic springtail *Megaphorura arctica* (Clark *et al.*, 2009) and the goldenrod gall fly *Eurosta solidaginis* (Philip *et al.*, 2008) and have been reported from an Antarctic nematode, *P. murrayi* (Adhikari *et al.*, 2009). In *P. davidi*, which survives intracellular freezing, cell membranes seem to present little barrier to ice propagation (Wharton and Ferns, 1995). Study of the aquaporins of this species should thus prove interesting.

Late embryogenesis abundant (LEA) proteins are implicated in anhydrobiotic survival in nematodes and other organisms (Browne *et al.*, 2004; see also Burnell and Tunnacliffe, Chapter 6, this volume) and may have cryoprotective functions (Honjoh *et al.*, 2000; Goyal *et al.*, 2005b). Freezing produces protein aggregation via freeze concentration effects and the resulting water stress. LEA proteins prevent this aggregation by acting as molecular chaperones or shields (Goyal *et al.*, 2005b).

Freezing-tolerant animals may need to tolerate anoxia when the freezing of their body fluids interrupts the delivery of oxygen to their tissues (Margesin *et al.*, 2007). For a small animal like a nematode, the problem may occur if its substrate is frozen and the animal is encased in ice. Antioxidant defences may also be required to survive freezing. Reactive oxygen species increase in frozen yeast cells and in other cells during cryopreservation, and antioxidant defences may be involved in the survival of freezing-tolerant frogs (Margesin *et al.*, 2007). Antioxidant systems are widespread in nematodes (see Barrett, Chapter 12, this volume).

8.4 Linking Mechanisms to Strategies

Arthropods appear to have relatively fixed cold tolerance strategies, adopting either a freeze-avoiding or a freezing-tolerant strategy. Examples of strategy switching or flexible strategies are rare and have only been observed in two species of beetle, which apparently switched strategy between freezing tolerance and freeze avoidance in different years (Duman, 1984; Horwarth and

Duman, 1984), and in the Antarctic midge *B. antarctica*, which can survive low temperatures by cryoprotective dehydration or by freezing tolerance (Lee *et al.*, 2006; Elnitsky *et al.*, 2008). By contrast, cold tolerance strategies in nematodes appear to be less immutable, and *P. davidi* can adopt cryoprotective dehydration, freezing tolerance or anhydrobiosis as cold tolerance strategies, depending on the circumstances. Do these strategies result from the activation of different cold tolerance mechanisms or do the cold tolerance mechanisms that nematodes possess allow them to survive by a variety of strategies?

The answer may be assisted by gene expression studies that investigate whether the upregulation or downregulation of particular genes are associated with particular strategies (freezing tolerance, cryoprotective dehydration, anhydrobiosis). Some progress in that direction has been achieved with *P. murrayi*, in which expression of a gene with homology to a type II antifreeze protein from fish was upregulated during freezing and downregulated during desiccation (Adhikari *et al.*, 2009; see Adhikari and Adams, Chapter 9, this volume). It is critical in such studies to determine the survival strategy being employed by the nematode at the time gene expression is being sampled.

Some cold tolerance mechanisms deal with the harmful effects of low temperatures per se and hence may be involved in all of the cold tolerance strategies that deal with the risk of freezing (Fig. 8.1). These potentially include changes in phospholipid saturation, Hsps/molecular chaperones produced in response to cold stress, cold-adapted enzymes and the regulation of ion homeostasis. Mechanisms that deal with freezing, or the risk of freezing, are likely to be more variable between strategies or to play rather different roles in the various strategies.

8.4.1 The role of trehalose

Trehalose is involved in the response to a variety of stresses. In nematodes in a state of anhydrobiosis, or under milder desiccation stress, trehalose is often synthesized and has various properties that protect membranes and proteins from the harmful effects of dehydration (Behm, 1997). In cold tolerance strategies which involve water loss (strategies 1 and 3, Fig. 8.1), trehalose plays a role related to its desiccation-protective properties. However, freezing also brings with it dehydration stress, since ice excludes solutes and concentrates them in the remaining unfrozen solution (the freeze concentration effect: Mazur, 1984), the dehydration-protective properties of trehalose could thus be important where the nematode is hydrated and frozen (strategy 4, Fig. 8.1). Since trehalose depresses the supercooling point of a solution via its colligative properties (Ring and Danks, 1998), it assists cold tolerance by freeze avoidance (strategy 2, Fig. 8.1). Trehalose plays a role in all nematode cold tolerance strategies (Table 8.1) and protects organisms from the effects of a variety of stresses, including desiccation, heat, cold and oxidation (Elbein *et al.*, 2003); thus, it appears to be part of a general stress response.

Table 8.1. The possible relationships between nematode cold tolerance mechanisms and strategies.

Mechanism	Cold tolerance strategy			
	1. Anhydrobiosis	2. Freeze avoidance	3. Cryoprotective dehydration	4. Freeze tolerance
Trehalose	+	+	+	+
Stress protein	LEA	AFP?	AFP? LEA?	RIP LEA?
Inoculative freezing	Prevent	Prevent	Prevent	Allow
Ice nucleators	–	–	–	–

+, present; –, absent; LEA, late embryogenesis abundant protein; AFP, antifreeze protein; RIP, recrystallization inhibiting protein; ?, role not demonstrated but possible.

A stress that results in the synthesis of trehalose will provide cross-tolerance to other stresses where trehalose plays a protective role. In the Antarctic midge *B. antarctica*, desiccation-induced trehalose synthesis also provides protection against cold and heat (Benoit *et al.*, 2009). In the Antarctic nematode *P. murrayi*, desiccation produces an upregulation of trehalose-6-phosphate synthase and provides cross-tolerance to freezing (see Adhikari and Adams, Chapter 9, this volume). However, there is not always a relationship between desiccation, cold tolerance and trehalose in nematodes. The desiccation survival abilities of *D. dipsaci* are better than those of *P. davidi* since it can survive immediate exposure to 0% relative humidity (Perry, 1977; Wharton and Aalders, 1999), whilst *P. davidi* requires slow desiccation at a high relative humidity before it will do so (Wharton and Barclay, 1993). By contrast, *P. davidi* will survive freezing to much lower temperatures than *D. dipsaci* (Smith *et al.*, 2008). *D. dipsaci* and *P. davidi* have been shown to synthesize trehalose in response to desiccation and low temperature, respectively (Womersley and Smith, 1981; Wharton *et al.*, 2000a).

8.4.2 Stress proteins in cold tolerance

In contrast to trehalose, there may be more differentiation in the role of stress proteins in nematode cold tolerance (Table 8.1). LEA proteins are involved in anhydrobiosis (strategy 1, Fig. 8.1) but they could also be involved in the other strategies that produce water stress (strategies 3 and 4, Fig. 8.1). Antifreeze proteins could play a role in freeze avoidance and cryoprotective dehydration, since they inhibit ice nucleation, and recrystallization-inhibiting proteins are involved in freezing tolerance.

8.5 Conclusions and Future Directions

Nematodes have a number of strategies to survive low temperatures and the risk of freezing. These include freeze avoidance, cryoprotective dehydration, anhydrobiosis and freezing tolerance, the latter involving surviving the formation of extracellular and/or intracellular ice. The relative importance of these cold tolerance strategies between different nematode species, or even the dominant strategy in a particular species, is unclear, but it seems likely to be a response to the particular sequence of changes in thermal and hydric conditions to which the nematode is exposed. Cold tolerance survival mechanisms may involve changes in membrane phospholipid saturation and the synthesis of organic osmolytes (particularly trehalose), stress proteins and ice-active proteins, with specific roles in controlling the formation and stability of ice in the nematode's body. These mechanisms are likely to contribute to several of the survival strategies that nematodes can employ.

8.6 References

Acker, J.P. and McGann, L.E. (2002) Innocuous intracellular ice improves survival of frozen cells. *Cell Transplantation* 11, 563–571.

Adhikari, B.N., Wall, D.H. and Adams, B.J. (2009) Desiccation survival in an Antarctic nematode: molecular analysis using expressed sequenced tags. *BMC Genomics* 10, 69.

Andrassy, I. (1998) Nematodes in the sixth continent. *Journal of Nematode Morphology and Systematics* 1, 107–186.

Ash, C.P.J. and Atkinson, H.J. (1983) Evidence for a temperature-dependent conversion of lipid reserves to carbohydrate in quiescent eggs of the nematode, *Nematodirus battus*. *Comparative Biochemistry and Physiology B* 76, 603–610.

Ash, C.P.J. and Atkinson, H.J. (1986) *Nematodirus battus:* development of cold hardiness in dormant eggs. *Experimental Parasitology* 62, 24–28.

Barrett, J. (1981) *Biochemistry of Parasitic Helminths*. Macmillan, London.

Behm, C.A. (1997) The role of trehalose in the physiology of nematodes. *International Journal for Parasitology* 27, 215–229.

Benoit, J.B., Lopez-Martinez, G., Elnitsky, M.A., Lee, R.E. and Denlinger, D.L. (2009) Dehydration-induced cross tolerance of *Belgica antarctica* larvae to cold and heat is facilitated by trehalose accumulation. *Comparative Biochemistry and Physiology A* 152, 518–523.

Brown, I.M. and Gaugler, R. (1996) Cold tolerance of steinernematid and heterorhabditid nematodes. *Journal of Thermal Biology* 21, 115–121.

Brown, I.M., Wharton, D.A. and Millar, R.B. (2004) The influence of temperature on the life history of the Antarctic nematode *Panagrolaimus davidi*. *Nematology* 6, 883–890.

Browne, J.A., Dolan, K.M., Tyson, T., Goyal, K., Tunnacliffe, A. and Burnell, A.M. (2004) Dehydration-specific induction of hydrophilic protein genes in the anhydrobiotic nematode *Aphelenchus avenae*. *Eukaryotic Cell* 3, 966–975.

Campbell, E.M., Ball, A., Hoppler, S. and Bowman, A.S. (2008) Invertebrate aquaporins: a review. *Journal of Comparative Physiology B* 178, 935–955.

Christner, B.C., Kvitko, B.H. and Reeve, J.N. (2003) Molecular identification of bacteria and eukarya inhabiting an Antarctic cryoconite hole. *Extremophiles* 7, 177–183.

Clark, M.S., Thorne, M.A.S., Purac, J., Burns, G., Hillyard, G., Popovic, Z.D., Grubor-Lajsic, G. and Worland, M.R.

(2009) Surviving the cold: molecular analyses of insect cryoprotective dehydration in the Arctic springtail *Megaphorura arctica* (Tullberg). *BMC Genomics* 10, 328.

Convey, P. and Worland, M.R. (2000) Survival of freezing by free-living Antarctic soil nematodes. *CryoLetters* 21, 327–332.

Cossins, A.R. and Bowler, K. (1987) *Temperature Biology of Animals*. Chapman & Hall, London and New York.

Coulson, S.J. and Birkemoe, T. (2000) Long-term cold tolerance in Arctic invertebrates: recovery after 4 years at below −20°C. *Canadian Journal of Zoology* 78, 2055–2058.

Davidson, R.K., Handeland, K. and Kapel, C.M.O. (2008) High tolerance to repeated cycles of freezing and thawing in different *Trichinella nativa* isolates. *Parasitology Research* 103, 1005–1010.

De Luca, F., Di Vito, M., Fanelli, E., Reyes, A., Greco, N. and De Giorgi, C. (2009) Characterization of the heat shock protein 90 gene in the plant parasitic nematode *Meloidogyne artiellia* and its expression as related to different developmental stages and temperature. *Gene* 440, 16–22.

Devine, K.J. (2010) Comparison of the effects of freezing and thawing on the cysts of the two potato cyst nematode species, *Globodera rostochiensis* and *G. pallida* using differential scanning calorimetry. *Nematology* 12, 81–88.

Dimander, S.O., Hoglund, J. and Waller, P.J. (1999) The origin and overwintering survival of the free living stages of cattle parasites in Sweden. *Acta Veterinaria Scandinavica* 40, 221–230.

Duman, J.G. (1984) Change in the overwintering mechanism in the Cucjus beetle *Cucjus clavipes*. *Journal of Insect Physiology* 30, 235–239.

Duman, J.G. (2001) Antifreeze and ice nucleator proteins in terrestrial arthropods. *Annual Review of Physiology* 63, 327–357.

Duman, J.G. (2002) The inhibition of ice nucleators by insect antifreeze proteins is enhanced by glycerol and citrate. *Journal of Comparative Physiology B* 172, 163–168.

Ehlers, R.-U., Oestergaard, J., Hollmer, S., Wingen, M. and Strauch, O. (2005) Genetic selection for heat tolerance and low temperature activity of the entomopathogenic nematode–bacterium complex *Heterorhabditis bacteriophora–Photorhabdus luminescens*. *Biocontrol* 50, 699–716.

Elbein, A.D., Pan, Y.T., Pastuszak, I. and Carroll, D. (2003) New insights on trehalose: a multifunctional molecule. *Glycobiology* 13, 17R–27R.

Elnitsky, M.A., Hayward, S.A.L., Rinehart, J.P., Denlinger, D.L. and Lee, R.E. (2008) Cryoprotective dehydration and the resistance to inoculative freezing in the Antarctic midge, *Belgica antarctica*. *Journal of Experimental Biology* 211, 524–530.

Forge, T.A. and MacGuidwin, A.E. (1990) Cold hardening of *Meloidogyne hapla* second-stage juveniles. *Journal of Nematology* 22, 101–105.

Forge, T.A. and MacGuidwin, A.E. (1992a) Impact of thermal history on tolerance of *Meloidogyne hapla* second-stage juveniles to external freezing. *Journal of Nematology* 24, 262–268.

Forge, T.A. and MacGuidwin, A.E. (1992b) Effects of water potential and temperature on survival of the nematode *Meloidogyne hapla* in frozen soil. *Canadian Journal of Zoology* 70, 1553–1560.

Gibson, D.M., Moreau, R.A., McNeil, G.P. and Brodie, B.B. (1995) Lipid composition of cyst stages of *Globodera rostochiensis*. *Journal of Nematology* 27, 304–311.

Gordon, R., Finney, J.R., Condon, W.J. and Rusted, T.N. (1979) Lipids in the storage organs of three mermithid nematodes and in the hemolymph of their hosts. *Comparative Biochemistry and Physiology B* 64, 369–374.

Goyal, K., Browne, J.A., Burnell, A.M. and Tunnacliffe, A. (2005a) Dehydration-induced *tps* gene transcripts from an anhydrobiotic nematode contain novel spliced leaders and encode atypical GT-20 family proteins. *Biochimie* 87, 565–574.

Goyal, K., Walton, L.J. and Tunnacliffe, A. (2005b) LEA proteins prevent protein aggregation due to water stress. *Biochemical Journal* 388, 151–157.

Gradinger, R.R. (2001) Adaptation of Arctic and Antarctic ice metazoa to their habitat. *Zoology – Analysis of Complex Systems* 104, 339–345.

Grewal, P.S. and Jagdale, G.B. (2002) Enhanced trehalose accumulation and desiccation survival of entomopathogenic nematodes through cold preacclimation. *Biocontrol Science & Technology* 12, 533–545.

Grewal, P.S., Bornstein-Forst, S., Burnell, A.M., Glazer, I. and Jagdale, G.B. (2006) Physiological, genetic, and molecular mechanisms of chemoreception, thermobiosis, and anhydrobiosis in entomopathogenic nematodes. *Biological Control* 38, 54–65.

Gross, M. (2004) Emergency services: a bird's eye perspective on the many different functions of stress proteins. *Current Protein & Peptide Science* 5, 213–223.

Grout, B.W.W. and Morris, G.J. (1987) *The Effects of Low Temperatures on Biological Systems*. Edward Arnold, London.

Hawes, T.C. and Bale, J.S. (2007) Plasticity in arthropod cryotypes. *Journal of Experimental Biology* 210, 2585–2592.

Hawes, T.C., Bale, J.S., Worland, M.R. and Convey, P. (2007) Plasticity and superplasticity in the acclimation potential of the Antarctic mite *Halozetes belgicae* (Michael). *Journal of Experimental Biology* 210, 593–601.

Hayward, S.A.L., Murray, P.A., Gracey, A.Y. and Cossins, A.R. (2007) Beyond the lipid hypothesis: mechanisms underlying phenotypic plasticity in inducible cold tolerance. In: Csermely, P. and Vigh, L. (eds) *Molecular Aspects of the Stress Response: Chaperones, Membranes and Networks*, Vol. 594. Springer, Berlin, pp. 132–142.

Hodson, A., Anesio, A.M., Tranter, M., Fountain, A., Osborn, M., Priscu, J., Laybourn-Parry, J. and Sattler, B. (2008) Glacial ecosystems. *Ecological Monographs* 78, 41–67.

Holmstrup, M. and Sømme, L. (1998) Dehydration and cold hardiness in the arctic collembolan *Onychiurus arcticus* Tullberg 1876. *Journal of Comparative Physiology B* 168, 197–203.

Holmstrup, M. and Westh, P. (1994) Dehydration of earthworm cocoons exposed to cold: a novel cold hardiness mechanism. *Journal of Comparative Physiology B* 164, 312–315.

Holmstrup, M. and Westh, P. (1995) Effects of dehydration on water relations and survival of lumbricid earthworm egg capsules. *Journal of Comparative Physiology B* 165, 377–383.

Honjoh, K., Matsumoto, H., Shimizu, H., Ooyama, K., Tanaka, K., Oda, Y., Takata, R., Joh, T., Suga, K., Miyamoto, T., Iio, M. and Hatano, S. (2000) Cryoprotective activities of group 3 late embryogenesis abundant proteins from *Chlorella vulgaris* C-27. *Bioscience Biotechnology & Biochemistry* 64, 1656–1663.

Horwarth, K.C. and Duman, J.G. (1984) Yearly variations in the overwintering mechanisms of the cold-hardy beetle, *Dendroides canadensis*. *Physiological Zoology* 57, 40–45.

Hoschitz, M. and Kaufmann, R. (2004) Soil nematode communities of Alpine summits – site differentiation and microclimatic influences. *Pedobiologia* 48, 313–320.

Huang, C.G., Lamitina, T., Agre, P. and Strange, K. (2007) Functional analysis of the aquaporin gene family in *Caenorhabditis elegans*. *American Journal of Physiology – Cell Physiology* 292, C1867–C1873.

Jagdale, G.B. and Gordon, R. (1997) Effect of temperature on the composition of fatty acids in total lipids and phospholipids of entomopathogenic nematodes. *Journal of Thermal Biology* 22, 245–251.

Jagdale, G.B. and Grewal, P.S. (2003) Acclimation of entomopathogenic nematodes to novel temperatures: trehalose accumulation and the acquisition of thermotolerance. *International Journal for Parasitology* 33, 145–152.

Jagdale, G.B., Grewal, P.S. and Salminen, S. (2005) Both heat-shock and cold-shock influence trehalose metabolism in an entomopathogenic nematode. *Journal of Parasitology* 91, 988–994.

Kelty, J.D. and Lee, R.E. (1999) Induction of rapid cold hardening by cooling at ecologically relevant rates in *Drosophila melanogaster. Journal of Insect Physiology* 45, 719–726.

Knight, C.A., Wen, D. and Laursen, R.A. (1995) Nonequilibrium antifreeze peptides and the recrystallization of ice. *Cryobiology* 32, 23–34.

Kostal, V., Vambera, J. and Bastl, J. (2004) On the nature of pre-freeze mortality in insects: water balance, ion homeostasis and energy charge in the adults of *Pyrrhocoris apterus. Journal of Experimental Biology* 207, 1509–1521.

Kuiper, M.J., Lankin, C., Gauthier, S.Y., Walker, V.K. and Davies, P.L. (2003) Purification of antifreeze proteins by adsorption to ice. *Biochemical and Biophysical Research Communications* 300, 645–648.

Lamitina, S.T., Morrison, R., Moeckel, G.W. and Strange, K. (2004) Adaptation of the nematode *Caenorhabditis elegans* to extreme osmotic stress. *American Journal of Physiology – Cell Physiology* 286, C785–C791.

Lee, R.E. (1991) Principles of insect low temperature tolerance. In: Lee, R.E. and Denlinger, D.L. (eds) *Insects at Low Temperatures*. Chapman & Hall, New York and London, pp. 17–46.

Lee, R.E., McGrath, J.J., Morason, R.T. and Taddeo, R.M. (1993) Survival of intracellular freezing, lipid coalescence and osmotic fragility in fat body cells of the freeze-tolerant gall fly *Eurosta solidaginis. Journal of Insect Physiology* 39, 445–450.

Lee, R.E., Elnitsky, M.A., Rinehart, J.P., Hayward, S.A.L., Sandro, L.H. and Denlinger, D.L. (2006) Rapid cold-hardening increases the freezing tolerance of the Antarctic midge *Belgica antarctica. Journal of Experimental Biology* 209, 399–406.

Lewis, S.C., Dyal, L.A., Hilburn, C.F., Weitz, S., Liau, W.S., LaMunyon, C.W. and Denver, D.R. (2009) Molecular evolution in *Panagrolaimus* nematodes: origins of parthenogenesis, hermaphroditism and the Antarctic species *P. davidi. BMC Evolutionary Biology* 9, 15.

Lyons, J.M., Keith, A.D. and Thomason, I.J. (1975) Temperature-induced phase transitions in nematode lipids and their influence upon respiration. *Journal of Nematology* 7, 98–104.

Mabbett, K. and Wharton, D.A. (1986) Cold tolerance and acclimation in the free-living nematode, *Panagrellus redivivus. Revue de Nématologie* 9, 167–170.

MacGuidwin, A.E. and Forge, T.A. (1991) Winter survival of *Pratylenchus scribneri. Journal of Nematology* 23, 198–204.

Margesin, R., Neuner, G. and Storey, K. (2007) Cold-loving microbes, plants, and animals – fundamental and applied aspects. *Naturwissenschaften* 94, 77–99.

Martinez, J., Perez-Serrano, J., Bernadina, W.E. and Rodriguez-Caabeiro, F. (2001) Stress response to cold in *Trichinella* species. *Cryobiology* 43, 293–302.

Martinez, J., Perez-Serrano, J., Bernadina, W.E. and Rodriguez-Caabeiro, F. (2002) Oxidative and cold shock cause enhanced induction of a 50 kDa stress protein in *Trichinella spiralis. Parasitology Research* 88, 427–430.

Mazur, P. (1984) Freezing of living cells: mechanisms and implications. *American Journal of Physiology* 247 (Cell Physiol. 16), C125–C142.

Murray, P., Hayward, S.A.L., Govan, G.G., Gracey, A.Y. and Cossins, A.R. (2007) An explicit test of the phospholipid saturation hypothesis of acquired cold tolerance in *Caenorhabditis elegans. Proceedings of the National Academy of Sciences of the USA* 104, 5489–5494.

Ogura, N. and Nakashima, T. (1997) Cold tolerance and preconditioning of infective juveniles of *Steinernema kushidai* (Nematoda, Steinernematidae). *Nematologica* 43, 107–115.

Pedersen, P.G. and Holmstrup, M. (2003) Freeze or dehydrate: only two options for the survival of subzero temperatures in the arctic enchytraeid *Fridericia ratzeli.*

Journal of Comparative Physiology B 173, 601–609.

Pellerone, F.I., Archer, S.K., Behm, C.A., Grant, W.N., Lacey, M.J. and Somerville, A.C. (2003) Trehalose metabolism genes in *Caenorhabditis elegans* and filarial nematodes. *International Journal for Parasitology* 33, 1195–1206.

Perry, R.N. (1977) Desiccation survival of larval and adult stages of the plant parasitic nematodes, *Ditylenchus dipsaci* and *D. myceliophagus*. *Parasitology* 74, 139–148.

Philip, B.N., Yi, S.X., Elnitsky, M.A. and Lee, R.E. (2008) Aquaporins play a role in desiccation and freeze tolerance in larvae of the goldenrod gall fly, *Eurosta solidaginis*. *Journal of Experimental Biology* 211, 1114–1119.

Pickup, J. (1990a) Strategies of cold-hardiness in three species of antarctic dorylaimid nematodes. *Journal of Comparative Physiology B* 160, 167–173.

Pickup, J. (1990b) Seasonal variation in the cold-hardiness of a free-living predatory nematode, *Coomansus gerlachei* (Mononchidae). *Polar Biology* 10, 307–315.

Pickup, J. (1990c) Seasonal variation in the cold hardiness of three species of free-living antarctic nematodes. *Functional Ecology* 4, 257–264.

Qiu, L.H. and Bedding, R. (1999) Low temperature induced cryoprotectant synthesis by the infective juveniles of *Steinernema carpocapsae*: biological significance and mechanisms involved. *CryoLetters* 20, 393–404.

Ramløv, H. (2000) Aspects of natural cold tolerance in ectothermic animals. *Human Reproduction* 15, 26–46.

Ramløv, H., Wharton, D.A. and Wilson, P.W. (1996) Recrystallization in a freezing tolerant Antarctic nematode, *Panagrolaimus davidi*, and an alpine weta, *Hemideina maori* (Orthoptera, Stenopelmatidae). *Cryobiology* 33, 607–613.

Reardon, W., Chakrabortee, S., Pereira, T.C., Tyson, T., Banton, M.C., Dolan, K.M., Culleton, B.A., Wise, M.J., Burnell, A.M. and Tunnacliffe, A. (2010) Expression profiling and cross-species RNA interference (RNAi) of desiccation-induced transcripts in the anhydrobiotic nematode *Aphelenchus avenae*. *BMC Molecular Biology* 11, 6.

Rinehart, J.P., Hayward, S.A.L., Elnitsky, M.A., Sandro, L.H., Lee, R.E. and Denlinger, D.L. (2006) Continuous up-regulation of heat shock proteins in larvae, but not adults, of a polar insect. *Proceedings of the National Academy of Sciences of the USA* 103, 14223–14227.

Ring, R.A. and Danks, H.V. (1998) The role of trehalose in cold-hardiness and desiccation. *CryoLetters* 19, 275–282.

Salinas-Flores, L., Adams, S.L., Wharton, D.A., Downes, M.F. and Lim, M.H. (2008) Survival of Pacific oyster, *Crassostrea gigas*, oocytes in relation to intracellular ice formation. *Cryobiology* 56, 28–35.

Scheiner, S.M. (1993) Genetics and evolution of phenotypic plasticity. *Annual Review of Ecology and Systematics* 24, 35–68.

Scholander, P.F., Flagg, W., Hock, R.J. and Irving, L. (1953) Studies on the physiology of frozen plants and animals in the arctic. *Journal of Cellular and Comparative Physiology* 42 (Supplement), 1–56.

Sinclair, B.J. and Renault, D. (2010) Intracellular ice formation in insects: unresolved after 50 years? *Comparative Biochemistry and Physiology A – Molecular & Integrative Physiology* 155, 14–18.

Sinclair, B.J., Vernon, P., Klok, C.J. and Chown, S.L. (2003) Insects at low temperatures: an ecological perspective. *Trends in Ecology & Evolution* 18, 257–262.

Smith, T., Wharton, D.A. and Marshall, C.J. (2008) Cold tolerance of an Antarctic nematode that survives intracellular freezing: comparisons with other nematode species. *Journal of Comparative Physiology B* 178, 93–100.

Storey, K.B. (1997) Organic solutes in freezing tolerance. *Comparative Biochemistry & Physiology A* 117, 319–326.

Stormo, S.K., Praebel, K. and Elvevoll, E.O. (2009) Cold tolerance in sealworm (*Pseudoterranova decipiens*) due to heat-shock adaptations. *Parasitology* 136, 1317–1324.

Surrey, M.R. (1996) The effect of rearing method and cool temperature acclimation on the cold tolerance of *Heterorhabditis zealandica* infective juveniles (Nematoda: Heterorhabditidae). *CryoLetters* 17, 313–320.

Tanaka, T., Ikita, K., Ashida, T., Motoyama, Y., Yamaguchi, Y. and Satouchi, K. (1996) Effects of growth temperature on the fatty acid composition of the free-living nematode *Caenorhabditis elegans*. *Lipids* 31, 1173–1178.

Tyson, T., Reardon, W., Browne, J.A. and Burnell, A.M. (2007) Gene induction by desiccation stress in the entomopathogenic nematode *Steinernema carpocapsae* reveals parallels with drought tolerance mechanisms in plants. *International Journal for Parasitology* 37, 763–776.

Weast, R.C. (1989) *Handbook of Chemistry and Physics*. CRC Press, Cleveland, Ohio.

Wharton, D.A. (1994) Freezing avoidance in the eggs of the Antarctic nematode *Panagrolaimus davidi*. *Fundamental and Applied Nematology* 17, 239–243.

Wharton, D.A. (1998) Comparison of the biology and freezing tolerance of *Panagrolaimus davidi*, an Antarctic nematode, from field samples and cultures. *Nematologica* 44, 643–653.

Wharton, D.A. (1999) Parasites and low temperatures. *Parasitology* 119, S7–S17.

Wharton, D.A. (2002) *Life at the Limits: Organisms in Extreme Environments*. Cambridge University Press, Cambridge, UK.

Wharton, D.A. (2003) The environmental physiology of Antarctic terrestrial nematodes: a review. *Journal of Comparative Physiology B* 173, 621–628.

Wharton, D.A. and Aalders, O. (1999) Desiccation stress and recovery in the anhydrobiotic nematode *Ditylenchus dipsaci* (Nematoda : Anguinidae). *European Journal of Entomology* 96, 199–203.

Wharton, D.A. and Aalders, O. (2002) The response of *Anisakis* larvae to freezing. *Journal of Helminthology* 76, 363–368.

Wharton, D.A. and Allan, G.S. (1989) Cold tolerance mechanisms of the free-living stages of *Trichostrongylus colubriformis* (Nematoda: Trichostrongylidae). *Journal of Experimental Biology* 145, 353–370.

Wharton, D.A. and Barclay, S. (1993) Anhydrobiosis in the free-living Antarctic nematode *Panagrolaimus davidi*. *Fundamental and Applied Nematology* 16, 17–22.

Wharton, D.A. and Block, W. (1993) Freezing tolerance of some Antarctic nematodes. *Functional Ecology* 7, 578–584.

Wharton, D.A. and Brown, I.M. (1991) Cold tolerance mechanisms of the Antarctic nematode *Panagrolaimus davidi*. *Journal of Experimental Biology* 155, 629–641.

Wharton, D.A. and Ferns, D.J. (1995) Survival of intracellular freezing by the Antarctic nematode *Panagrolaimus davidi*. *Journal of Experimental Biology* 198, 1381–1387.

Wharton, D.A. and Ramløv, H. (1995) Differential scanning calorimetry studies on the cysts of the potato cyst nematode *Globodera rostochiensis* during freezing and melting. *Journal of Experimental Biology* 198, 2551–2555.

Wharton, D.A. and Surrey, M.R. (1994) Cold tolerance mechanisms of the infective larvae of the insect parasitic nematode, *Heterorhabditis zealandica* Poinar. *CryoLetters* 25, 749–752.

Wharton, D.A. and Worland, M.R. (1998) Ice nucleation activity in the freezing tolerant Antarctic nematode *Panagrolaimus davidi*. *Cryobiology* 36, 279–286.

Wharton, D.A. and Worland, M.R. (2001) Water relations during desiccation of cysts of the potato-cyst nematode *Globodera rostochiensis*. *Journal of Comparative Physiology B* 171, 121–126.

Wharton, D.A., Young, S.R. and Barrett, J. (1984) Cold tolerance in nematodes. *Journal of Comparative Physiology B* 154, 73–77.

Wharton, D.A., Perry, R.N. and Beane, J. (1993) The role of the eggshell in the cold tolerance mechanisms of the unhatched juveniles of *Globodera rostochiensis*. *Fundamental and Applied Nematology* 16, 425–431.

Wharton, D.A., Judge, K.F. and Worland, M.R. (2000a) Cold acclimation and cryoprotectants in a freeze-tolerant Antarctic nematode, *Panagrolaimus davidi. Journal of Comparative Physiology B* 170, 321–327.

Wharton, D.A., Rolfe, R.N. and Perry, R.N. (2000b) Electrophysiological activity during recovery from anhydrobiosis in fourth stage juveniles of *Ditylenchus dipsaci. Nematology* 2, 881–886.

Wharton, D.A., Goodall, G. and Marshall, C.J. (2003) Freezing survival and cryoprotective dehydration as cold tolerance mechanisms in the Antarctic nematode *Panagrolaimus davidi. Journal of Experimental Biology* 206, 215–221.

Wharton, D.A., Barrett, J., Goodall, G., Marshall, C.J. and Ramløv, H. (2005a) Ice-active proteins from the Antarctic nematode *Panagrolaimus davidi. Cryobiology* 51, 198–207.

Wharton, D.A., Downes, M.F., Goodall, G. and Marshall, C.J. (2005b) Freezing and cryoprotective dehydration in an Antarctic nematode (*Panagrolaimus davidi*) visualised using a freeze substitution technique. *Cryobiology* 50, 21–28.

Womersley, C. (1981) Biochemical and physiological aspects of anhydrobiosis. *Comparative Biochemistry and Physiology B* 70, 669–678.

Womersley, C. and Smith, L. (1981) Anhydrobiosis in nematodes – 1. the role of glycerol, myo-inositol and trehalose during desiccation. *Comparative Biochemistry and Physiology B* 70, 579–586.

Worland, M.R. and Convey, P. (2001) Rapid cold hardening in Antarctic microarthropods. *Functional Ecology* 15, 515–524.

Worland, M.R., Gruborlajsic, G. and Montiel, P.O. (1998) Partial desiccation induced by sub-zero temperatures as a component of the survival strategy of the Arctic collembolan *Onychiurus arcticus* (Tullberg). *Journal of Insect Physiology* 44, 211–219.

Wu, Z., Nagano, I., Boonmars, T. and Takahashi, Y. (2007) Thermally induced and developmentally regulated expression of a small heat shock protein in *Trichinella spiralis. Parasitology Research* 101, 201–212.

Yancey, P.H. (2005) Organic osmolytes as compatible, metabolic and counteracting cytoprotectants in high osmolarity and other stresses. *Journal of Experimental Biology* 208, 2819–2830.

Yen, J.H., Niblack, T.L., Karr, A.L. and Wield, W.J. (1996) Seasonal biochemical changes in eggs of *Heterodera glycines* in Missouri. *Journal of Nematology* 28, 442–450.

Zhang, C. and Guy, C.L. (2006) In vitro evidence of Hsc70 functioning as a molecular chaperone during cold stress. *Plant Physiology and Biochemistry* 44, 844–850.

9 Molecular Analyses of Desiccation Survival in Antarctic Nematodes

Bishwo N. Adhikari and Byron J. Adams

Department of Biology, Brigham Young University, Provo, Utah, USA

9.1 Introduction

The natural habitat of terrestrial nematodes, the soil, is a harsh environment that poses significant challenges to survival and persistence. As such, nematodes have evolved a myriad of strategies that enable them to respond to environmental stress, including their gross morphology, cell wall constituents and the expression of a number of genes that encode metabolites, cryoprotectants and signalling compounds. Suites of adaptive traits such as these enable soil-dwelling nematodes to withstand the rigors of fluctuating environmental extremes or survive unfavourable environmental conditions in a dormant state that considerably prolongs their lifespan. For example, some nematodes can survive freezing conditions and/or exposure to desiccation for long periods of time. The latter are capable of 'anhydrobiosis' or 'life without water' and can survive in a desiccated state for many years (see Perry and Moens, Chapter 1, this volume).

Different nematode species can also vary in their responses to environmental changes, and this is most likely a function of the habitat in which they have evolved. For example, nematodes that have evolved in desert-like habitats may respond to desiccation stress differently from nematodes now optimally fit to mesic environments. Terrestrial Antarctic nematodes are part of a community of organisms that are living in one of the harshest environments on Earth. These nematodes have to face the problems of exposure to desiccation, freezing, high radiation, higher soil salinity and frequent freeze–thaw events. Nematodes exposed to such extreme environmental conditions typically show two broad responses. Nematodes such as *Plectus antarcticus* are adapted to grow and reproduce under conditions that are constantly extreme (extremophiles, displaying capacity adaptation), whilst others such as *Scottnema lindsayae* survive extreme conditions in a dormant state, only growing and reproducing when conditions are favourable (cryptobiotes, displaying resistance adaptation).

Our understanding of stress tolerance strategies of Antarctic nematodes has undergone a number of significant shifts. Substantial information has accumulated on physiological and biochemical mechanisms of stress survival by nematodes. However, research programmes addressing the molecular mechanisms and gene functions involved in desiccation and freeze survival of nematodes are only now beginning to emerge. Because environmental stresses can have a profound effect on the developmental programmes of nematodes, it is expected that the timing and order of gene expression will be the major regulatory mechanisms by which nematodes respond to these external cues. In the present chapter we will characterize the molecular mechanisms of Antarctic nematode stress survival and describe these mechanisms in relation to desiccation survival. We will discuss the inducible as well as constitutive mechanisms of stress response, along with signal transduction mechanisms. The genetic response of nematodes to stress-induced metabolic changes and oxidative stress response will also be discussed. Finally, we will point out interesting conserved mechanisms and cross-talk between stress responses that may be used by highly divergent organisms to mediate transcriptional responses to stress. In addition, some of the future areas of research that may be important for our understanding of molecular mechanisms relative to stress survival will be presented. Because of the limited amount of data that have been generated in the area of molecular stress survival by Antarctic nematodes, the foundation of this chapter will be the work we have done on stress survival of the Antarctic nematode *Plectus murrayi*.

9.2 Molecular Anhydrobiology of Antarctic Nematodes

Desiccation tolerance is one of the most fundamental features of many terrestrial organisms. Most organisms are homeohydric (ability to restrict cellular water loss regardless of environmental conditions) and avoid desiccation by preventing water loss under dry conditions. However, some organisms (Collembola, nematodes, resurrection plants) are extremely well adapted to dehydration; they are able to survive extended periods of drought in a biological state called anhydrobiosis (Watanabe, 2006). Anhydrobiosis is a survival strategy employed by some species of nematodes, rotifers and tardigrades in response to desiccation (Crowe and Madin, 1974). Nematodes in anhydrobiosis lose all their detectable body water content and cease metabolic activity (Crowe and Madin, 1975). While in an anhydrobiotic state, nematodes are capable of surviving extreme cold (Wharton and Brown, 1991) as well as desiccation (Pickup and Rothery, 1991). Laboratory studies of nematodes and other organisms indicate that anhydrobiosis typically is accompanied by the production of large quantities of non-reducing sugars, such as trehalose, which stabilize molecules (proteins, membrane lipids) within the cells of anhydrobiotes (Higa and Womersley, 1993; Crowe *et al.*, 2002; see Burnell and Tunnacliffe, Chapter 6, and Barrett, Chapter 12, this volume). Recent research suggests anhydrobiotes synthesize many other compounds (primarily proteins) that are essential to survival (Solomon *et al.*, 2000; Browne *et al.*, 2002; Oliver *et al.*, 2002; Tunnacliffe and Lapinski, 2003), and our understanding of this complex process is increasing. Several reviews of the physiology and biochemistry of desiccation survival and anhydrobiosis in nematodes have been published (Wharton, 1986; Barrett, 1991; Crowe *et al.*, 1992; Womersley *et al.*, 1998; Perry, 1999).

Despite recent work on behavioural, biochemical and molecular stress response mechanisms (Liu and Glazer, 2000; Browne *et al.*, 2004; Gal *et al.*, 2005), the molecular mechanisms governing anhydrobiosis in nematodes are not fully understood. Studies on desiccation-responsive compounds in nematodes have resulted in the identification of many genes that play important roles in stress pre-treatment and cross-tolerance. The molecular mechanism of anhydrobiosis is surprisingly similar in unrelated species ranging from nematodes to plants. An important process seems to be the formation of intracellular filamentous structures by folding of late embryogenesis abundance (LEA) proteins into superhelical structures (Wise and Tunnacliffe, 2004). Furthermore, the dehydration process is accompanied by induction of genes coding for organic osmolytes (trehalose, glycerol, sorbitol), which prevent membrane damage, scavenge free radicals and avoid oxidation of proteins and phospholipids (Watanabe, 2006). Heat shock proteins (Hsps) serve as chaperones to prevent denaturation of proteins. Induced as well as constitutive expression of these Hsp genes has also been reported (Adhikari *et al.*, 2009). Decreased water content is thought to be the first signal to trigger anhydrobiosis, followed by the activation of signal transduction cascades, pathways activated by the binding of an extracellular signal molecule (such as a mitogen) to a receptor protein on the cell membrane.

To understand the molecular mechanisms activated during anhydrobiosis, a condition induced by slow dehydration, Adhikari *et al.* (2009) initiated a genomic-level analysis of gene expression during anhydrobiosis of *P. murrayi* (an Antarctic nematode capable of surviving desiccation as well as freezing conditions). Accordingly, an expressed sequence tag (EST; a short subsequence of a transcribed complementary DNA, or cDNA) library was generated from nematodes that had been slowly desiccated. Desiccation-induced transcripts were sequenced and transcripts differentially expressed during desiccation stress were identified using suppressive subtractive hybridization (SSH; an approach that allows for polymerase chain reaction (PCR)-based amplification of only cDNA fragments that differ between a control and an experimental transcriptome). In that study 2486 ESTs were generated from a cDNA library and the unique transcripts from *P. murrayi* were compared with known sequences. Of these, 56% were considered to have homologues in *Caenorhabditis elegans* and other organisms (including other nematodes), while 44% did not closely resemble any known gene sequences. The breakdown of functions was assessed by Gene Ontology (Ashburner *et al.*, 2000) and by assignment to metabolic pathways using the Kyoto Encyclopaedia of Genes and Genomes (KEGG) (Kanehisa and Goto, 2000) database. These analyses showed that the desiccation-induced transcriptome encompasses a wide range of functions associated with the EST transcripts, representing many familiar functions that might be expected of a eukaryotic organism. Analyses of the transcripts that were abundantly expressed suggested that metabolic genes and those associated with the processing of environmental information, mainly environmental stresses, were highly expressed. KEGG analysis indicated that gene transcription related to metabolic activity, protein folding, sorting and degradation, membrane transport and signal transduction-related activities was high.

Using subtractive hybridization (hybridization between cDNAs from desiccated and hydrated nematodes), a library of transcripts present in desiccated nematodes was made. Sequencing of the subtracted library of *P. murrayi* produced 80 sequences specific to the desiccated sample. A large portion of these were considered to be associated with metabolism, followed by environmental information processing, genetic information processing and novel proteins. To validate the differential expression of genes identified by subtractive hybridization, 14 of the genes were further examined by real-time PCR. Many of these showed a significant increase in mRNA expression level after desiccation, including LEA protein, trehalose-6-phosphate synthase (TPS), aldehyde dehydrogenase (ALH) and glycerol kinase (GK). Genes encoding glycogen synthase (GS) and an antifreeze protein (AFP) homologue showed a decrease in expression. The Hsps of 70 and 90 kDa (Hsp70 and Hsp90) did not alter expression during desiccation, indicating that these Hsps are not induced by environmental challenge but instead are constitutively expressed at high levels. The expressed genome of *P. murrayi* showed that anhydrobiotic survival in nematodes involves a suite of genes from diverse functional areas, such as hormone signalling transduction, transcription regulation, reactive oxygen species (ROS) scavenging, re-establishment of homeostasis, molecular chaperoning and transcriptional regulation of ribosomal proteins and other genes. In the following sections, each of these processes will be discussed in detail, focusing on the findings of our work on the Antarctic nematode *P. murrayi*.

9.3 Stress Response System

All organisms must deploy stress response systems with more or less specific tasks to cope with environmental insults and restore normal physiological conditions after each disturbance. These mechanisms are found in almost all cells, and several systems are conserved among all living things. The widespread occurrence of stress response systems indicates that mediating environmental stress was a crucial problem to solve very early in the evolution of life. Stress response is thus closely linked to the idea of homeostasis, the tendency to regulate the internal state at a level independent from the changeable environment. At the cellular level, stress response is a defensive reaction to a strain imposed by environmental force(s) on macromolecules. Such strain commonly results in deformation of or damage to proteins, DNA or other essential macromolecules (Kültz, 2005). Stress response systems assess and counteract stress-induced damage, temporarily increase tolerance of such damage and/or remove terminally damaged cells by programmed cell death (apoptosis). The overall capacity of stress response systems to mediate environmental insults depends on the gene expressed in a cell at a particular time and therefore can be unique to particular cell types and species.

Desiccation is one of the most prevalent and extreme environmental stress factors encountered by nematodes in Antarctic terrestrial environments. The capacity of these nematodes to withstand significant desiccating conditions requires a wide array of molecular responses involving a number of genes from different functional areas (Adhikari *et al.*, 2009). These stress-related genes can be either constitutively expressed (not induced by stress) or expressed upon exposure to desiccation. Stress-induced genes are typically enriched for antioxidant functions as well as for carbohydrate metabolism and energy generation functions, whereas most stress-repressed genes have growth-related functions, such as translation and ribosome biogenesis, reflecting a redirection of resources from rapid proliferation and repair to stress protection.

9.3.1 Constitutively expressed genes

Stress genes are ubiquitously present in the genomes of organisms from bacteria to humans and are generally highly conserved. One of the most common and well-studied classes of stress proteins are often termed heat shock proteins, because heat is one of the best-known inducers of stress gene expression. These proteins, which commonly show enhanced synthesis after cessation of stressful events, are called molecular chaperones because some of them are involved in the folding of newly synthesized proteins or in the repair of misfolded proteins (Feder and Hofmann, 1999). Chaperones are also involved in the transport of proteins, prevent the aggregation of proteins and serve to stabilize certain protein conformations. During stress, such as exposure to heat or chemical stressors, several members of the stress and chaperone gene families increase their expression and thus the amount of chaperones in the cell.

Hsps are well known to respond to high temperature and other environmental stresses in a wide range of organisms (Feder and Hofmann, 1999; see Devaney, Chapter 10, this volume). Typically, the genes encoding these proteins are not expressed under normal conditions but are quickly turned on in response to stress and are quickly turned off again when the stress is removed. Concurrent with this upregulation of Hsps is the cessation of synthesis of most other proteins. Interestingly, recent evidence suggests that some Antarctic species may respond somewhat differently. Adhikari *et al.* (2009) showed that in the Antarctic nematode *P. murrayi* the transcripts of 70 and 90 kDa Hsps and small heat shock proteins (sHsps) were abundantly expressed following desiccation stress. Hsps have been implicated in response to desiccation in many nematodes (Browne *et al.*, 2004; Gal *et al.*, 2005), but were constitutively expressed in *P. murrayi* (Fig. 9.1). Although Hsps may contribute to enhanced stress resistance overall, there is no evidence to suggest that expression levels of these Hsps were altered by desiccation. It has also been shown that other Antarctic organisms constitutively express Hsp70 and Hsp90, indicating that a significant upregulation of these genes is not a necessary component of a response to thermal stress (La Terza *et al.*, 2001; Buckley *et al.*, 2004; Rinehart *et al.*, 2006). Characteristically, Hsps are expressed not at the environmental optimum

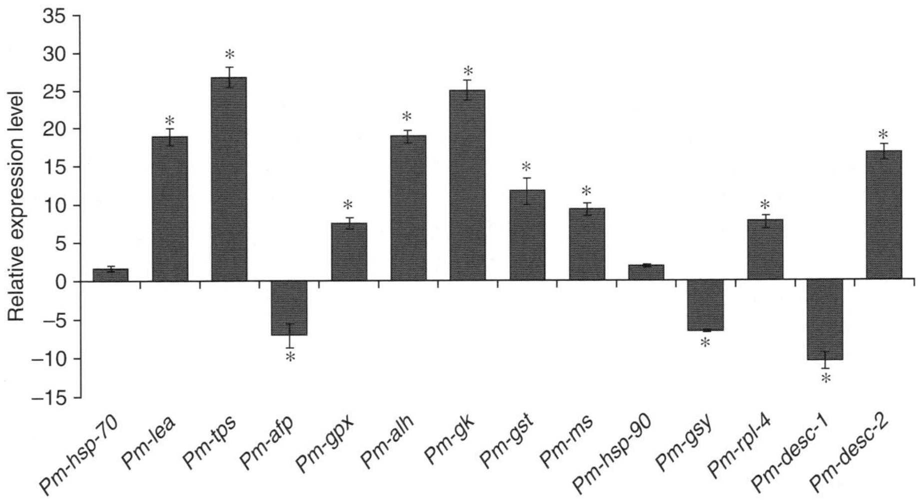

Fig. 9.1. Quantitative real-time PCR analysis of gene expression in *Plectus murrayi* in response to desiccation. Values were determined using qRT-PCR and represent relative expression of genes from desiccated to undesiccated nematodes (control). The relative expression of the target gene (*Pm-alh*: aldehyde dehydrogenase; *Pm-tps*: trehalose-6-phosphate synthase; *Pm-gpx*: glutathione peroxidase; *Pm-afp*: antifreeze protein; *Pm-hsp-70*: heat shock protein 70; *Pm-lea*: late embryogenesis abundant protein; *Pm-gk*: glycerol kinase; *Pm-ms*: malate synthase; *Pm-gsy*: glycogen synthase; *Pm-hsp-90*: heat shock protein 90; *Pm-rpl-4*: ribosomal protein-4; *Pm-desc-1*: novel protein I; *Pm-desc-2*: novel protein II; *Pm-gst*: glutathione *S*-transferase 1), normalized to *Pm-18s*:18S rRNA and relative to the expression of control, was calculated for each sample using the $2^{-\Delta\Delta CT}$ method (Livak and Schmittgen, 2001). Gene expression was determined in each sample using three independent technical replicates. A transcript with a relative abundance of 1 is equivalent to the abundance of 18S rRNA. Bars represent standard errors calculated from three replicates of each experiment. *Significant difference (P<0.05) from control. (Adapted from Adhikari *et al.*, 2009.)

of the animal but only as part of the organism's response to a suboptimal environment. In fact, the expression of these genes is thought to be incompatible with ongoing protein synthesis and the progression of development (Feder *et al.*, 1992; Krebs and Feder, 1997). Is it possible that *P. murrayi* has evolved a way to maintain Hsp function without disrupting normal metabolism and growth? Consistent with the observed constitutive expression, it is possible that desiccation stress does not activate these genes and that the mild desiccation failed to boost Hsp expression. However, an alternative explanation for the constitutive expression of Hsps could be that, because Antarctic nematodes are frequently, although unpredictably, exposed to a variety of environmental stressors such as desiccation, high pH, extreme osmotic shock, freezing and anoxia, as well as temperature (Treonis and Wall, 2005), their survival may be optimized by maintaining continuous expression of molecular chaperones. Because of the unpredictability and the potential rapidity of exposure to diverse environmental stresses, the continuous production of these molecular chaperones may be energetically justified.

In the same study, Adhikari *et al.* (2009) reported downregulation of transcripts encoding type II AFP during desiccation stress (Fig. 9.1). Many overwintering organisms, including insects, fish, bacteria, fungi and plants, accumulate AFPs, which bind to the faces of ice crystals during freezing and inhibit their growth (Duman, 2001). An ice-active protein that is thought to play an important role in freezing tolerance, particularly intracellular freezing, has been reported in the Antarctic nematode *Panagrolaimus davidi* (Wharton *et al.*, 2005). Adhikari *et al.* (2009) showed that when *P. murrayi* was exposed to freezing (−10°C) there was significant upregulation of the AFP gene. Although our experiments in this area are still preliminary, we think it is possible that *Pm-afp* is continuously upregulated during freezing and that once the nematodes are exposed to desiccation, other stress response mechanisms are activated by downregulating AFP. As AFPs can prevent the potential injurious process of recrystallization even at very low concentrations (Knight *et al.*, 1984), we hypothesize that even the slightest expression of AFP genes during freezing could be sufficient to prevent ice nucleation and ultimately protect individuals from further mechanical damage to their cells.

Taken together, these findings suggest that adaptation to multiple stresses may, at times, involve the production of molecular chaperones at a constant level, coupled with the loss of desiccation-induced changes in the genes that encode Hsps. These findings suggest that the Hsp genes are constitutively expressed, which might be due to an alteration in the transcriptional apparatus controlling Hsp genes. If this is the case, the constant production of Hsps, while energetically costly, must be relatively advantageous. It is possible that constitutive expression of genes like Hsps is a common phenomenon in many Antarctic organisms. Similarly, continuous induction of AFP during freezing might be a consequence of adaptation to highly stable, cold temperatures and could therefore be a general phenomenon that extends to other metazoans that are able to persist in cold environments.

9.3.2 Stress-induced genes

An organism's response to stress can take many forms (behavioural, biochemical, molecular, etc.), most of which are species- and stress-specific. However, so far the most well-known universal response is the production of stress proteins (Gross, 2004) and induction of several stress-related genes. Inducible stress tolerance is increasingly understood to result from numerous molecular mechanisms, of which Hsps are collectively only one. Other mechanisms include synthesis of osmotic stress protectants, such as polyols and trehalose, modifications of the saturation of cell membrane lipids (homeoviscous adaptation), radical scavengers (superoxide dismutase, glutathione system, cytochrome P450), and so on. The most common stress-inducible genes expressed in Antarctic nematodes in response to different environmental stress are discussed below.

9.3.2.1 Late embryogenesis abundant proteins

Survival of desiccation and dehydration is one of the most intriguing phenomena in nature but is far from fully understood. Various non-reducing disaccharides, such as trehalose in animals and fungi, accumulate in diverse anhydrobiotic organisms before desiccation and are thought to have a protective function as either water replacement molecules or vitrification agents in the dry state (Clegg, 2001; see Barrett, Chapter 12, this volume). However, none of these compounds is an absolute requirement for anhydrobiosis, being absent from some organisms (Lapinski and Tunnacliffe, 2003) and apparently unnecessary in others (Ratnakumar and Tunnacliffe, 2006). Therefore, discovery programmes have been launched in a number of anhydrobiotic organisms to determine key genes and proteins important in desiccation tolerance. One recurring theme is the presence of a number of abundant, highly hydrophilic proteins in desiccation-tolerant organisms (Tyson *et al.*, 2007), chief among which are the LEA proteins (Tunnacliffe and Wise, 2007; see Burnell and Tunnacliffe, Chapter 6, this volume). LEA proteins have been known for many years to accumulate in maturing plant seeds as they acquire desiccation tolerance (Tunnacliffe and Wise, 2007), but their discovery in invertebrates (Goyal *et al.*, 2005a) suggests that similar mechanisms govern anhydrobiosis in both animals and plants. Although LEA proteins are widely held to protect cells against water loss, their precise role has been a puzzle since they were first described. Recently, evidence supporting possible functions has been obtained (Tunnacliffe and Wise, 2007), including *in vitro* data suggesting that some LEA proteins act to prevent other proteins from aggregating during water loss. Probably because of their hydrophilic, unstructured nature, LEA proteins themselves are not susceptible to aggregation upon desiccation. Their anti-aggregation activity could mean they represent a novel form of dehydration-specific molecular chaperone. An alternative, perhaps simpler, explanation is that LEA proteins behave as molecular shields, preventing the approach and interaction of aggregation-prone protein species by steric or electrostatic repulsion, analogous to polymer stabilization of colloidal suspensions (Tunnacliffe and Wise, 2007).

Protein aggregation during desiccation is likely to be a major potential hazard for anhydrobiotes; LEA proteins acting as molecular chaperones or molecular shields play an important role in preventing this aggregation (Goyal *et al.*, 2005b). Adhikari *et al.* (2009) reported that the presence of transcripts similar to plant LEA-related family members of *C. elegans* was upregulated during desiccation stress of *P. murrayi* (*Pm-lea*) (Fig. 9.1). The *C. elegans* genome encodes three LEA genes (Crowe, 2002), and silencing of the *lea-1* gene by RNA interference caused a marked reduction in desiccation resistance in dauer larvae (Gal *et al.*, 2004). It is likely that LEA proteins also contribute to protection and recovery from desiccation stress in anhydrobiotic Antarctic terrestrial nematodes.

LEA proteins are widespread in anhydrobiotes and function as molecular chaperones. Transcripts encoding the LEA gene were upregulated in response to desiccation stress in dauer larvae of the entomopathogenic nematode *Steinernema carpocapsae*, and a hydrophilic heat-stable protein isolated from *Steinernema feltiae* was identified as a group 3 LEA protein (Solomon *et al.*, 2000). An EST transcript identified as a member of the LEA group 3 was also found to be upregulated in *S. feltiae* in response to desiccation (Gal *et al.*, 2003). An LEA group 3 gene, *Aav-lea-1*, is also strongly induced in *Aphelenchus avenae* during the induction of anhydrobiosis (Browne *et al.*, 2002; Oliver *et al.*, 2002). The *C. elegans* genome encodes three LEA genes; there are four LEA genes in *Caenorhabditis briggsae* and the *Drosophila melanogaster* genome encodes one LEA gene (Browne *et al.*, 2004). A desiccation-induced LEA-like protein has also been detected in anhydrobiotic bdelloid rotifers (Tunnacliffe *et al.*, 2005). Group 3 LEA genes have recently been isolated in the larvae of the African chironomid, *Polypedium vanderplanki*, an anhydrobiotic arthropod (Kikiwada *et al.*, 2006). These authors found that the mRNA levels of three *P. vanderplanki* LEA genes increased in response to desiccation and salinity stress and that the expression of *Pv-LEA* proteins also increased in response to these stresses.

9.3.2.2 Small heat shock proteins

Upon exposure to stress, both prokaryotic and eukaryotic cells produce a group of proteins with a molecular mass of 15 to 42 kDa, designated small heat shock proteins (sHsps; Vierling, 1997). sHsps are ubiquitous, being found in the cytosol of eukaryotes as well as prokaryotes. In Antarctic nematodes, sHsps are among the most abundant desiccation-induced proteins (Adhikari *et al.*, 2009; Table 9.1). The production of abundant sHsps in Antarctic nematodes may reflect their need to adapt quickly to ever-changing environmental conditions (such as temperature and humidity). The correlation between the abundant expression of sHsps and desiccation led to the hypothesis that sHsps protect cells from the detrimental effects of desiccation stress. There is evidence to support the idea that sHsps function as molecular chaperones that bind to partially folded or denatured substrate proteins and thereby prevent irreversible aggregation or promote correct substrate folding (Scharf *et al.*, 2001).

Table 9.1. The most abundantly represented transcripts in a *Plectus murrayi* cDNA library exposed to desiccation. (Adapted from Adhikari *et al.*, 2009.)

Contig no.	Tentative annotation[a]	Number of ESTs	Percentage[b] (%)
Contig_57	Ribosomal protein	37	1.48
Contig_132	Cytochrome *c* oxidase	33	1.32
Contig_9	Small heat shock protein family	27	1.08
Contig_312	Rab-family member	25	1.00
Contig_87	Aquaporin	24	0.96
Contig_231	DNA-binding protein	23	0.92
Contig_32	Y-box family member	22	0.88
Contig_198	Zinc finger protein	18	0.72
Contig_54	Cu/Zn superoxide dismutase	17	0.68
Contig_56	Translation initiation factor	17	0.68
Contig_67	Chaperonin-containing subunit	16	0.64
Contig_301	Glutamate synthase	13	0.52
Contig_287	NADH dehydrogenase	13	0.52
Contig_93	Cathespin B-like cysteine proteinase	13	0.52
Contig_74	Elongation factor 1α	13	0.52
Contig_63	Glutathione *S*-transferase	12	0.48
Contig_152	Heat shock 70 kDa protein	11	0.44
Contig_242	60S ribosomal protein	11	0.44
Contig_131	Elongation factor family member	11	0.44
Contig_29	Aldehyde dehydrogenase	7	0.28
Contig_149	Heat shock 90 kDa protein	7	0.28
Contig_313	ATP synthase subunit	7	0.28
Contig_112	Nuclear hormone protein family	7	0.28

[a]Annotation based on most significant BLAST alignment for each cluster.
[b]Percentage based on total number of high quality sequences.

Some sHsps have been demonstrated to act as molecular chaperones. Molecular chaperones are formally defined as 'proteins that bind to and stabilize an otherwise unstable conformer of another protein and by controlled binding and release, facilitate its correct fate *in vivo*, be it folding, oligomeric assembly, transport to a particular sub-cellular compartment, or disposal by degradation' (Hendrick and Hartl, 1995). The ability to recognize and bind unfolded proteins, to suppress protein aggregation and to influence the yield of protein folding are taken into account in this definition of molecular chaperone activities (Lee, 1995). The current model for the function of sHsps is that sHsps can bind selectively non-native proteins, prevent their aggregation and maintain them in a state competent for ATP-dependent refolding by other chaperones (Veinger *et al.*, 1998; Lee *et al.*, 2000). Experiments performed with sHsps from diverse organisms have demonstrated that sHsps are particularly effective in preventing thermal aggregation of other proteins by an ATP-independent mechanism (Lee *et al.*, 1997). sHsps can undergo temperature-dependent conformational changes, resulting in either dissociation of the oligomeric complex or alteration of the surface hydrophobicity (Lee *et al.*, 1997; Torok *et al.*, 2001) and an increase in the affinity for substrate binding.

9.3.2.3 Ubiquitin

The ubiquitin-dependent pathway is one of the putative repair mechanisms. It participates in DNA repair and selective removal of damaged or obsolete proteins (Shang and Taylor, 1995; Bregman *et al.*, 1996). Ubiquitin genes may be either constitutively expressed or stress-inducible. Stress induction of ubiquitin genes occurs in some organisms, including Antarctic nematodes (Adhikari *et al.*, 2009). However, in *C. elegans*, ubiquitin is not a stress-induced protein. In nematodes, the ubiquitin-dependent pathway is required to withstand oxidative stress and heavy metal toxicity. In addition, the ubiquitin-dependent pathway also plays roles in regulating diverse cellular processes, such as signal transduction, cell cycle control, differentiation and apoptosis (Shang *et al.*, 1997). The hallmark of the ubiquitin-dependent pathway is the covalent attachment of ubiquitin to proteins to form ubiquitin–protein conjugates in a process termed ubiquitinylation. The best-known role of ubiquitinylation is selectively to target proteins for degradation, but ubiquitinylation of some proteins, such as calmodulin, histones H2A and H2B, actin and some membrane receptors, serves a regulatory function without targeting them for cytosolic degradation (Jennissen, 1995).

Nematodes rely on proteomic plasticity to remodel themselves during periods of developmental change, to endure varying environmental conditions and to respond to biotic and abiotic stresses. Regulated ubiquitin- and proteasome-mediated degradation therefore plays a crucial role in enabling nematodes to alter their proteome to maximize their chances of survival under many different environmental circumstances. In nematodes, as in all eukaryotes, the ubiquitin/proteasome system (UPS) typically targets proteins for degradation. The terrestrial Antarctic nematodes represent a special case of stress tolerance because the environments they experience are both extreme and extremely unpredictable. Two different genes are expressed; a gene homologue to *C. elegans ubq-2* was expressed in a desiccated *P. murrayi* library and a ubiquitin-specific protease was shown to be differentially expressed in desiccated nematodes (Adhikari *et al.*, 2009). In the entomopathogenic nematode *S. feltiae* IS-6, a homologue of *ubq-2* from *C. elegans* is upregulated after 8 h of desiccation. In *C. elegans*, *ubq-2* encodes the fusion of ubiquitin and 52 amino acid ribosomal proteins (Jones *et al.*, 1995) and is involved in the heat shock response, which leads to the activation of heat shock promoters (Muller-Taubenberger *et al.*, 1988; Jentsch, 1992). In *C. elegans*, *ubq-2* is expressed in all life stages of the worm, and the abundance of its transcript is unaffected by heat stress (Jones *et al.*, 1995). By contrast, the observed differential expression of *Pm-ubq-2* transcripts upon desiccation suggests that in *P. murrayi* it may be stress related and plays a somewhat different role than it does in *C. elegans*.

9.4 Signal Transduction System

In order for nematodes to respond to their environment, signal transduction mechanisms have evolved to transfer environmental information to the cell, resulting in alterations in gene expression. Many different component transduction pathways may be required to achieve alterations in gene expression. In the case of signals related to cell growth, differentiation, proliferation and stress, this often results in a transcriptional activation of genes related to these processes. Receptors in the plasma membrane, including secondary messengers, such as cyclic adenosine monophosphate (cAMP), and protein kinase systems, such as mitogen-activated protein kinase (MAPK), belong to these signal transduction systems. Components of the major stress response systems are constitutively present to maintain cellular homeostasis and to contribute to cellular processes such as growth, proliferation and apoptosis. In these processes, the actions of the response system are combined with those emanating from signals from, for example, growth factors and hormones via the signal transduction systems. For this reason, these signal transduction systems are also activated under stressful conditions, thus contributing to the stress response.

Neuronal signal transduction in response to desiccation stress is required to initiate the coiling response of desiccating *P. murrayi*. Transcripts with putative roles in neuronal signal transduction are strongly upregulated in *P. murrayi* in response to desiccation stress. A number of desiccation-responsive *P. murrayi* ESTs encode putative signalling molecules or transcription factors. One of the *P. murrayi* ESTs was most similar to the *unc-16* gene of *C. elegans*, which encodes a c-Jun N-terminal kinase (JNK)-interacting protein (Byrd *et al.*, 2001), and one of the members of JNK kinase family (*jkk-1*) is highly expressed during desiccation (Fig. 9.2). JNK (also known as stress-activated MAPK or SAPK) is a member of the MAPKs that regulate cellular responses to a variety of extracellular signals, including desiccation stress (Davis, 2000). The JNK pathway is activated by the exposure of cells to stress. However, the role of JNK in response to stress is unclear. It is possible that JNK may mediate some of the effects of stress on cells. Alternatively, JNK activation may represent a protective response that is initiated by the exposure to stress. The specific role of JNK may therefore depend upon the cellular context. Indeed, the JNK pathway has been implicated in both apoptosis and survival signalling (Ip and Davis, 1998).

Adhikari *et al.* (2009) demonstrated that the levels of transcripts for JNK kinase increased markedly when nematodes were exposed to desiccation (Fig. 9.2). These results suggest that some of the MAPK cascades in nematodes function as transducing signals in the presence of environmental stress and that MAPK cascades are regulated at the transcriptional level in nematodes. Tyson *et al.* (2007) reported transcripts encoding JNK-interacting protein in a desiccation-induced entomopathogenic nematode, *S. carpocapsae*. Under stressful environmental conditions the elevated levels of the mRNAs of these putative signal transducers may increase their protein level, which is likely to amplify the signal transduction efficiency of the cascade. Analyses

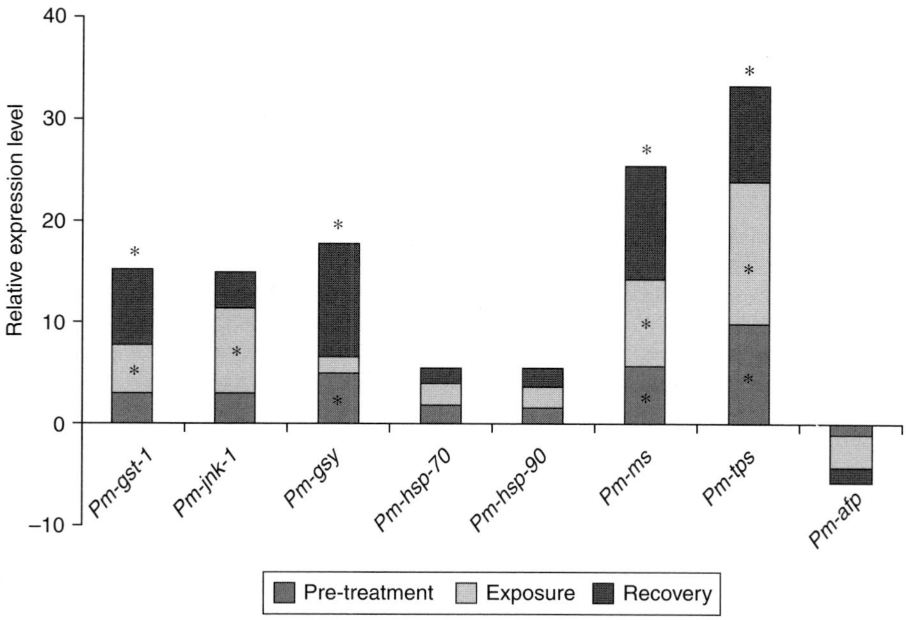

Fig. 9.2. Relative expression of eight transcripts in *Plectus murrayi* during pre-treatment, exposure to and recovery from desiccation. The relative expression of the target gene (*Pm-gst-1*: glutathione *S*-transferase 1; *Pm-jnk-1*: c-Jun N-terminal kinase; *Pm-gsy*: glycogen synthase; *Pm-hsp-70*: heat shock protein 70; *Pm-hsp-90*: heat shock protein 90; *Pm-ms*: malate synthase; *Pm-tps*: trehalose-6-phosphate synthase; *Pm-afp*: antifreeze protein), normalized to *Pm-18s*:18S rRNA and relative to the expression of control, was calculated for each sample. Note the *y*-axis, where values >1 indicate upregulation and values <1 indicate downregulation of the transcript. Asterisks indicate values that are significantly different from 1 ($P<0.05$). Mean value presented, where $N = 3$ in all cases.

of nematodes that over-express or repress genes for JNK will help to elucidate the molecular mechanisms of stress response and adaptation to environmental changes in Antarctic nematodes.

9.5 Metabolic System

Changes in metabolism are a general response to stress in nematodes. Evidence suggests that Antarctic nematodes activate metabolic pathways in response to desiccation within hours of exposure to dehydrating conditions. Transcriptional analyses of Antarctic nematodes exposed to desiccation have been carried out by sequencing cDNA libraries of desiccated nematodes and assigning the transcripts into different biochemical pathways based on KEGG (Table 9.2). The expressed genome of desiccated *P. murrayi* reveals a large number of genes related to metabolism (Adhikari *et al.*, 2009). Pathways well represented by the *P. murrayi* clusters were carbohydrate metabolism,

Table 9.2. KEGG biochemical mappings for *Plectus murrayi* clusters obtained from sequencing a desiccated *P. murrayi* cDNA library. (Adapted from Adhikari *et al.*, 2009.)

KEGG categories represented	Unique sequences (number of enzymes)	Percentage[a]
Metabolism	84 (52)	11
Carbohydrate metabolism	29 (18)	4
Amino acid metabolism	14 (9)	2
Lipid metabolism	13 (8)	2
Xenobiotics biodegradation and metabolism	8 (5)	1
Biosynthesis of secondary metabolites	6 (3)	<1
Energy metabolism	5 (2)	<1
Nucleotide metabolism	3 (2)	<1
Metabolism of other amino acids	3 (2)	<1
Glyoxylate and dicarboxylate metabolism	3 (3)	<1
Genetic information processing	83 (48)	11
Folding, sorting and degradation	42 (25)	6
Transcription	17 (9)	2
Translation	16 (8)	2
Replication and repair	8 (6)	1
Environmental information processing	95 (47)	12
Membrane transport	30 (15)	4
Ligand-receptor interaction	28 (15)	4
Signal transduction	14 (8)	2
Signalling molecules and interaction	13 (9)	2
Cellular processes	19 (11)	3
Cell growth and death	8 (5)	1
Cell communication	6 (4)	<1
Cell motility	3 (1)	<1
Development	2 (1)	<1
Unassigned[b]	385	49
Hypothetical	116	15

[a]Percentage based on total unique transcripts (782) with significant similarity to sequences in public databases.
[b]Unassigned sequences are those that have significant similarity to known sequences whose functions are unclear.

amino acid metabolism, lipid metabolism, biosynthesis of secondary metabolites, and glyoxylate and decarboxylation metabolism. The lipid metabolism pathway in anhydrobiotes is one of the most active pathways as lipids are the main reservoir of energy and the most likely source of carbon for the synthesis of trehalose, a disaccharide implicated in anhydrobiosis (Wharton, 2003; see Barrett, Chapter 12, this volume). A number of metabolism-related transcripts encoding cytochrome *c* oxidase subunit II, glutamate synthase, NADH dehydrogenase subunit I and aldehyde dehydrogenase are also abundantly expressed during desiccation stress. Adhikari *et al.* (2009) used subtractive hybridization to show that 28% of the genes differentially expressed in desiccated nematodes are metabolism related. These genes include aldehyde dehydrogenase, TPS, thymidylate synthase, GK, GS, ATP synthase, ADP/ATP translocase and malate dehydrogenase. Expression levels of some of these genes are highly induced during desiccation stress (Fig. 9.1).

Interestingly, transcripts encoding a bifunctional glyoxylate cycle protein, malate synthase (*Pm-ms*), a distinct and anaplerotic variant of the tricarboxylic acid cycle, are also highly expressed during desiccation. Analyses of gene expression during different stages of desiccation and freezing stress show that malate synthase is implicated during both stresses. Nematodes are unique among animals in utilizing the glyoxylate cycle to generate carbohydrates from the β-oxidation of fatty acids (Barrett and Wright, 1998). The glyoxylate pathway, generally found in plants and microorganisms, is similar to the citrate cycle but relies on two critical enzymes, malate synthase and isocitrate lyase, to bypass two decarboxylation steps. The glyoxylate cycle provides a means for the conversion of two-carbon acetyl-CoA units derived from the β-oxidation of fatty acids into the four-carbon molecules succinate and malate, which can then be used as precursors in the biosynthesis of carbohydrates and other cellular components. Analyses of gene expression during different stages of desiccation and freeze stress show that Antarctic nematodes use the glyoxylate cycle, not only during induction but also during survival and recovery from stress (Figs 9.2 and 9.3). Interestingly, the anhydrobiotic nematode *A. avenae* has been reported to use the glyoxylate cycle during induction of anhydrobiosis (Madin *et al.*, 1985). The abundant expression of malate synthase transcripts in the EST collection and its upregulation upon desiccation and freezing stress provide experimental support for an active role of glyoxylate cycle proteins during induction of anhydrobiosis by *P. murrayi*. Analyses of mRNA expression data show that desiccation induced upregulation of several anaerobic pathways, including the 'malate dismutation' and fermentation pathways. In nematodes, the malate dismutation process enables mitochondria to function anaerobically and requires the use of a specialized electron transport chain.

9.6 Oxidative Stress Response and Detoxification System

Upon exposure to environmental stressors including desiccation, extra amounts of ROS may be generated, which can accumulate to toxic levels. Desiccation stress induces the generation of ROS in nematodes, and therefore it is important for nematodes to have effective ROS-scavenging mechanisms. At these elevated concentrations, ROS can lead to oxidative stress, resulting in cytotoxicity and damage of cellular structures, such as membranes. These effects may initiate a series of events that culminate in apoptosis or necrosis. The molecular events that increase ROS during some types of stress, including desiccation, are a direct consequence of the stress. However, the molecular basis for oxidative stress is poorly understood in nematodes. A number of ESTs encoding proteins that detoxify ROS like superoxide dismutase (SOD), Ras-related protein, and glutathione *S*-transferase (GST) are expressed in

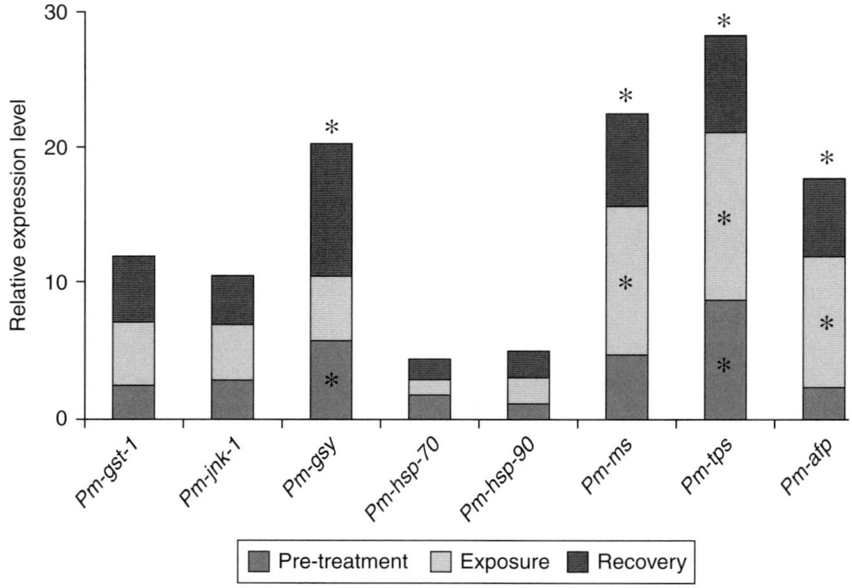

Fig. 9.3. Relative expression of eight transcripts in *Plectus murrayi* during pre-treatment, exposure to and recovery from freezing. The relative expression of the target gene (*Pm-gst-1*: glutathione S-transferase 1; *Pm-jnk-1*: c-Jun N-terminal kinase; *Pm-gsy*: glycogen synthase; *Pm-hsp-70*: heat shock protein 70; *Pm-hsp-90*: heat shock protein 90; *Pm-ms*: malate synthase; *Pm-tps*: trehalose-6-phosphate synthase; *Pm-afp*: antifreeze protein), normalized to *Pm-18s*:18S rRNA and relative to the expression of control, was calculated for each sample. Note the *y*-axis, where values >1 indicate upregulation and values <1 indicate downregulation of the transcript. Asterisks indicate values that are significantly different from 1 ($P<0.05$). Mean value presented, where $N = 3$ in all cases.

P. murrayi in response to desiccation. Expression of ESTs similar to *C. elegans sod-1*, which encodes a copper/zinc superoxide dismutase, is also abundant. Three transcripts encoding GST-1 were expressed in the ESTs of *P. murrayi*, and one of them was upregulated following desiccation stress. GSTs are a diverse superfamily of multifunctional proteins that play prominent roles in detoxification metabolism in nematodes (Lindblom and Dodd, 2006). More than a dozen different GSTs have been isolated from *C. elegans* (Van Rossum *et al.*, 2001) and these detoxifying enzymes are reported to be involved in several functions, including xenobiotic detoxification and oxidative stress tolerance (Leiers *et al.*, 2003; Lindblom and Dodd, 2006). Differential expression of a number of detoxifying enzymes and upregulation of *Pm-gst-1* (Figs 9.1–9.3) suggest that *P. murrayi* has efficient ROS-scavenging mechanisms under desiccation stress.

One of the most deleterious effects of dehydration in the cell is oxidative damage (Hermes-Lima and Zenteno-Savin, 2002; Franca *et al.*, 2007). Although the origin of the excess oxygen radicals is not fully understood, a tenfold increase in oxidation as a result of dehydration has been recorded in yeast cells (Pereira *et al.*, 2001). It is suspected that greater oxidation also plays a major role in death caused by dehydration in nematodes. Thus, protection against oxidation by enzymatic breakdown of oxygen radicals is crucial during water loss. A universal primary defence against oxygen toxicity is SOD, which catalyses the breakdown of the superoxide anion into hydrogen peroxide and oxygen. Hydrogen peroxide is then broken down to water and oxygen by several enzymes, the main one being catalase. Catalase is involved in the dehydration response of *P. murrayi* (Adhikari *et al.*, 2009) as well as in the Antarctic midge *Belgica antarctica* (Lopez-Martinez *et al.*, 2008).

Several broad-spectrum detoxification gene families are also present in *P. murrayi*. These include UDP-glucuronosyltransferases (UGTs) and GSTs (Adhikari *et al.*, 2009). UGTs add a glucuronic acid group to a wide range of structurally diverse lipophilic molecules, thereby increasing their aqueous solubility and allowing them to be partitioned into intra- and extracellular aqueous compartments. GSTs catalyse the detoxification of a wide variety of endobiotic and xenobiotic electrophilic substrates by conjugation with glutathione, rendering the glutathione adducts less toxic and more water-soluble. Additionally, GSTs play a role in oxidative stress resistance. These detoxification reactions are metabolically costly – each UGT and GST reaction consumes, respectively, a molecule of glucose or a molecule of glutathione. Since anhydrobiotic nematodes are non-feeding, these detoxification reactions are likely to be targeted towards degrading or metabolizing endogenous compounds. Some of these reactions may be involved in the biosynthesis or degradation of lipophilic hormones required for the maintenance of, or recovery from, the anhydrobiotic state, but many of the compounds that this system targets are likely to be potentially toxic endogenous metabolites.

9.7 Cryoprotectant

Desiccation involves two universal reactions: production of organic osmolytes such as trehalose and glycerol and changes in membrane composition. These two processes are linked, as removal of water profoundly affects the physical properties of membrane phospholipids and leads to destructive events such as fusion, phase transitions and increased permeability. The sugar-based cryoprotectants prevent damage from dehydration by inhibiting fusion between adjacent vesicles during desiccation and also by maintaining the lipids in a fluid state in the absence of water (Crowe *et al.*, 1987). This protective process is substantiated in *P. murrayi* by elevations in the expression levels of *Pm-tps* (Adhikari *et al.*, 2009). Sugar-based cryoprotectants are well known to reduce osmotic outflow from the cell and thus reduce injury due to cell dehydration (Storey and Storey, 1988). Trehalose has been implicated in anoxia tolerance by reducing protein aggregation and maintaining proteins in a partially folded position, aiding chaperone refolding, enhancing recovery from anoxia exposure and possibly as a carbon source under extraordinary conditions (Chen *et al.*, 2002). Trehalose has also been demonstrated to prevent damage caused by oxygen radicals when combined with heat shock (Benaroudj *et al.*, 2001; Pereira *et al.*, 2001).

Our work reveals a significant change in the expression of a gene encoding T6PS (*Pm-tps*) during both desiccation and freezing, with significant influence of pre-treatment on the relative expression: acclimated samples had higher levels of mRNA transcripts than those without pre-treatment (Figs 9.2 and 9.3). A characteristic feature of anhydrobiotic organisms is their synthesis of high concentrations of non-reducing sugars during the induction of anhydrobiosis (Goyal *et al.*, 2005c), which protects membranes and proteins from desiccation damage by replacing structural water (Crowe *et al.*, 1992), and formation of an intracellular organic glass (Crowe *et al.*, 1998) to stabilize the cell's contents. In nematodes, cold and desiccation pre-treatment results in the accumulation of trehalose (Solomon *et al.*, 2000; Grewal and Jagdale, 2002; see Grewal *et al.*, Chapter 7, and Wharton, Chapter 8, this volume) and plays an important role in enhanced freezing (Wharton *et al.*, 2000) and desiccation tolerance (Solomon *et al.*, 2000; Grewal and Jagdale, 2002). Elevated levels of trehalose, which enhance freezing survival, have also been reported from a coastal Antarctic bacterial-feeding nematode, *P. davidi* (Wharton *et al.*, 2000). In our experiments, expression levels of *Pm-tps* were higher in both acclimated and non-acclimated treatments, suggesting the nematode's ability to produce significant amounts of trehalose without prior pre-treatment. Since trehalose is reported to be the major compatible solute during multiple stresses (Santos and de Costa, 2001; Yancey, 2005), increased transcription of genes encoding trehalose could be a mechanism to counter multiple stresses in addition to desiccation and freezing. On the other hand, the increased *Pm-tps* mRNA levels observed in desiccated nematodes is in keeping with a role for trehalose in anhydrobiosis; they are of the same order as, and therefore could be largely responsible for, the concomitant increase in trehalose concentration (Browne *et al.*, 2004). However, to a similar extent, *Pm-tps* is also upregulated by freezing stress, which might indicate a more general role for trehalose in alleviating the effects of reduced water activity and perhaps other stresses. Certainly, our data do not support an exclusive role for trehalose in anhydrobiosis and, indeed, further evidence is required to demonstrate that it is actually necessary for desiccation tolerance.

The process of anhydrobiosis is accompanied by a concomitant increase in the levels of the carbohydrate trehalose, which is presumed to also act as a cryoprotectant in many organisms, including Antarctic nematodes (Womersley and Smith, 1981; Shannon *et al.*, 2005; Adhikari *et al.*, 2009). However, there are also a number of examples where species undergo anhydrobiosis in the absence of disaccharide accumulation, including bdelloid rotifers (Caprioli *et al.*, 2004) and some tardigrades (Hengerr *et al.*, 2008). Also, the model species of baker's yeast, *Saccharomyces cerevisiae*, and the nematode *C. elegans* accumulate trehalose in response to stressful conditions, but knockout experiments using T6PS genes have shown very little, if any, effect (Pellerone *et al.*, 2003). These conflicting data suggest that trehalose may not be the single most important molecule in determining whether an animal survives dehydration but is one of a series of mechanisms, several of which may act synergistically to improve survival (Hengerr *et al.*, 2008).

Additional metabolic pathways associated with desiccation in *P. murrayi* are GK (*Pm-gk*) and GS (*Pm-gsy*), but these compounds also act as cryoprotectants. GK is the key enzyme in regulating glycerol catabolism and catalyses the conversion of glycerol to glycerol-3-phosphate and dihydroxy-acetone phosphate as part of the process of entering into either the glycolysis or the gluconeogenesis pathway (Nelson and Cox, 2000). It is likely that gluconeogenesis is the preferred pathway under desiccation stress because the glucose produced by gluconeogenesis may be metabolized to the osmoprotectant trehalose. An increase in glycerol content has been demonstrated to reduce rates of water loss (Yoder *et al.*, 2006) and to protect membranes and proteins during dehydration (Tang and Pikal, 2005). GS (*Pm-gsy*) is a rate-limiting enzyme in the synthesis of glycogen and the primary storage form of glucose in higher eukaryotes, including nematodes (Behm, 1997). A number of studies have suggested a metabolic shift from glycogen to trehalose production during dehydration of nematodes (Behm, 1997; Solomon *et al.*, 1999; Gal *et al.*, 2001). We also observe an inverse relationship between the transcription of *Pm-tps* and *Pm-gsy*. It could be that we are seeing glycogen being converted to trehalose, which would explain the observation of significantly lower expression of *Pm-gsy* during exposure, while exhibiting a dramatic increase at the recovery phase (Figs 9.2 and 9.3). Such a reduction in GS transcription levels suggests a shift from glycogen to trehalose synthesis during exposure and, perhaps, a shift from trehalose to glycogen synthesis during recovery. It is possible that the shift from glycogen to trehalose synthesis occurs during both stresses and is regulated, at least in part, by suppression of *Pm-gsy* transcription. Adjustments of gene expression levels and changes in kinetics may play a major role in the induction and maintenance of stress tolerance, allowing nematodes to persist in some of the Earth's harshest environments.

9.8 Cross-tolerance and Stress-hardening

Environmental stress tolerance varies widely depending on the species (genome) and on cell type and differentiation state (proteome). The latter is a function of gene–environment interactions during development and of pre-exposure to stress over the lifetime of the organism. Stress-hardening (increased tolerance of a stress after preconditioning at low doses of that stress) and cross-tolerance (increased tolerance of one stress after preconditioning by another) are common and significant. Environmental stresses can result in the modification of gene expression by epigenetic events through an alteration of the epigenome (Jirtle and Skinner, 2007). Epigenetically mediated changes in gene expression due to environmental stress can not only trigger the activation of genes involved in stress resistance but also persist over several generations (Boyko and Kovalchuk, 2008). However, no research has been done to determine if epigenetics play any role in the survival of nematodes under environmental stresses and their subsequent inheritance.

Adhikari *et al.* (2009) showed that exposure to mild stress is accompanied by increased transcription of a number of genes which code for several osmolytes (e.g. trehalose and glycerol). We also showed that *P. murrayi* undergoes significant transcriptional change and accumulates osmolytes if desiccation occurs at slower rates of water loss and under high relative humidity. This point is exemplified by the finding that *P. murrayi* accumulates significantly higher transcripts of a number of genes when exposed to desiccation and freezing with pre-treatment as compared to no pre-treatment (Figs 9.2 and 9.3). Our study also supports the idea that pre-exposure to desiccation stress promotes enhanced cold tolerance and provides evidence that gradual desiccation can enhance the lower limit of freeze tolerance in this Antarctic nematode (Adhikari *et al.*, 2010). Although limited data exist regarding molecular mechanisms of the cross-stress tolerance of Antarctic nematodes, the observed enhanced tolerance could be due partly to induced upregulation of gene transcript levels, including *Pm-gst-1*, *Pm-jnk-1*, *Pm-gsy*, *Pm-ms* and *Pm-tps*, during both stresses (Figs 9.2 and 9.3). Furthermore, increased induction of some of these genes during exposure to one stress could be contributing to the cross-tolerance of this Antarctic nematode. The reprogramming of gene expression and post-transcriptional mechanisms are essential components of acclimatory mechanisms. In view of the large number of inducible stress genes responding during desiccation stress and their induction during freezing, we postulate that the induction of stress-related genes might not be sufficient to confer desiccation-pre-treatment-induced tolerance to nematodes. Thus, we hypothesize that desiccation pre-treatment involves the activation of more than one molecular cryoprotective network and additional genes, including those not currently among the consensus of the desiccation-inducible set of genes. Preconditioning involves the activation of a number of genes from many different functional areas. We hypothesize that the observed profound change in the expression of these genes reinforces inherent coordinated cryoprotective mechanisms and contributes to the cryoprotective aspects of the cross-tolerance phenomenon.

Work done by Adhikari *et al.* (2009, 2010) and other studies have shown that many genes that are induced by cold are also induced by desiccation, probably because many cold-inducible genes encode proteins to protect the nematodes from the consequences of freezing stress, which include dehydration (Chen *et al.*, 2005). It is highly likely that the activation and induction of a common set of stress proteins is the molecular basis of both cross-tolerance and stress-hardening. After the initial stress, these proteins remain active/elevated for a period that varies depending on species, cell type, history of prior stress exposure, gene–environment interactions during development and stress severity. During this period, activated/elevated stress proteins confer resistance to many different types of stress because of their involvement in general aspects of cellular protection such as protein stabilization, DNA repair and free radical scavenging. Differences in constitutive levels of critical stress proteins are also responsible for species-specific variation in tolerance thresholds within multicellular organisms (Beck *et al.*, 2000; Kültz, 2005).

Work done on *P. murrayi* gene transcription during different stages of desiccation and freezing stress and the effect of pre-treatment on gene transcription suggests that many stress-related genes, signal transduction pathways and transcription factors are involved in desiccation as well as freeze pre-treatment (Adhikari *et al.*, 2010). It is interesting to note that almost all stress factors switch on more than one stress-responsive pathway. The molecular mechanisms underlying interactions between different pathways may involve the requirement of more than one type of transcription factor for gene activation and the complicated interactions between different kinases in signal transduction pathways. Also, the stress factor itself may activate more than one response pathway, for example, when triggering several metabolic disturbances. For nematodes, it is not yet possible to identify an exhaustive set of genes induced by all stress factors at the same time.

9.9 Conclusions and Future Directions

Anhydrobiosis is broadly distributed among dissimilar taxa in the biological world, suggesting that this is a deeply rooted, ancient physiological process. The finding of a plant desiccation gene in a nematode genome, a putative homologue, supports this idea (Browne *et al.*, 2002). Alternatively, but perhaps less parsimoniously, anhydrobiosis has evolved through repeated episodes of strong selection and convergent evolution. Mechanisms of desiccation tolerance such as anhydrobiosis are likely to have accompanied the spread of life from sediments to soils, from oceans to the land (Oliver *et al.*, 2002). The soils of the earliest terrestrial environments on Earth were probably very similar in some ways to those of the Antarctic dry valleys: desert-like, poorly weathered, with low water-holding capacity and low organic content. Nematodes are one of the most desiccation-tolerant animals on Earth, and their ability to employ an anhydrobiotic survival strategy may help to explain this phylum's almost ubiquitous distribution. Studies of nematode anhydrobiosis in the Antarctic dry valleys and other arid environments should help us to reconstruct the evolutionary history of adaptations that facilitated the success of life's invasion of terrestrial environments.

Antarctic nematodes exhibit an interesting case of cross-tolerance as well as stress-hardening. Many genes that are induced by cold are also induced by desiccation, probably because at the cellular level the challenges faced through cold or desiccation are very similar, particularly in terms of increase in solute concentration and osmotic stress, reduction in super-cooling point and membrane shrinkage (Convey and Williams, 2002). Desiccation and cold tolerance are considered to be overlapping adaptations (Ring and Danks, 1994) eliciting similar responses to limit injury during periods of extreme environmental conditions. The induction of genes that code for sugars and polyols like trehalose and glycerol seems to be a common trait of both cold- and drought-tolerant ectotherms. But it should be noted that although large portions of the tolerance programme against desiccation are also turned on in defence against freezing, additional genes are activated to deal with each individual stress. A gene encoding AFP was only activated during freezing; desiccation resulted in downregulating its expression. It is possible that different suites of genes will be transcribed when nematodes are exposed to both desiccation and freezing.

Studying abiotic stress-signalling pathways in isolation is valuable but it can be misleading because they typically form only small parts of very complex networks. In the future, the onus will be on taking this fact into account, both intellectually and in terms of technology development. A perfect example of this is the utility of microarray technology. This approach enables researchers to examine the expression of not only all their particular stress-induced genes of interest but also thousands of others, without prejudice and without extra efforts. Approaches such as this may lead to a greater understanding of the effect that abiotic stress pathways have on each other as well as on pathways and processes that were not previously known to be connected. Large-scale genome or transcriptome sequencing efforts can enable the microarray projects and inform work on specificity as well as cross-talk between different stress factors and the proteins associated with those events. Finally, signalling components whose genes are induced by abiotic stress, and hence implied in abiotic stress signalling, can be directly tested for specificity or cross-talk (problems of redundancy aside) by phenotypic analyses of knockout mutants.

The study of stress biology of organisms that can survive environmental extremes offers plenty of challenges to biologists to understand the mechanisms by which they manage to survive under conditions that would be fatal to most other organisms. It also provides an opportunity to use this knowledge to improve the stress tolerance of other beneficial organisms (e.g. entomopathogenic nematodes) using molecular genetics tools. Proteins and enzymes isolated from stress-tolerant organisms could have considerable biotechnological as well as industrial potential. Antifreezes from polar fish, and other cold-tolerant organisms, such as nematodes and tardigrades, may improve our ability to cryopreserve biological materials. Understanding the mechanisms of anhydrobiosis may improve our ability to store biological materials (Wharton, 2002). Proteins like proteases and key enzymes of fundamental biological processes such as DNA replication, DNA repair and RNA maturation could have considerable biotechnological potential (Marsh *et al.*, 2001). Biosynthetic pathways and enzymes like alcohol dehydrogenase active in desiccated and cold environments could be promising biocatalysts in industrial processes. The molecular analysis of the desiccation response has arrived at a stage where research can build upon a large collection of characterized genes. The use of novel approaches combining genes and biochemical and molecular techniques should provide exciting results in the near future.

9.10 Acknowledgements

We thank Diana H. Wall and numerous other McMurdo Dry Valley Wormherders for useful advice and support. Chin-Yo Lin, Josh Udall, Jeff Maughan, Ed Wilcox,

Chris Cheng-Devries and David Wharton provided helpful discussions of the subject over the years. Fieldwork related to this project was supported by the National Science Foundation McMurdo Long Term Ecological Research Project (OPP #98-10219) to Diana H. Wall, and a Brigham Young University Mentored Environment Grant to Byron J. Adams.

9.11 References

Adhikari, B.N, Wall, D.H. and Adams, B.J. (2009) Desiccation survival in an Antarctic nematode: molecular analysis using expressed sequenced tags. *BMC Genomics* 10, 69.

Adhikari, B.N., Wall, D.H. and Adams, B.J. (2010) Effect of slow desiccation and freezing on gene transcription and stress survival of an Antarctic nematode. *Journal of Experimental Biology* 213, 1803–1812.

Ashburner, M., Ball, C.A., Blake, J.A. *et al.* (2000) Gene Ontology: tool for the unification of biology. *Nature Genetics* 25, 25–29.

Barrett, J. (1991) Anhydrobiotic nematodes. *Agricultural Zoology Reviews* 4, 161–176.

Barrett, J. and Wright, D.J. (1998) Intermediary metabolism. In: Perry, R.N and Wright, D.J. (eds) *The Physiology and Biochemistry of Free-living and Plant-parasitic Nematodes*. CAB International, Wallingford, UK, pp. 331–354.

Beck, F.X., Grunbein, R., Lugmayr, K. and Neuhofer, W. (2000) Heat shock proteins and the cellular response to osmotic stress. *Cell Physiology Biochemistry* 10, 303–306

Behm, C.A. (1997) The role of trehalose in the physiology of nematodes. *International Journal for Parasitology* 27, 215–229.

Benaroudj, N., Lee, D.H. and Goldberg, A.L. (2001) Trehalose accumulation during cellular stress protects cells and cellular proteins from damage by oxygen radicals. *Journal of Biological Chemistry* 26, 24261–24267.

Boyko, A. and Kovalchuk, I. (2008) Epigenetic control of plant stress response. *Environmental and Molecular Mutagenesis* 49, 61–72.

Bregman, D.B., Halaban, R., van Gool, A.J., Henning, K.A., Friedberg, E.C. and Warren, S.L. (1996) UV-induced ubiquitination of RNA polymerase II: a novel modification deficient in Cockayne syndrome cells. *Proceedings of National Academy of Sciences of the USA* 93, 11586–11590.

Browne, J.A., Tunnacliffe, A. and Burnell, A. (2002) Anhydrobiosis – plant desiccation gene found in a nematode. *Nature* 416, 38.

Browne, J.A., Dolan, K.M., Tyson, T., Goyal, K., Tunnacliffe, A. and Burnell, A.M. (2004) Dehydration-specific induction of hydrophilic protein genes in the anhydrobiotic nematode *Aphelenchus avenae. Eukaryotic Cell* 3, 966–975.

Buckley, B.A., Place, S.P. and Hofmann, G.E. (2004) Regulation of heat shock genes in isolated hepatocytes from an Antarctic fish, *Trematomus bernacchii. Journal of Experimental Biology* 207, 3649–3656.

Byrd, D.T., Kawasaki, M., Walcoff, M., Hisamoto, N., Matsumoto, K. and Jin, Y.S. (2001) UNC-16, a JNK-signaling scaffold protein, regulates vesicle transport in *Caenorhabditis elegans. Neuron* 32, 787–800.

Caprioli, M., Krabbe Katholm, A., Melone, G., Ramlov, H., Ricci, C. and Santo, N. (2004) Trehalose in desiccated rotifers: a comparison between a bdelloid and a monogonont species. *Comparative Biochemistry and Physiology A – Molecular and Integrative Physiology* 139, 527–532.

Chen, Q., Ma, E., Behar, K.L, Xu, T. and Haddad, G.G. (2002) Role of trehalose phosphate synthase in anoxia tolerance and development in *Drosophila melanogaster. Journal of Biological Chemistry* 277, 3274–3279.

Chen, S., Golop, N. and Glazer, I. (2005) Cross-stress tolerance and expression of stress-related proteins in osmotically desiccated entomopathogenic *Steinernema feltiae* IS-6. *Parasitology* 131, 695–703.

Clegg, J.S. (2001). Cryptobiosis – a peculiar state of biological organization. *Comparative Biochemistry and Physiology B – Molecular and Integrative Physiology* 128, 613–624.

Convey, P. and Williams, D.D. (2002) Antarctic soil nematode response to artificial climate amelioration. *European Journal of Soil Biology* 38, 255–259.

Crowe, J.H. and Madin, K.A.C. (1974) Anhydrobiosis in tardigrades and nematodes. *Transactions of the American Microscopical Society* 93, 513–524.

Crowe, J.H. and Madin, K.A.C. (1975) Anhydrobiosis in nematodes – evaporative water-loss survival. *Journal of Experimental Zoology* 193, 323–333.

Crowe, J.H., Crowe, L.M., Carpenter, J.F. and Aurell, C.W. (1987) Stabilization of dry phospholipid bilayers and proteins by sugars. *Biochemistry Journal* 242, 1–10.

Crowe, J.H., Hoekstra, F.A. and Crowe, L.M. (1992) Anhydrobiosis. *Annual Review of Physiology* 54, 579–599.

Crowe, J.H., Carpenter, J.F. and Crowe, L.M. (1998) The role of vitrification in anhydrobiosis. *Annual Review of Physiology* 60, 73–103

Crowe, J.H., Oliver, A.E. and Tablin, F. (2002) Is there a single biochemical adaptation to anhydrobiosis? *Integrative and Comparative Biology* 42, 497–503.

Crowe, L.M. (2002) Lessons from nature: the role of sugars in anhydrobiosis. *Comparative Biochemistry and Physiology A – Molecular and Integrative Physiology* 131, 505–513.

Davis, R.J. (2000) Signal transduction by the JNK group of MAP kinases. *Cell* 103, 239–252.

Duman, J.G. (2001) Antifreeze and ice nucleator proteins in terrestrial arthropods. *Annual Review of Physiology* 63, 327–357.

Feder, M.E. and Hofmann, G.E. (1999) Heat-shock proteins, molecular chaperones, and the stress response: evolutionary and ecological physiology. *Annual Review of Physiology* 61, 243–282.

Feder, J.H., Rossi, J.M., Solomon, J., Solomon, N. and Lindquist, S. (1992) The consequences of expressing hsp70 in *Drosophila* cells at normal temperatures. *Genes and Development* 6, 1402–1413.

Franca, M.B., Panek, A.D. and Eleutherio, E.C.A. (2007) Oxidative stress and its effects during dehydration. *Comparative Biochemistry and Physiology A – Molecular and Integrative Physiology* 146, 621–631.

Gal, T.Z., Solomon, A., Glazer, I. and Koltai, H. (2001) Alterations in the levels of glycogen and glycogen synthase transcripts during desiccation in the insect-killing nematode *Steinernema feltiae* IS-6. *Journal of Parasitology* 87, 725–732.

Gal, T.Z., Glazer, I. and Koltai, H. (2003) Differential gene expression during desiccation stress in the insect-killing nematode *Steinernema feltiae* IS-6. *Journal of Parasitology* 89, 761–766.

Gal, T.Z., Glazer, I. and Koltai, H. (2004) An LEA group 3 family member is involved in survival of *Caenorhabditis elegans* during exposure to stress. *FEBS* 577, 21–26.

Gal, T.Z., Glazer, I. and Koltai, H. (2005) Stressed worms: responding to the post-genomics era. *Molecular and Biochemical Parasitology* 143, 1–5.

Goyal, K., Walton, L.J., Browne, J.A., Burnell, A.M. and Tunnacliffe, A. (2005a) Molecular anhydrobiology: identifying molecules implicated in invertebrate anhydrobiosis. *Integrative and Comparative Biology* 45, 702–709.

Goyal, K., Pinelli, C., Maslen, S.L., Rastogi, R.K., Stephens, E. and Tunnacliffe, A. (2005b) Dehydration-regulated processing of late embryogenesis abundant protein in a desiccation-tolerant nematode. *FEBS Letters* 579, 4093–4098.

Goyal, K., Browne, J.A., Burnell, A.M. and Tunnacliffe, A. (2005c) Dehydration-induced *tps* gene transcripts from an anhydrobiotic nematode contain novel spliced leaders and encode atypical GT-20 family proteins. *Biochimie* 87, 565–574.

Grewal, P.S. and Jagdale, G.B. (2002) Enhanced trehalose accumulation and desiccation survival of entomopathogenic nematodes through cold pre-acclimation. *Biocontrol Science and Technology* 12, 533–545.

Gross, M. (2004) Emergency services: a bird's eye perspective of the many different functions of stress proteins. *Current Protein Peptide Science* 5, 213–223.

Hendrick, J.P. and Hartl, F.U. (1995) The role of molecular chaperones in protein folding. *FASEB Journal* 9, 1559–1569.

Hengerr, S., Heyer, A.G. and Schill, R.O. (2008) Trehalose and anhydrobiosis in tardigrades – evidence for divergence in response to dehydration. *FEBS Journal* 275, 281–288.

Hermes-Lima, M. and Zenteno-Savin, T. (2002) Animal response to drastic changes in oxygen availability and physiological oxygen stress. *Comparative Biochemistry and Physiology C* 133, 537–556.

Higa, L.M. and Womersley, C.Z. (1993) New insights into the anhydrobiotic phenomenon: the effects of trehalose content and differential rates of evaporative water loss on the survival of *Aphelenchus avenae. Journal of Experimental Biology* 267, 120–129.

Ip, Y.T. and Davis, R.J. (1998) Signal transduciton by the c-Jun N-terminal kinase (JNK) – from inflamation to development. *Current Opinion in Cell Biology* 10, 205–219.

Jennissen, H.P. (1995) Ubiquitin and the enigma of intracellular protein degradation. *European Journal of Biochemistry* 231, 1–30.

Jentsch, S. (1992) The ubiquitin-conjugation system. *Annual Review of Genetics* 26, 179–207.

Jirtle, R.L. and Skinner, M. (2007) Environmental epigenomics and disease susceptibility. *Nature Reviews Genetics* 8, 253–262.

Jones, D., Stringham, E.V., Graham, R.W. and Candido, P.M. (1995) A portable regulatory element directs specific expression of the *Caenorhabditis elegans* ubiquitin gene *ubq-2* in the somatic gonad. *Developmental Biology* 171, 60–72.

Kanehisa, M. and Goto, S. (2000) KEGG: Kyoto Encyclopedia of Genes and Genomes. *Nucleic Acids Research* 28, 27–30.

Kikiwada, T., Nakahara, Y., Kanamori, Y., Iwata, K.I., Watanabe, M., McGee, B., Tunnacliffe, A. and Okuda, T. (2006) Dehydration-induced expression of LEA proteins in an anhydrobiotic chironomid. *Biochemical and Biophysical Research Communications* 348, 56–61.

Knight, C.A., DeVries, A.L. and Oolman, L.D. (1984) Fish antifreeze protein and the freezing and recrystallization of ice. *Nature* 308, 295–296.

Krebs, R.A. and Feder, M.E. (1997) Negative consequences of Hsp70 over-expression in *Drosophila melanogaster* larvae. *Cell Stress and Chaperones* 2, 60–71.

Kültz, D. (2005) Molecular and evolutionary basis of the cellular stress response. *Annual Review of Physiology* 67, 225–257.

Lapinski, J. and Tunnacliffe, A. (2003) Anhydrobiosis without trehalose in bdelloid rotifers. *FEBS Letters* 553, 387–390.

La Terza, A., Papa, G., Miceli, C. and Luporini, P. (2001) Divergence between two Antarctic species of the ciliate *Euplotes*, *E. focardii* and *E. nobilii*, in the expression of heat-shock protein 70 genes. *Molecular Ecology* 10, 1061–1067.

Lee, B.H., Won, S.H., Lee, H.S., Miyao, M., Chung, W., Kim, I.J. and Jo, J. (2000) Expression of the chloroplast-localized small heat shock protein by oxidative stress in rice. *Gene* 245, 283–290.

Lee, G.J. (1995) Assaying proteins for molecular chaperone activity. *Methods in Cell Biology* 50, 325–334.

Lee, G.J., Roseman, A.M., Saibil, H.R. and Vierling, E. (1997) A small heat shock protein stably binds heat-denatured model substrates and can maintain a substrate in a folding-competent state. *EMBO Journal* 16, 659–671.

Leiers, B.R., Kampkotter, A., Grevelding, C.G., Link, C.D., Johnson, T.E. and Henkle-Duhrsen, K. (2003) A stress-responsive glutathione S-transferase confers resistance to oxidative stress in *Caenorhabditis elegans. Free Radical Biology and Medicine* 34, 1405–1415.

Lindblom, T.H. and Dodd, A.K. (2006) Xenobiotic detoxification in the nematode *Caenorhabditis elegans. Journal of Experimental Zoology Part A – Ecological Genetics and Physiology* 305A, 720–729.

Liu, Q.Z. and Glazer, I. (2000) Desiccation survival of entomopathogenic nematodes of the genus *Heterorhabditis*. *Phytoparasitica* 28, 331–340.

Livak, K.J. and Schmittgen, T.D. (2001) Analysis of relative gene expression data using real-time quantitative PCR and the $2^{-\Delta\Delta CT}$ method. *Methods* 25, 402–408.

Lopez-Martinez, G., Elnitsky, M.A., Benoit, J.B., Lee, R.E. and Denlinger, D.L. (2008) High resistance to oxidative damage in the Antarctic midge *Belgica antarctica*, and developmentally linked expression of genes encoding superoxide dismutase, catalase and heat shock proteins. *Insect Biochemistry and Molecular Biology* 38, 796–804.

Madin, K.A.C., Loomis, S.H. and Crowe, J.H. (1985) Anhydrobiosis in nematodes – control of carbon flow through the glyoxylate cycle. *Journal of Experimental Zoology* 234, 341–350.

Marsh, A.G., Maxson, R.E. and Manahan, D.T. (2001) High macromolecular synthesis with low metabolic cost in Antarctic sea urchin embryos. *Science* 291, 1950–1952.

Muller-Taubenberger, A., Hagmann, J., Noegel, A. and Gerisch, G. (1988) Ubiquitin gene expression in *Dictyostelium* is induced by heat and cold shock, cadmium and inhibitors of protein synthesis. *Journal of Cell Science* 90, 51–58.

Nelson, D.L. and Cox, M.M. (2000) Structure and catalysis. In: Nelson, D.L. and Cox, M.M. (eds) *Lehninger Principles of Biochemistry*. Worth Publishers, New York, pp. 45–421.

Oliver, A.E., Hincha, D.K. and Crowe, J.H. (2002) Looking beyond sugars: the role of amphiphilic solutes in preventing adventitious reactions in anhydrobiotes at low water contents. *Comparative Biochemistry and Physiology A – Molecular and Integrative Physiology* 131, 515–525.

Pellerone, F.L., Archer, S.K., Behm, C.A., Grant, W.N., Lacey, M.J. and Somerville, A.C. (2003) Trehalose metabolism genes in *Caenorhabditis elegans* and filarial nematodes. *International Journal for Parasitology* 33, 1195–1206.

Pereira, M.D., Eleutherio, E.C.A. and Panek, A.D. (2001) Acquisition of tolerance against oxidative damage in *Saccharomyces cerevisiae*. *BMC Microbiology* 1, 11–18.

Perry, R.N. (1999) Desiccation survival of parasitic nematodes. *Parasitology* 119, S19–S30.

Pickup, J. and Rothery, P. (1991) Water-loss and anhydrobiotic survival in nematodes of Antarctic fellfields. *Oikos* 61, 379–388.

Ratnakumar, S. and Tunnacliffe, A. (2006) Intracellular trehalose is neither necessary nor sufficient for desiccation tolerance in yeast. *FEMS Yeast Research* 6, 902–913.

Rinehart, J.P., Hayward, S.A.L., Elnitsky, M.A., Sandro, L.H., Lee, R.E. and Denlinger, D.L. (2006) Continuous up-regulation of heat shock proteins in larvae, but not adults, of a polar insect. *Proceedings of the National Academy of Sciences of the USA* 103, 14223–14227.

Ring, R.A. and Danks, H.V. (1994) Desiccation and cryoprotection – overlapping adaptations. *Cryo-Letters* 15, 181–190.

Santos, H. and de Costa, M.S. (2001) Organic solutes from thermophiles and hyperthermophiles. *Methods in Enzymology* 334, 302–315.

Scharf, K.D., Siddique, M. and Vierling, E. (2001) The expanding family of *Arabidopsis thaliana* small heat stress proteins and new family of proteins containing α-crystallin domains (Acd proteins). *Cell Stress and Chaperones* 6, 225–237.

Shang, F. and Taylor, A. (1995) Oxidative stress and recovery from oxidative stress are associated with altered ubiquitin conjugating and proteolytic activities in bovine lens epithelial cells. *Biochemistry Journal* 307, 297–303.

Shang, F., Gong, X. and Taylor, A. (1997) Activity of ubiquitin dependent pathway in response to oxidative stress: ubiquitin activating enzyme (E1) is transiently up-regulated. *Journal of Biological Chemistry* 272, 23086–23093.

Shannon, A.J., Browne, J.A., Boyd, J., Fitzpatrick, D.A. and Burnell, A.M.

(2005) The anhydrobiotic potential and molecular phylogenetics of species and strains of *Panagrolaimus* (Nematoda, Panagrolaimidae). *Journal of Experimental Biology* 208, 2433–2445.

Solomon, A., Paperna, I. and Glazer, I. (1999) Desiccation survival of the entomopathogenic nematode *Steinernema feltiae*: induction of anhydrobiosis. *Nematology* 1, 61–68.

Solomon, A., Salomon, R., Paperna, I. and Glazer, I. (2000) Desiccation stress of entomopathogenic nematodes induces the accumulation of a novel heat-stable protein. *Parasitology* 121, 409–416.

Storey, K.B. and Storey, J.M. (1988) Freeze tolerance in animals. *Physiological Review* 68, 27–84.

Tang, X. and Pikal, M.J. (2005) The effects of stabilizers and denaturants on the cold denaturation temperatures of proteins and implications for freeze-drying. *Pharmacological Research* 22, 1167–1175.

Torok, Z., Goloubinoff, P. and Horvath, I. (2001) Synechocystis HSP17 is an amphitropic protein that stabilizes heat-stressed membranes and binds denatured proteins for subsequent chaperone-mediated refolding. *Proceedings of National Academy of Sciences of the USA* 98, 3098–3103.

Treonis, A.M. and Wall, D.H. (2005) Soil nematodes and desiccation survival in the extreme arid environment of the Antarctic dry valleys. *Integrative and Comparative Biology* 45, 741–750.

Tunnacliffe, A. and Lapinski, J. (2003) Resurrecting Van Leeuwenhoek's rotifers: a reappraisal of the role of disaccharides in anhydrobiosis. *Philosophical Transactions of the Royal Society of London Series B – Biological Sciences* 358, 1755–1771.

Tunnacliffe, A. and Wise, M.J. (2007) The continuing conundrum of the LEA proteins. *Naturwissenschften* 94, 791–812.

Tunnacliffe, A., Lapinski, J. and McGee, B. (2005) A putative LEA protein, but no trehalose, is present in anhydrobiotic bdelloid rotifers. *Hydrobiologia* 546, 315–321.

Tyson, T., Reardon, W., Browne, J.A. and Burnell, A.M. (2007) Gene induction by desiccation stress in the entomopathogenic nematode *Steinernema carpocapsae* reveals parallels with drought tolerance mechanisms in plants. *International Journal for Parasitology* 37, 763–776.

Van Rossum, A.J., Brophy, P.M., Tait, A., Barrett, J. and Jefferies, J.R. (2001) Proteomic identification of glutathione S-transferases from the model nematode *Caenorhabditis elegans*. *Proteomics* 1, 1463–1468.

Veinger, L., Diamant, S., Buchner, J. and Goloubinoff, P. (1998) The small heat-shock protein IbpB from *Escherichia coli* stabilizes stress-denatured proteins for subsequent refolding by a multichaperone network. *Journal of Biological Chemistry* 273, 11032–11037.

Vierling, E. (1997) Plant HSP101/ClpB. In: Gething M.J. (ed.) *Guidebook to Molecular Chaperones and Protein-folding Catalysts*. Oxford University Press, Oxford, UK, pp. 253–261.

Watanabe, M. (2006) Anhydrobiosis in invertebrates. *Applied Entomology and Zoology* 41, 15–31.

Wharton, D.A. (1986) *A Functional Biology of Nematodes*. The John Hopkins University Press, Baltimore, Maryland.

Wharton, D.A. (2002) *Life at the Limits: Organisms in Extreme Environments*. Cambridge University Press, Cambridge, UK.

Wharton, D.A. (2003) The environmental physiology of Antarctic terrestrial nematodes: a review. *Journal of Comparative Physiology B – Biochemical Systemic and Environmental Physiology* 173, 621–628.

Wharton, D.A. and Brown, I. (1991) Cold tolerance mechanisms of the Antarctic nematode *Panagrolaimus davidi*. *Journal of Experimental Biology* 155, 629–641.

Wharton, D.A., Judge, K.F. and Worland, M.R. (2000) Cold acclimation and cryoprotectants in a freeze-tolerant Antarctic nematode, *Panagrolaimus davidi*. *Journal of Comparative Physiology B – Biochemical Systemic and Environmental Physiology* 170, 321–327.

Wharton, D.A., Barrett, J., Goodall, G., Marshall, C.J. and Ramlov, H. (2005) Ice-active proteins from the Antarctic nematode

Panagrolaimus davidi. *Cryobiology* 51, 198–207.

Wise, M.J. and Tunnacliffe, A. (2004) POPP the question: what do LEA proteins do? *Trends in Plant Science* 9, 13–17.

Womersley, C. and Smith, L. (1981) Anhydrobiosis in nematodes. 1. The role of glycerol, myoinositol and trehalose during desiccation. *Comparative Biochemistry and Physiology B – Biochemistry & Molecular Biology* 70, 579–586.

Womersley, C., Wharton, D.A. and Higa, L.M. (1998) Survival biology. In: Perry R.N. and Wright, D.J. (eds) *The Physiology and Biochemistry of Free-living and Plant-parasitic Nematodes*. CAB International, Wallingford, UK, pp. 271–302.

Yancey, P.H. (2005) Organic osmolytes as compatible, metabolic and conteracting cytoprotectants in high osmolarity and other stresses. *Journal of Experimental Biology* 208, 2819–2830.

Yoder, J.A., Benoit, J.B., Denlinger, D.L. and Rivers, D.B. (2006) Stress-induced accumulation of glycerol in the flesh fly, *Sarcophaga bullata*: evidence indicating anti-desiccant cryoprotectant functions of this polyol and a role for the brain in coordinating the response. *Journal of Insect Physiology* 52, 202–214.

10 Thermobiotic Survival

Eileen Devaney

Parasitology Group, Veterinary Infection and Immunity, Institute of Comparative Medicine, University of Glasgow, Glasgow, UK

10.1 Introduction

As is clear from the preceding chapters in this volume, nematodes are extremely successful animals, with free-living species invading almost all known ecological niches, while other species are parasites of humans, animals or plants. In part, this success relates to the ability of nematodes to adapt to changing environmental conditions. In this chapter I focus on the molecular mechanisms by which nematodes withstand thermal stress and discuss the integration of the stress response with developmental pathways. In all organisms, tolerance of elevated temperature is mediated by the heat shock response, an evolutionary conserved mechanism that promotes cell survival under changing environmental conditions. Our understanding of the heat shock response in nematodes comes largely from genetic and molecular studies using the free-living model nematode *Caenorhabditis elegans*. However, it is probable that much of that information can be applied to understanding how parasitic species adapt to thermal stress. Temperature affects all aspects of the biology of *C. elegans*, and the ability to integrate temperature and development, while important for

 Molecular and Physiological Basis of Nematode Survival (eds R.N. Perry and D.A. Wharton)

the survival of free-living nematodes, may be key for the adaptation to parasitism. The life cycle of many animal-parasitic species involves a switch from ambient temperature to mammalian body temperature and the resulting heat shock may be one signal for the resumption of development.

10.2 Temperature Regulates Development in Nematodes

The life cycle of *C. elegans* and that of many parasitic nematodes is known to be temperature dependent, i.e. the speed of development increases with increasing temperature over a particular range, known as the temperature niche breadth (Anderson and Coleman, 1982). *Caenorhabditis elegans* wild-type N2 worms develop 2.1 times faster at 25°C than at 16°C and 1.3 times faster at 20°C than at 16°C (Stiernagle, 2006). However, deleterious effects are observed at extremes of temperature; for example, both the lifespan and fecundity of *C. elegans* N2 are significantly decreased at very high or low temperatures (Klass, 1977). The reduction in lifetime fecundity at elevated temperature may relate to the production of decreased numbers of viable sperm (Harvey and Viney, 2007). The optimal temperature for development varies from species to species within a genus and depends upon the normal growth habitat of the organism. For example, some species of *Caenorhabditis*, such as those found in close association with birds or mammals, thrive at much higher temperatures than does *C. elegans*. One such species, *Caenorhabditis bovis*, was isolated from the ears of zebu cattle, an environment rich in the bacterial food source of *Caenorhabditis* spp. This ability to tolerate elevated temperatures has been suggested to predispose *Caenorhabditis* spp. to a pseudo-parasitic lifestyle (Kiontke and Sudhaus, 2006). It is interesting to note that *C. bovis* may be deposited in the bovine ear by a fly carrying dauer stages, a life cycle with parallels to that of some parasitic nematodes.

Caenorhabditis elegans has the ability to 'remember' the temperature at which it was cultivated, so that when worms cultured at a particular temperature are placed on a temperature gradient, they migrate back to their original growth temperature. This migratory behaviour is known as thermotaxis (Hedgecock and Russell, 1975; Mori and Ohshima, 1995). In *C. elegans*, temperature also modulates the neuroendocrine signalling pathway that controls the decision between reproductive development and dauer formation (Ailion and Thomas, 2000) (see Grant and Viney, Chapter 5, this volume, for a full discussion of the dauer larva). The choice of developmental pathway is made at the first larval (L1) stage, in response to environmental factors such as population density and food availability, and is modulated by temperature. Entry into and exit from the dauer pathway is controlled by neurones situated in the amphids at the head of the worm. These sense population density by detecting a pheromone, a mixture of so-called ascarosides, which are modified derivatives of the dideoxysugar ascarylose (reviewed in Edison, 2009). Temperatures greater than 20°C promote entry of N2 wild-type *C. elegans* into the dauer pathway as long as worms are exposed to elevated temperatures prior to the first larval moult (Golden and Riddle, 1984). Likewise, recovery from the dauer stage is influenced by a temperature downshift, with a reduction in temperature amplifying the number of worms recovering in the presence of a food signal.

The parasitic nematode *Strongyloides ratti* provides an excellent 'half-way house' between nematodes that are free-living and those that are parasitic. *Strongyloides ratti* can undergo one of two alternative life cycles in response to differing environmental conditions. The homogonic life cycle gives rise to infective third-stage larvae (L3) that invade the host to develop into parasitic females, while in the heterogonic cycle, free-living adult worms are produced (Viney, 1996). Developmental switching in *S. ratti* is also influenced by temperature, with elevated temperature favouring the heterogonic pathway (Harvey *et al.*, 2000). Although the molecular mechanisms by which larvae respond to temperature are not defined for *S. ratti*, by analogy with other *Strongyloides* spp. it is likely that thermosensitive neurones play a role in developmental switching (see below).

10.3 How Does *Caenorhabditis elegans* Sense Temperature?

The observation that temperature regulates the speed of development and lifespan in *C. elegans* indicates that the worm has a mechanism for detecting and responding to changes in ambient temperature. This is indeed the case: laser ablation of specific neurones together with the availability of genetic mutants has allowed the neural circuits and downstream signalling pathways involved in thermal sensing to be defined (reviewed in Mori *et al.*, 2007; Kuhara *et al.*, 2008). The bilateral AFD neurones positioned in the amphids of *C. elegans* are the major thermosensory neurones and can detect temperature differences as small as 0.1°C or less. Laser ablation of these cells deletes the thermotactic response (Mori and Ohshima, 1995). Studies in which the levels of intracellular calcium were monitored under different conditions suggested that the AFD responds to increases in temperature rather than to warm temperatures per se (Kimura *et al.*, 2004). The thermosensory response in AFD is mediated by the opening and closing of a cGMP-gated ion channel, which bears a notable similarity to the signalling mechanisms involved in vertebrate photoreception (Ramot *et al.*, 2008). The AFD neurones are wired to the AIY and AIZ interneurones, which signal to the RIA interneurone (Fig. 10.1a). RIA integrates the input from AIY (thermophilic) and AIZ (cryophilic) and connects to the motor neurones that control worm movement, thus affecting behaviour (Mori and Ohshima, 1995; Mori *et al.*, 2007). In addition, the AWC neurone, the major role of which is in olfaction, also acts in thermal sensing (Kuhara *et al.*, 2008).

As well as laser ablation, mutations in genes such as *ttx-3* (abnormal thermotaxis-3), which encodes a LIM homeodomain protein essential for the function of the AIY interneurone, have been used to dissect the thermosensory pathway (Hobert *et al.*, 1997; Ramot *et al.*, 2008). *ttx-3* mutants are defective in thermotaxis and show a similar phenotype to animals in which AIY has been killed. Using mutants defective in AFD or AIY function, Prahlad *et al.* (2008) investigated the relationship between temperature sensing and the heat shock response. In *ttx-3* mutants or in *gcy-8* mutants (*gcy-8* encodes a guanylyl

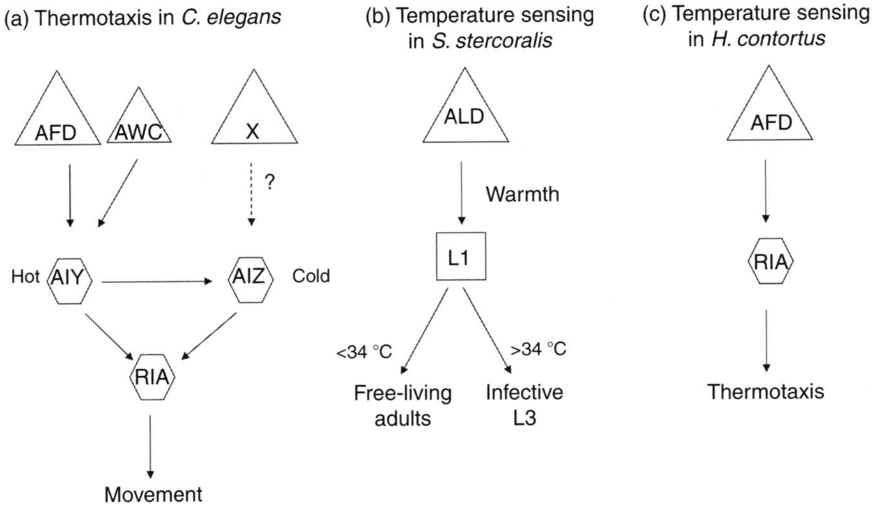

Fig. 10.1. Thermal sensing in *Caenorhabditis elegans*, *Strongyloides stercoralis* and *Haemonchus contortus*. (a) *C. elegans* pathway adapted from Kuhara *et al.* (2008). In *C. elegans* the major thermosensory neurone pair is AFD, although recent data suggest that the olfactory neurone AWC also responds to changes in temperature (Kuhara *et al.*, 2008). AFD and AWC synapse with the AIY interneurone to signal movement towards higher temperatures, while the AIZ interneurone mediates the response to cold. RIA integrates the input from AIY and AIZ and signals to motor neurones in the body wall muscle to regulate movement. (b) The major temperature-sensitive neurone in the amphids of *S. stercoralis* L1 is the ALD pair (Lopez *et al.*, 2000). These may be equivalent to AFD in *C. elegans* but also share some similarity with AWC of *C. elegans*. Ablation of ALD in the L1 disrupts thermotaxis behaviour in the L3. The major neurones involved in developmental switching in *S. stercoralis* are ASF and ASI, but ALD controls the temperature-sensitive choice of developmental pathway (Nolan *et al.*, 2004). (c) The L3 of *H. contortus* migrate to the temperature at which they were grown, in a similar manner to that described for *C. elegans* thermotaxis. Ablation of the AFD pair in the L1 stage of the parasite ablates the thermotaxis response, as does killing of the proposed RIA-analogous interneurone (Li *et al.*, 2000).

cyclase, which regulates thermotaxis in the AFD neurone), the accumulation of heat shock proteins (Hsps) was defective and the mutant worms showed lower rates of survival than did N2 worms following heat shock. Most interestingly, a decrease in *hsp70* expression was apparent in the somatic cells of *gcy-8* mutant worms following heat shock, an observation that was mimicked in worms in which heat shock factor (*hsf-1*) had been silenced by RNA interference (RNAi). As the AFD neurones do not directly innervate any of the tissues in which the heat shock response was observed, these observations provide further evidence of the involvement of a neuroendocrine pathway that coordinates temperature sensing and the induction of the stress response in *C. elegans*. Lee and Kenyon (2009) identified a steroid signalling pathway by which the thermosensory neurones influence lifespan in *C. elegans*. They examined the interaction between elevated temperature and reduced lifespan and showed that in animals in which AFD was killed or in various thermosensory mutant worms, lifespan was further compromised at 25°C but not at 15°C. Sensing of elevated temperature by the AFD neurone in wild-type animals induced the expression of *daf-9*, a cytochrome P450 homologue. DAF-9 is required for the biosynthesis of dafachronic acids, the ligands for DAF-12, a nuclear hormone receptor that regulates development and longevity in *C. elegans* (Antebi *et al.*, 2000). DAF-12 signalling appears to be conserved in other nematodes (Ogawa *et al.*, 2009), including some parasitic species (Wang *et al.*, 2009).

10.4 Temperature Sensing in Parasitic Nematodes

While thermotaxis may contribute to the sensing of nutrient-rich organic matter by *C. elegans*, for parasitic nematodes the ability to move along a thermal gradient is likely to be an important aspect of host finding. For example, the behaviour of the dog hookworm, *Ancylostoma caninum*, changes in response to warming and other stimuli associated with the presence of a mammalian host (Granzer and Haas, 1991). The elegant studies of Schad and co-workers used a similar laser ablation approach to that described above for *C. elegans* to identify neurones involved in thermosensing in parasitic nematodes (reviewed in Ashton *et al.*, 1999). Ablation of the presumed AFD-analogous neurone in *A. caninum* L1 resulted in L3 that had a significantly reduced thermotactic response (Bhopale *et al.*, 2001). Similar laser ablation studies have been carried out in *Strongyloides stercoralis*; like *S. ratti* described above, *S. stercoralis* can also undergo an alternative developmental pathway, the choice of which is partially influenced by temperature. However, in contrast to *S. ratti*, culturing the L1 stage of *S. sterocoralis* at temperatures of 34°C and above resulted in the majority of larvae developing to infective L3, while at lower temperatures (15–31°C) the majority of the larvae developed to free-living adult worms, a developmental choice made during the L1 stage (Nolan *et al.*, 2004). The ALD neurone pair is the major thermosensitive neurone in *S. stercoralis* larvae, equivalent to the AFD neurones of *C. elegans*, and ablation of these neurones reduced the thermotactic response (Lopez *et al.*, 2000). In addition, killing the ALD neurones in the L1 stage resulted in a reduction in the numbers of worms developing to infective L3 (Fig. 10.1b), demonstrating the link between temperature sensing and developmental choice in *S. stercoralis* (Nolan *et al.*, 2004). Temperature sensing has also been studied in the L3 of *Haemonchus contortus* (Li *et al.*, 2000), a parasite that infects its host by passive ingestion, in contrast to *Ancylostoma* and *Strongyloides* spp., both of which can migrate through the skin. *Haemonchus contortus* L3 were shown to display a similar thermotaxis response to *C. elegans*, with larvae migrating on a thermal gradient to the temperature at which they had been cultivated (Fig. 10.1c). Laser ablation of the AFD neurone or the likely RIA interneurone in L1 larvae destroyed the thermotaxis response in the L3 stage (Li *et al.*, 2000).

10.5 Heat Shock Factor – the Master Regulator of the Heat Shock Response

When exposed to elevated temperatures, most organisms respond with a heat shock response and nematodes are no exception. Indeed many parasitic nematodes undergo heat shock as part of their normal life cycle during infection of a mammalian host. The heat shock response is an ancient mechanism that permits the maintenance of homeostasis in cells when exposed to conditions that compromise protein integrity. Ritossa (1962) first described the heat shock response in *Drosophila* spp. upon the exposure of flies to elevated temperatures, which apparently occurred when the incubator temperature was altered in error (Ritossa, 1996). The large polytene chromosomes of fly larvae were observed to 'puff' in response to heat shock, the puffs being loci of active gene transcription. Despite the terminology, the heat shock response is also induced by a variety of conditions that do not involve elevated temperature but which constitute cellular stress or insult; these include oxidative stress, metabolic inhibition, exposure to toxins and various pathophysiological states (reviewed in Morimoto, 1998; see Barrett, Chapter 12, this volume). The common feature here is the accumulation of misfolded or unfolded proteins, elevated levels of which are responsible for the induction of the heat shock response.

During a heat shock response, normal protein synthesis is repressed in favour of expression of a complex of proteins known collectively as heat shock proteins (Hsps). These protect the cell by chaperoning misfolded or denatured proteins and targeting them for refolding or degradation. The heat shock response has to be initiated rapidly but tightly controlled to allow the cellular environment to return to homeostasis. This is accomplished through the activity of the heat shock transcription factor, HSF-1. *Caenorhabditis elegans* expresses only one HSF, HSF-1 (Walker *et al.*, 2003), equivalent to the major stress-related HSF of vertebrates, and heat shock of *C. elegans* N2 worms results in the transcription of a full range of Hsps (Snutch and Baillie, 1983). The molecular mechanisms regulating the activation of HSF-1 are extremely complex in the intracellular environment and still not completely understood. The transcriptional competence of HSF-1 is negatively regulated through both intra- and inter-molecular interactions. In most eukaryotes the molecular architecture of HSF-1 is similar, with conservation of the DNA-binding domain, an oligomerization domain (HR-A/B) that is involved in the stress-induced trimerization of HSF-1 and a region of hydrophobic heptad repeats (HR-C) at the carboxy terminus, which functions in the suppression of trimerization. Comparison of the sequence of *C. elegans* HSF-1 with other eukaryotic heat shock factors reveals a high degree of conservation over the DNA-binding domain. The HR-A/B domain is also conserved, although the spacing and the sequence of the heptad repeats are not well conserved. By contrast, the amino acid residues in the region corresponding to HR-C are poorly conserved in *C. elegans*, although this region contains an equivalent number of hydrophobic residues to that of higher eukaryotes. In response to stress, HSF-1 monomers trimerize and bind to heat shock response elements (HSEs) (Perisic *et al.*, 1989), inverted repeats of nGAAn located in the upstream region of Hsp genes. A minimum of three repeats of the 5 bp unit are required for stable binding of HSF-1, but longer arrays of the pentamer result in much enhanced binding of HSF-1, particularly at heat shock temperatures (Xiao *et al.*, 1991). Many Hsp genes contain several HSEs, presumably facilitating the binding of multiple trimers of HSF-1.

Under non-stress conditions, HSF-1 is maintained in a transcriptionally inert monomeric state (reviewed in Pirkkala *et al.*, 2001) in the cytoplasm and/or nucleus. The localization of HSF-1 under non-stress conditions has been the subject of debate, with some studies showing a cytoplasmic and others a nuclear location. However, recent studies in living *Drosophila* cells have shown that HSF-1 is localized to the nucleoplasm under non-stress conditions and rapidly translocates to chromosomal loci following heat shock (Yao *et al.*, 2006). In *C. elegans*, staining of freeze-cracked worms with an affinity-purified antibody to HSF-1 supported a nuclear localization for HSF-1, with a redistribution within the nucleus following heat shock (Devaney, unpublished data). The *C. elegans* genome contains a large series of tandem repeats of the HSF-1 binding motif, which, although not transcriptionally active *in situ*, can drive transcription in a heterologous system following heat shock (La Volpe *et al.*, 1988). Searching the *C. elegans* genome for sequences containing five tandem repeats composed of a 10 bp unit (nGAAnnTTCn) demonstrated that clusters of this sequence are distributed over the six chromosomes, with chromosome IV showing the highest number of repeats, although the X chromosome shows a particularly dense area of HSE repeats (Emes and Devaney unpublished data; see Fig. 10.2 for a graphical representation). Whether these may represent sites at which HSF-1 is sequestered in the absence of heat shock has yet to be determined.

RNA polymerase II is paused on heat shock promoters (Rasmussen and Lis, 1993) and the recruitment of HSF-1 to the promoter releases the polymerase, resulting in the rapid transcription of Hsp genes. The transcriptional competence of HSF-1 is further enhanced by a range of post-translational modifications, including phosphorylation on specific serine residues (Holmberg *et al.*, 2001). While the heat shock response must be rapidly activated, its termination is equally important. Here studies in *C. elegans* have proved to be illuminating. Yeast two-hybrid studies identified a protein, heat shock binding protein 1 (HSBP-1), which binds to the heptad repeats (HR-A/B) in the trimerization domain of HSF-1. Over-expression of HSBP-1 in *C. elegans* blocked the heat shock response, resulting in reduced thermotolerance, while loss of HSBP-1 resulted in increased thermotolerance (Satyal *et al.*, 1998), suggesting that HSBP-1 functions as a negative regulator of HSF-1.

In vitro studies have implied that HSF-1 may have the ability to sense temperature, oxidative stress and pH directly. Exposure of recombinant *Drosophila* HSF-1 monomers to elevated temperatures resulted in trimer formation, as assessed by size exclusion chromatography and acquisition of DNA binding (Zhong *et al.*, 1998, 1999). However, other studies have shown that the temperature at which HSF-1 becomes activated is a feature of the cellular environment in which it is expressed. For example, expression of human HSF-1 in *Drosophila* cells results in activation at *Drosophila* heat shock

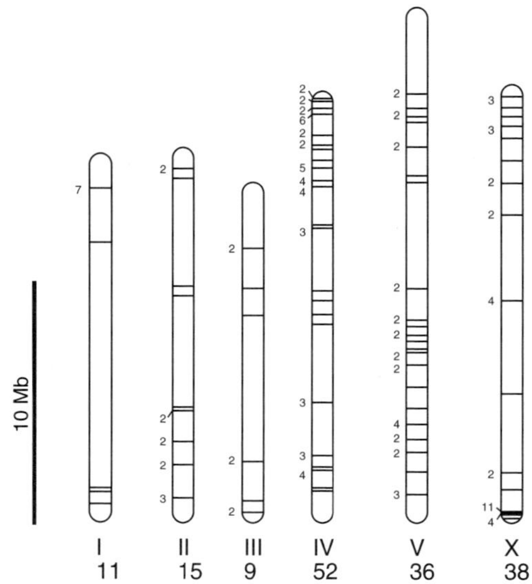

Fig. 10.2. The *Caenorhabditis elegans* genome contains multiple copies of the heat shock response element (HSE) sequence. The *C. elegans* genome was searched for repeats of the HSE sequence nGAAnnTTCn. The diagram maps the location of sequences containing five tandem repeats of the 10 bp unit. Each horizontal bar on the chromosome represents a single five tandem repeat; numbers next to the horizontal bars show where the number of repeats is >1. The total number of five tandem repeats of the 10 bp HSE is shown under each chromosome. Analysis and diagram provided by Dr R. Emes.

temperatures some 10°C less than normal (Clos *et al.*, 1993), suggesting that cell autonomous factors are important in regulating the activity of HSF-1.

10.6 Integration of the Stress Response and Developmental Pathways

The first evidence that HSF-1 might have additional functions related to development came from studies on *Drosophila melanogaster*. Using antibody labelling, it was shown that HSF-1 bound at multiple puff sites on the polytene chromosomes in addition to known *hsp* loci. Binding was observed at loci that were normally repressed during heat shock, suggesting that HSF-1 may act as a negative regulator of gene expression during heat shock and thus inhibit developmental programmes until homeostasis is restored (Westwood *et al.*, 1991). Direct evidence of a role for HSF-1 in development came from additional studies on *Drosophila*, where analysis of mutants suggested two separate roles for HSF-1 in development, both of which were independent of *hsp* induction (Jedlicka *et al.*, 1997). In that study, all four *Drosophila hsf-1* mutants arrested development at the first or second larval instar, while analysis of germline mosaics demonstrated a further role for *hsf-1* in oogenesis. In addition, mice deficient in *hsf-1* survive to adulthood and show normal basal Hsp levels but display a variety of developmental phenotypes, including growth retardation and female infertility, demonstrating that HSF-1 has additional roles in mammals as well as in invertebrates (Xiao *et al.*, 1999).

In *C. elegans*, *hsf-1* is essential both for thermotolerance and for normal development of the animal, as defined by RNAi (Walker *et al.*, 2003) and mutagenesis studies (Hajdu-Cronin *et al.*, 2004). An *hsf-1* mutant (*sy441*) carries a single point mutation in the seventh exon of *hsf-1*, resulting in a stop codon and truncation of the final 85 amino acids (Hajdu-Cronin *et al.*, 2004). These animals do not express *hsp16.2* following heat shock, have defects in egg laying and arrest at L2–L3 when grown at 25°C. RNAi studies have confirmed the complexity of *hsf-1* function and demonstrated a link between the stress response and the insulin-like signalling (ILS) pathway, which is an important regulator of the dauer pathway and ageing (Ogg *et al.*, 1997; Lin *et al.*, 2001). Reduction of levels of *hsf-1* by RNAi results in arrest of many worms at the L2/L3 stage, suggesting that HSF-1 is required for progression beyond these stages. Those worms that do develop to adulthood are small and scrawny and show significant reductions in thermotolerance and fertility (Walker *et al.*, 2003). Most notably, *hsf-1(RNAi)* worms have a significantly reduced lifespan (Garigan *et al.*, 2002; Walker *et al.*, 2003), while, conversely, over-expression of *hsf-1* is reported to extend lifespan (Hsu *et al.*, 2003), with the most significant effect observed when *hsf-1* is expressed in body wall muscle or in neurones (Morley and Morimoto, 2004). HSF-1 also has a role in the dauer pathway, which, given the sensitivity of the dauer pathway to elevated temperature and other stresses, is not unexpected. However, these studies have provided additional evidence of the cross-talk between the stress response and ILS pathways. *hsf-1(RNAi)* carried out on a variety of *C. elegans* mutants that are dauer-constitutive (daf-c) at 25°C (i.e. form dauer larvae under favourable conditions) resulted in a significant reduction in worms entering the dauer pathway, an effect that was particularly pronounced in mutants in the ILS pathway (Walker *et al.*, 2003; Morley and Morimoto, 2004).

It has long been known that exposure to a short burst of elevated temperature protects the cell from more extreme temperatures. This so-called 'induced thermotolerance' results in the synthesis of Hsps, and studies in *C. elegans* have demonstrated the role of ILS in this process (Lithgow *et al.*, 1995). DAF-16, the downstream target of the ILS pathway, is a forkhead transcription factor that functions in dauer formation and affects lifespan and reproduction. Ligation of DAF-2, the insulin-like receptor, initiates a signalling cascade resulting in the phosphorylation of DAF-16 and its sequestration in the cytoplasm, where it is inactive. When ILS is disrupted, or in response to environmental stresses such as heat, oxidative stress or starvation, DAF-16 translocates to the nucleus, where it acquires DNA binding activity (Henderson and Johnson, 2001; Lee *et al.*, 2001). Over-expression of *daf-16* results in worms that are long-lived and stress-resistant (Henderson and Johnson, 2001), as does over-expression of the small Hsp, *hsp16.2* (Walker and Lithgow, 2003). Indeed, *daf-16* is essential for the lifespan extension conferred by over-expression of *hsp16.2* (Walker and Lithgow, 2003). Subsequent studies showed that the *hsp16* genes contained DAF-16 binding sites in the promoter region and thus are likely to be regulated by DAF-16 as well as HSF-1 (Hsu *et al.*, 2003). Recent studies have described *daf-16* orthologues from the parasitic nematodes *S. stercoralis* (Massey *et al.*, 2006) and *H. contortus* (Hu *et al.*, 2010) and shown that a single isoform of *daf-16* from both parasites can complement a *C. elegans daf-16* mutant. Thus, aspects of the ILS pathway are conserved in parasitic species.

In *hsf-1(RNAi)* animals, the gut is blocked by proliferating bacteria (Garigan *et al.*, 2002), a phenomenon that contributes to the accelerated ageing phenotype. This observation suggested that *hsf-1* might function in a pathway that acts to restrict bacterial proliferation, and a number of recent studies have shown that *hsf-1* and ILS are required for innate immunity to bacterial infection in *C. elegans*. Heat-shocked *C. elegans* display higher levels of resistance to infection with *Pseudomonas aeruginosa* (Singh and Aballay 2006), an effect dependent upon HSF-1. By knocking down individual Hsps, a role for Hsp90 and Hsp16 in immunity to infection was demonstrated. Long-lived ILS mutants showed an increased resistance to infection that was dependent on *hsf-1*, and animals over-expressing *hsf-1* were more resistant to infection than wild-type worms (Singh and Aballay, 2006). In a different pathogen system, infection with *Enterococcus faecalis* was shown to cause protein aggregation (a sign of damage) in the intestine, an effect that could be overcome by the expression of *hsf-1* in the cells of the intestine (Mohri-Shiomi and Garsin, 2008). Here the protective effects of HSF-1 were, at least in part, mediated by the expression of a range of Hsps including Hsp90 and a number of small Hsps (sHsps). Interestingly, the role of Hsp90 in resistance of plants to various pathogens is well documented (reviewed in Shirasu, 2009). Here, Hsp90 collaborates with other proteins such as SGT-1 to regulate the signalling pathway that results in the expression of plant resistance genes. The resistance of tomato plants to infection with the root-knot nematode *Meloidogyne javanica* was shown to require an intact Hsp90 signalling pathway (Bhattarai *et al.*, 2007).

10.7 Heat Shock Protein Families

The major role of HSF-1 is in the induction of Hsps. Hsps are classified into families on the basis of their approximate molecular mass on SDS-PAGE. Under normal conditions, Hsps act as molecular chaperones during protein synthesis or degradation. Protein structure is compromised as a consequence of elevated temperature and other cellular insults and Hsps are an important mechanism by which unfolded/misfolded proteins are targeted and removed from the cell via the ubiquination and proteasome pathways.

Like other eukaryotes, nematodes express a full range of Hsps, most of which are rather poorly characterized in parasitic species. Many Hsps are highly antigenic in the mammalian host, a feature that probably relates to their function as chaperones in the immune system, where they play an important role in antigen presentation (Tsan and Gao, 2009). Consequently, several parasitic nematode Hsps were first identified as major antigens, such as Hsp70 from the filarial nematodes *Brugia malayi* (Selkirk *et al.*, 1989), *Onchocerca volvulus* (Rothstein *et al.*, 1989) and *Wuchereria bancrofti* (Ravi *et al.*, 2004). A notable feature of Hsp genes in parasitic nematodes is that many are expressed under normal conditions of growth and show only a minimal induction in response to elevated temperature (Devaney, 2006), emphasizing the fact that Hsps have important roles under non-stress conditions. However, given the lack of a consistent system for RNAi in parasitic nematodes (Knox *et al.*, 2007), the precise function of most Hsps is unclear, except by analogy with known functions in *C. elegans* and other eukaryotes. A full profile of Hsps has been reported for *C. elegans* (Snutch and Baillie, 1983; see Table 10.1 for a summary of Hsp families in *C. elegans*), but information on parasitic nematodes relates mainly to individual Hsps. In *C. elegans*, the most consistently heat-inducible protein is Hsp70, with a range of other proteins observed, including those migrating on SDS-polyacrylamide gels at approximately 81, 41, 38, 29, 19, 18 and 16kDa. All developmental stages of *C. elegans* displayed a classic heat shock response (Snutch and Baillie, 1983). Here we briefly summarize current knowledge of the major Hsp families (Hsp90, Hsp70 and the sHsps) in nematodes.

10.7.1 Hsp90

Hsp90 (often referred to as Hsp83) is a relatively well-characterized Hsp in nematodes. It is encoded by a single gene, *daf-21*, in *C. elegans*. In early studies in *C. elegans*, *daf-21* (*hsp90*) mRNA was shown to be 10- to 15-fold enriched in dauer larvae compared to other life cycle stages. Levels of *daf-21* mRNA decreased rapidly upon stimulation to emerge from the dauer stage (Dalley and Golomb, 1992). Subsequent studies using a variety

Table 10.1. The major heat shock protein (Hsp) families in *Caenorhabditis elegans*.

Hsp family	Gene name	Localization	Function
Hsp90	*daf-21* (C47E8.5)	Cytoplasm	Development, dauer, lifespan
Hsp70	Multi-gene family, e.g. *hsp-70* (C12C8.1)	Cytoplasm, ER, mitochondria	Stress response, lifespan
	hsp-4 (F43E2.8)	ER (BiP homologue), gut, hypodermis	ER UPR, lifespan
	hsp-1 (F26D10.3)	Most tissues	Stress response, lifespan
Hsp60	*hsp-60* (Y22D7AL.5)	Mitochondria	Mitochondrial UPR
Hsp40	Multi-gene family, e.g. *dnj-19* (T05C3.5)	Probable cytoplasm, ER	May interact with Hsp70 as molecular chaperone
sHsps	Multi-gene family, e.g. *hsp-16-1* (T27E4.8)	Many tissues, some tissue-specific	Stress response, lifespan, dauer
	hsp-25 (C09B8.6)	Muscle, gonad	Molecular chaperone, not strongly heat-inducible
	hsp-12.6 (F38E11.2)	Predominantly L1, dauer, most tissues	Not stress-inducible, lifespan, dauer, developmental arrest

Many Hsps in *C. elegans* belong to large multi-gene families. Only examples are given in the table, with localization and function where known. See text for additional detail. The full descriptions are available on wormbase (www.wormbase.org). ER, endoplasmic reticulum; BIP, binding immunoglobulin protein; UPR, unfolded protein response.

of differential screening methods have broadly confirmed these results and shown that *daf-21* is upregulated in dauer larvae or in long-lived mutants in the ILS pathway (Cherkasova *et al.*, 2000; McElwee *et al.*, 2003). Conversely, knock down of *daf-21* by RNAi results in a small, but significant, decrease in lifespan (Morley and Morimoto, 2004). In mammalian cells, Hsp90 chaperones a specific subset of client proteins that tend to be unstable and regulatory in function, such as kinases, transcription factors and steroid receptors (see list at http://www.picard.ch). A large number of likely client proteins of DAF-21 have been identified in yeast two-hybrid screens or predicted from genetic interactions (see http://www.wormbase.org/db/gene/interaction?list= WBGene00000915) in *C. elegans* and many of these are proteins with regulatory function. Two well-characterized *daf-21* mutants are available (Birnby *et al.*, 2000). The first of these carries a point mutation in *daf-21* (E292K) that results in a daf-c phenotype at 25°C. This phenotype is suppressed in the presence of a cGMP analogue, suggesting that DAF-21 might be required to regulate proteins involved in cGMP signalling. The second mutant is a loss-of-function mutation resulting in worms that fail to progress beyond the L2/L3 stage. *daf-21(RNAi)* has identified additional phenotypes, including sterility, a blocked gut phenotype and embryonic lethality (Gillan *et al.*, 2009). Thus, *daf-21* is clearly an essential gene in *C. elegans*, presumably because of the nature of its client proteins. In our studies, Hsp90 was not highly heat inducible in either *Brugia* or *C. elegans* (Thompson *et al.*, 2001; Devaney *et al.*, 2005) and knock-down of HSF-1 in *C. elegans* had minimal effect upon Hsp90 levels, in contrast to Hsp16, the level of which was significantly decreased in heat-shocked *hsf-1(RNAi)* worms (Walker *et al.*, 2003). However, other studies using different approaches have reported increased expression of *hsp90* following heat shock in *C. elegans* (Inoue *et al.*, 2003).

Specific inhibitors, such as the well-characterized geldanamycin (GA), have been invaluable for defining the cellular functions of Hsp90, although not all nematode Hsp90s bind GA (David *et al.*, 2003; Him *et al.*, 2009). Exposure of *Brugia pahangi* to nanomolar concentrations of GA is lethal to microfilariae and to adult worms (Devaney *et al.*, 2005). Hsp90 is reportedly secreted by adult female *B. malayi in vitro* (Kumari *et al.*, 1994), although its extracellular function is not defined. Interestingly, some tumour cells express an extracellullar form of Hsp90, which is required for the activity of a matrix metalloproteinase that functions in metastasis (Eustace *et al.*, 2004). To investigate whether *Brugia hsp90* was functionally conserved with *Ce-daf-21*, we attempted to rescue the developmental arrest phenotype of a *C. elegans daf-21* mutant. This was not successful, although a partial rescue was obtained using *hsp90* from the clade V nematode *H. contortus* (Gillan *et al.*, 2009). *Brugia* and *C. elegans* are known to differ in their affinity for GA (Him *et al.*, 2009) and this feature may reflect differences in Hsp90 function/client proteins in the two species, which renders complementation impossible. *hsp90* has been cloned from a variety of other parasitic nematodes (Him *et al.*, 2009) and has been characterized in more detail from the plant-parasitic nematode *Meloidogyne* (Skantar and Carta, 2004; De Luca *et al.*, 2009) and from *Trichinella spiralis* (Martinez *et al.*, 1999).

10.7.2 The small heat shock protein family

The sHsps are also relatively well characterized in nematodes. These range in size from approximately 12 to 43 kDa, depending upon species. sHsps are characterized by the presence of a C-terminal α-crystallin domain and have the ability to form large oligomers *in vitro*. In the original analysis of the *hsp16* family in *C. elegans*, four very similar genes were identified that were organized in two clusters, one of which was much more heat inducible than the other (Candido *et al.*, 1989; Dixon *et al.*, 1990). However, with the availability of the full *C. elegans* genome, the sHsps are now known to constitute a large multi-gene family with 18 members (Aevermann and Waters, 2008), eight of which code for the Hsp16 family. The heat sensitivity of *hsp16* in *C. elegans* has been exploited as an inducible system to drive expression of transgenes in the worm (Stringham *et al.*, 1992). By placing green fluorescent protein under the control of the *hsp16* promoter, a transgenic line of *C. elegans* was generated, which has proved to be very useful for studying factors that induce the stress response in live animals (Link *et al.*, 1999).

Mild heat shock is known to increase *C. elegans* lifespan (Lithgow *et al.*, 1995), and in various long-lived mutants, Hsps are upregulated. Overexpression of *hsp16.2* resulted in increased longevity and resistance to a variety of stresses, an effect that was dependent upon an intact DAF-16 signalling pathway (Walker and Lithgow, 2003). In keeping with these findings, genes of the *hsp16* family were shown to be very highly upregulated in long-lived stress-resistant *daf-2* mutants (Halaschek-Wiener *et al.*, 2005). These findings presumably relate to the activation of DAF-16 in *daf-2* mutants and the induction of the sHsps by DAF-16 (Hsu *et al.*, 2003). In a differential screen of genes expressed in dauer larvae versus mixed stages of *C. elegans*, the most abundant transcript identified in the dauer stage was another sHsp, *hsp12.6* (Jones *et al.*, 2001). It was hypothesized that the accumulation of this mRNA in the dauer larva may be related to developmental arrest, as *hsp12.6* is also abundant in the L1 of *C. elegans* arrested by starvation following hatching. Interestingly, Hsp12.6 protein was reported to be expressed only by the L1 stage of *C. elegans* and was not induced by heat or oxidative stress (Leroux *et al.*, 1997), although dauer larvae were not analysed in that study. In other eukaryotes, sHsps have been shown to accumulate in cell-cycle-arrested cells (Pauli *et al.*, 1990) and are also abundantly expressed in the microfilariae of *B. pahangi*, which is essentially an arrested L1 stage (Devaney *et al.*, 1992). As described previously, the *hsp16* genes in *C. elegans* are induced by the forkhead transcription factor, DAF-16, as well as HSF-1. This dual regulation of sHsps is not without precedent. In *Drosophila*, ecdysterone, as well as heat shock, induces the transcription of the sHsps (Ireland *et al.*, 1982). Analysis of the *Drosophila hsp27* promoter identified separate ecdysone response and heat shock elements (Riddihough and Pelham, 1986). A variety of other sHsps have been characterized from *C. elegans*, but the precise function of most of these is unknown. Many show tissue-specific expression patterns and some have been shown to act as molecular chaperones *in vitro* (e.g. Ding and Candido, 2000).

A large number of genes belonging to the sHsp family have been cloned from parasitic nematodes including *Brugia* spp. (Thompson *et al.*, 1996; Raghavan *et al.*, 1999), *Dirofilaria immitis* (Lillibridge *et al.*, 1996), *Nippostrongylus brasiliensis* (Tweedie *et al.*, 1993), *H. contortus* (Hartman *et al.*, 2003), *Ostertagia ostertagi* (Vercauteren *et al.*, 2006), *T. spiralis* (Wu *et al.*, 2007) and *S. ratti* (Tazir *et al.*, 2009). Although these are all referred to as sHsps, they differ significantly in the level of sequence similarity and in their temporal expression profiles. As in *C. elegans*, they are frequently members of multi-gene families (Thompson *et al.*, 1996). In our studies on the sHsps of *B. pahangi*, we observed constitutive expression of the sHsp mRNA and protein only in microfilariae cultured at 37°C, although expression could be induced in adult worms following heat shock at 41°C (Devaney *et al.*, 1992; Thompson *et al.*, 1996). By contrast, the *B. malayi* sHsp cloned by Raghavan *et al.* (1999) was expressed only in fourth-stage larvae (L4) and adult worms and could be further induced by exposure of adult worms to 42°C. The L3 of filarial nematodes undergo a heat shock as part of their normal development upon transmission from mosquito to mammalian host, and a range of Hsps are expressed coincident with that transition. The sHsps were the most dominant group of heat-inducible proteins in *B. pahangi* L3 (Jecock and Devaney, 1992) and were similarly inducible in the L3 of *B. malayi* when shifted from 28°C to 37°C or 42°C (Raghavan *et al.*, 1999). The precise function of sHsps in parasitic nematodes remains unknown. A small Hsp, Hsp12.6, cloned from *B. malayi* was shown to bind interleukin (IL)-10 and may have IL-10-like activity *in vitro* (Gnanasekar *et al.*, 2008), whilst the *S. ratti* sHsp was detected in secretory products of the parasite and thus has the potential to interact with the host (Tazir *et al.*, 2009).

In plants, a sHsp gene, *Hahsp17.7G4*, was shown to be expressed in giant cells, specialized feeding cells that are induced by infection with root-knot nematodes. Studies in transgenic tobacco plants identified a minimal promoter element of only 83 bp, which contained HSEs and was sufficient to drive expression of the *Hahsp17.7G4* gene. Mutation of the HSE sequences in the promoter reduced the nematode-driven expression of *Hahsp17.7G4* (Escobar *et al.*, 2003). Whether this sHsp is expressed in giant cells as part of a generalized stress response induced by nematode infection has yet to be determined.

10.7.3 Hsp70

The *hsp70* genes constitute a multi-gene family in *C. elegans* (Heschl and Baillie, 1989), with different temporal and spatial expression patterns described for some family members. Some *hsp70* genes are heat inducible while others are constitutively expressed. More recently, 13 members of the *hsp70* family have been identified in the *C. elegans* genome (ten *hsp70* and three related *hsp110* genes) (Nikolaidis and Nei, 2004). Hsp70 members are important mediators of the unfolded protein response. For example, in *C. elegans*, *hsp4* is highly inducible in the endoplasmic reticulum, while *hsp6* is expressed in the mitochondria during the unfolded protein response (Yoneda *et al.*, 2004). Over-expression of *hsp6* (otherwise known as *mot-2*) resulted in a significant extension in lifespan in *C. elegans* (Yokoyama *et al.*, 2002).

Hsp70 has been characterized from a range of parasitic nematodes, including *B. malayi* (Selkirk *et al.*, 1989) and *Parastrongyloides trichosuri* (Newton-Howes *et al.*, 2006). Both parasitic Hsp70s were initially identified on the basis of their immunogenicity with serum from their respective hosts. In both *Brugia* and *P. trichosuri*, the *hsp70* mRNA appeared to be constitutively expressed in all life cycle stages examined and was further inducible by heat shock. Comparison of *Pt-hsp70* with the *hsp70* gene family in *C. elegans* suggested that *Pt-hsp70* was most similar to *hsp1* of *C. elegans*, an *hsp70* family member that is expressed in all life cycle stages and mildly upregulated upon heat shock. Analysis of the *Pt-hsp70* promoter sequence in *C. elegans* demonstrated a similar expression pattern to *Ce-hsp1*, with both genes being expressed in the cells of the intestine (Newton-Howes *et al.*, 2006). However, the precise function of the Hsp70 gene family in parasitic nematodes remains unclear.

10.8 Conclusions and Future Directions

The molecular mechanisms by which cells respond to temperature changes are increasingly well understood. The role of HSF-1 in the induction of Hsps and the function of Hsps in protecting cells from thermal stress are well defined. While a heat shock response can be induced in isolated cells in culture, studies in *C. elegans* have shown that the stress response is coordinated by a neuroendocrine signalling pathway. For metazoans such as worms, a mechanism that results in the coordination of the stress response in different tissues may be essential to ensure an orchestrated response to stress, as discussed by Prahlad and Morimoto (2009). Although HSF-1 is best described as an inducer of Hsps, studies analysing HSF-1 binding sites have thrown up some surprising results. For example, one study used chromatin immunoprecipitation in a genome-wide analysis of human HSF-1 binding sites under heat shock and non-heat shock conditions. An enrichment of G-protein-coupled olfactory receptor genes was observed under non-heat shock conditions (Page *et al.*, 2006). Although these genes were shown to contain HSEs, it remains unclear whether they are actually regulated by HSF-1. Given the role of neuroendocrine signalling in the stress response in *C. elegans*, it is tempting to speculate that such HSF-1 targets may reflect that evolutionary history. In *C. elegans*, a bioinformatics analysis identified HSEs in a range of genes other than conventional Hsps that are upregulated in response to heat shock (GuhaThakurta *et al.*, 2002). Thus, defining the full range of HSF-1 targets and determining whether HSF-1 acts to repress or induce transcription at these sites will be informative in terms of mapping the complete network of genes regulated by HSF-1 under normal and stress conditions.

Many nematodes are important parasites of animals and plants. Parasitism has arisen on multiple occasions during the evolution of nematodes (Blaxter *et al.*, 1998), with parasitic species adopting a diversity of life cycles. As demonstrated in this chapter, free-living nematodes, such as *C. elegans*, are able to respond and adapt to a range of environmental stresses, including alterations in ambient temperature. The ability to integrate development with the stress response makes good evolutionary sense: when conditions are unfavourable for reproduction, *C. elegans* and other free-living nematodes enter a dauer pathway and, indeed, in the environment *C. elegans* is usually found as a dauer stage (Kiontke and Sudhaus, 2006). The ability to acclimatize to temperature differences may be a key factor in the adaptation to parasitism, at least for those nematodes that utilize mammalian hosts. Many parasites have free-living larval stages in the environment, which, at least in temperate climates, results in exposure to significant increases in temperature upon infection (e.g. from <20 to 37°C). The heat shock response presumably underpins the ability to tolerate such extreme temperature shifts.

The infective L3 of those animal-parasitic nematodes found in the environment are often considered to be equivalent to the stress-resistant dauer stage (Hotez *et al.*, 1993). In parallel with entry into the dauer pathway, temperature also influences the choice of developmental pathway in parasitic species such as *Strongyloides* spp. The similarity between the sensory structures of *C. elegans* and some parasitic nematodes indicates that mechanisms for sensing environmental changes, such as temperature fluctuations, may be conserved. For parasitic nematodes that switch between a mammalian host and the environment, temperature fluctuation may be one factor that allows the parasite to monitor its changing location and implement the appropriate gene expression profile (Hunter *et al.*, 2001). In *C. elegans*, a neuroendocrine pathway is important in transmitting the signals from thermosensory neurones to other cells in the animal, and it is possible that components of this pathway are conserved in parasitic nematodes. With the availability of genome sequence from parasitic nematodes (e.g. the animal-parasitic nematodes *Brugia* and *Haemonchus* and the plant-parasitic nematodes *Meloidogyne incognita* and *Meloidogyne hapla*), it is possible to identify orthologues of *C. elegans* genes, such as *daf-12*, *daf-16* and *hsf-1*, which play important roles in integrating development and the stress response. Indeed, aspects of the neuroendocrine signalling pathways that control development in *C. elegans* are conserved in some parasitic nematodes, as demonstrated by the complementation of *C. elegans* mutants with orthologous genes from parasitic nematodes (e.g. *daf-16*). Recent studies have provided evidence that additional aspects of the dauer pathway are conserved in parasitic nematodes: e.g. DAF-12 from *Ancylostoma* spp. and from *S. stercoralis* can be activated by dafachronic acids from *C. elegans*, resulting in the re-initiation of development in the L3 stage of the parasite (Wang *et al.*, 2009).

For the many parasitic nematodes whose life cycles involve a switch between environmental temperatures and a mammalian host, the infection event presents additional challenges. Not only must the parasite adapt to a completely novel environment but it must simultaneously confront the complexities of the host immune system and also reinitiate the developmental cycle. While it seems likely that these processes will be linked, much remains to be learned about the mechanisms by which parasitic nematodes integrate environmental and intrinsic signals to regulate development. However, dissecting these pathways may be particularly fruitful as these transitions in the life cycle could represent points of vulnerability at which parasites are susceptible to attack. The wealth of information available on the basic biology of *C. elegans* is already informing studies on parasitic nematodes, with the promise of novel therapeutics to combat these infections (see for example Wang *et al.*, 2009).

10.9 Acknowledgements

I would like to thank MRC, Wellcome Trust and BBSRC for funding, Dr Collette Britton (University of Glasgow) and Professor Mark Viney (University of Bristol) for their helpful comments on the manuscript, Dr R. Emes (University of Nottingham) for permission to use unpublished data, and past and present staff and students of the lab for their input into the work.

10.10 References

Aevermann, B.D. and Waters, E.R. (2008) A comparative genomic analysis of the small heat shock proteins in *Caenorhabditis elegans* and *briggsae*. *Genetica* 133, 307–319.

Ailion, M. and Thomas, J.H. (2000) Dauer formation induced by high temperatures in *Caenorhabditis elegans*. *Genetics* 156, 1047–1067.

Anderson, R.V. and Coleman, D.C. (1982) Nematode temperature responses: a niche dimension in populations of bacterial-feeding nematodes. *Journal of Nematology* 14, 69–76.

Antebi, A., Yeh, W.H., Tait, D., Hedgecock, E.M. and Riddle, D.L. (2000) *daf-12* encodes a nuclear receptor that regulates the dauer diapause and developmental age in *C. elegans*. *Genes & Development* 14, 1512–1527.

Ashton, F.T., Li, J. and Schad, G.A. (1999) Chemo- and thermosensory neurons: structure and function in animal parasitic nematodes. *Veterinary Parasitology* 84, 297–316.

Bhattarai, K.K., Li, Q., Liu, Y., Dinesh-Kumar, S.P. and Kaloshian, I. (2007) The *Mi-1*-mediated pest resistance requires *Hsp90* and *Sgt1*. *Plant Physiology* 144, 312–323.

Bhopale, V.M., Kupprion, E.K., Ashton, F.T., Boston, R. and Schad, G.A. (2001) *Ancylostoma caninium*: the finger cell neurons mediate thermotactic behavior by infective larvae of the dog hookworm. *Experimental Parasitology* 97, 70–76.

Birnby, D.A., Link, E.M., Vowels, J.J., Tian, H., Colacurcio, P.L. and Thomas, J.H. (2000) A transmembrane guanylyl cyclase (DAF-11) and Hsp90 (DAF-21) regulate a common set of chemosensory behaviors in *Caenorhabditis elegans*. *Genetics* 155, 85–104.

Blaxter, M.L., De Ley, P., Garey, J.R. *et al.* (1998) A molecular evolutionary framework for the phylum Nematoda. *Nature* 392, 71–75.

Candido, E.P., Jones, D., Dixon, D.K., Graham, R.W., Russnak, R.H. and

Kay, R.J. (1989) Structure, organization, and expression of the 16-kDa heat shock gene family of *Caenorhabditis elegans*. *Genome* 31, 690–697.

Cherkasova, V., Ayyadevara, S., Egilmez, N. and Reis, R.M. (2000) Diverse *Caenorhabditis elegans* genes that are upregulated in dauer larvae also show elevated transcript levels in long-lived, aged, or starved adults. *Journal of Molecular Biology* 300, 433–448.

Clos, J., Rabindran, S., Wisniewski, J. and Wu, C. (1993) Induction temperature of human heat shock factor is reprogrammed in a *Drosophila* cell environment. *Nature* 364, 252–255.

Dalley, B.K. and Golomb, M. (1992) Gene expression in the *Caenorhabditis elegans* dauer larva: developmental regulation of Hsp90 and other genes. *Developmental Biology* 151, 80–90.

David, C.L., Smith, H.E., Raynes, D.A., Pulcini, E.J. and Whitesell, L. (2003) Expression of a unique drug-resistant Hsp90 ortholog by the nematode *Caenorhabditis elegans*. *Cell Stress & Chaperones* 8, 93–104.

De Luca, F., Di Vito, M., Fanelli, E., Reyes, A., Greco, N. and De Giorgi, C. (2009) Characterization of the heat shock protein 90 gene in the plant parasitic nematode *Meloidogyne artiellia* and its expression as related to different developmental stages and temperature. *Gene* 440, 16–22.

Devaney, E. (2006) Thermoregulation in the life cycle of nematodes. *International Journal for Parasitology* 36, 641–649.

Devaney, E., Egan, A., Lewis, E., Warbrick, E.V. and Jecock, R.M. (1992) The expression of small heat shock proteins in the microfilaria of *Brugia pahangi* and their possible role in development. *Molecular and Biochemical Parasitology* 56, 209–218.

Devaney, E., O'Neill, K.M., Whitesell, L. and Kinnaird, J. (2005) Hsp90 is essential in the filarial nematode *Brugia*. *International Journal for Parasitology* 35, 625–636.

Ding, L. and Candido, E.P. (2000) HSP25, a small heat shock protein associated with dense bodies and M-lines of body wall muscle in *Caenorhabditis elegans*. *Journal of Biological Chemistry* 275, 9510–9517

Dixon, D.K., Jones, D. and Candido, E.P. (1990) The differentially expressed 16-kD heat shock genes of *Caenorhabditis elegans* exhibit differential changes in chromatin structure during heat shock. *DNA and Cell Biology* 9, 177–191.

Edison, A.S. (2009) *Caenorhabditis elegans* pheromones regulate multiple complex behaviours. *Current Opinion in Neurobiology* 19, 378–388.

Escobar, C., Barcala, M., Portillo, M., Almoguera, C., Jordano, J. and Fenoll, C. (2003) Induction of the *Hahsp17.7G4* promoter by root-knot nematodes: involvement of heat shock elements in promoter activity in giant cells. *Molecular Plant Microbe Interactions* 16, 1062–1068.

Eustace, B.K., Sakurai, T., Stewart, J.K. *et al.* (2004) Functional proteomic screens reveal an essential extracellular role for hsp90α in cancer cell invasiveness. *Nature Cell Biology* 6, 507–514

Garigan, D., Hsu, A.L., Fraser, A.G., Kamath, R.S., Ahringer, J. and Kenyon, C. (2002) Genetic analysis of tissue aging in *Caenorhabditis elegans*: a role for heat-shock factor and bacterial proliferation. *Genetics* 161, 1101–1112

Gillan, V., Maitland, K., McCormack, G., Nik Him, N.A.I.I. and Devaney, E. (2009) Functional genomics of *hsp-90* in parasitic and free-living nematodes. *International Journal for Parasitology* 39, 1071–1081.

Gnanasekar, M., Anandharaman, V., Anand, S.B., Nutman, T.B. and Ramaswamy, K. (2008) A novel heat shock protein 12.6 (HSP12.6) from *Brugia malayi* functions as a human IL-10 receptor binding protein. *Molecular and Biochemical Parasitology* 159, 98–103.

Golden, J.W. and Riddle, D.L. (1984) A pheromone-induced developmental switch in *Caenorhabditis elegans*: temperature-sensitive mutants reveal a wild-type temperature-dependent process. *Proceedings of the National Academy of Sciences of the USA* 81, 819–823.

Granzer, M. and Haas, W. (1991) Host-finding and host recognition of infective *Ancylostoma caninum* larvae. *International Journal for Parasitology* 21, 429–440.

GuhaThakurta, D., Palomar, L., Stormo, G.D., Tedesco, P., Johnson, T.E., Walker, D.W., Lithgow, G., Kim, S. and Link, C.D. (2002) Identification of a novel *cis*-regulatory element involved in the heat shock response in *Caenorhabditis elegans* using microarry gene expression and computational methods. *Genome Research* 12, 701–712.

Hajdu-Cronin, Y.M., Chen, W.J. and Sternberg, P.W. (2004) The L-type cyclin CYL-1 and the heat shock factor HSF-1 are required for heat-shock-induced protein expression in *Caenorhabditis elegans*. *Genetics* 168, 1937–1949.

Halaschek-Wiener, J., Khattra, J.S., McKay, S., Pouzyrev, A., Stott, J.M., Yang, G.S., Holt, R.A., Jones, S.J., Marra, M.A., Brooks-Wilson, A.R. and Riddle, D.L. (2005) Analysis of long-lived *C. elegans daf-2* mutants using serial analysis of gene expression. *Genome Research* 15, 603–615.

Hartman, D., Cottee, P.A., Savin, K.W., Bhave, M., Presidente, P.J., Fulton, L., Walkiewicz, M. and Newton, S.E. (2003) *Haemonchus contortus*: molecular characterisation of a small heat shock protein. *Experimental Parasitology* 104, 96–103.

Harvey, S.C. and Viney, M.E. (2007) Thermal variation reveals natural variation between isolates of *Caenorhabditis elegans*. *Journal of Experimental Zoology* 308B, 409–416.

Harvey, S.C., Gemmill, A.W., Read, A.F. and Viney, M.E. (2000) The control of morph development in the parasitic nematode *Strongyloides ratti*. *Proceedings of the Royal Society of London* 267, 2057–2063.

Hedgecock, E.M. and Russell, R.L (1975) Normal and mutant thermotaxis in the nematode *Caenorhabditis elegans*. *Proceedings of the National Academy of Sciences of the USA* 72, 4061–4065.

Henderson, S.T. and Johnson, T.E. (2001) *daf-16* integrates developmental and environmental inputs to mediate aging in the nematode *Caenorhabditis elegans*. *Current Biology* 11, 1975–1980.

Heschl, M.F. and Baillie, D.L. (1989) Characterization of the hsp70 multigene family of *Caenorhabditis elegans*. *DNA* 8, 233–243.

Him, N.A.I.I.N., Gillan, V., Emes, R.D., Maitland, K. and Devaney, E. (2009) Hsp-90 and the biology of nematodes. *BMC Evolutionary Biology* 9, 254.

Hobert, O., Mori, I., Yamashita, Y., Honda, H., Ohshima, Y., Liu, Y. and Ruvkun, G. (1997) Regulation of interneuron function in the *C. elegans* thermoregulatory pathway by the *ttx-3* LIM homeobox gene. *Neuron* 19, 345–357.

Holmberg, C.I., Hietakangas, V., Mikhailov, A., Rantanen, J.O., Kallio, M., Meinander, A., Hellma, J., Morrice, N., MacKintosh, C., Morimoto, R.I., Eriksson, J.E. and Sistonen, L. (2001) Phosphorylation of serine 230 promotes inducible transcriptional activity of heat shock factor 1. *EMBO Journal* 20, 3800–3810.

Hotez, P., Hawdon, J. and Schad, G.A. (1993) Hookworm larval infectivity, arrest and amphiparatenesis: the *Caenorhabditis elegans* Daf-c paradigm. *Parastiology Today*, 9, 23–26.

Hsu, A-L, Murphy, C.T. and Kenyon, C. (2003) Regulation of aging and age-related disease by DAF-16 and heat-shock factor. *Science* 300, 1142–1145.

Hu, M., Lok, J.B., Ranjit, N., Massey, H.C. Jr, Sternberg, P.W. and Gasser, R.B. (2010) Structural and functional characterisation of the fork head transcription factor-encoding gene, *Hc-daf-16*, from the parasitic nematode *Haemonchus contortus* (Strongylida). *International Journal for Parasitology* 40, 405–415.

Hunter, S.J., Thompson, F.J., Tetley, L. and Devaney, E. (2001) Temperature is a cue for gene expression in the L3 of the filarial nematode *Brugia pahangi*. *Molecular and Biochemical Parasitology* 112, 1–9.

Inoue, T., Takamura, K., Yamae, H., Ise, N., Kawakami, M., Tabuse, Y., Miwa, J. and Yamaguchi, Y. (2003) *Caenorhabditis elegans* DAF-21 (HSP90) is characteristically and predominantly expressed in germline cells: spatial and temporal analysis. *Development, Growth and Differentiation* 45, 369–376.

Ireland, R.C., Berger, E., Sirotkin, K., Yund, M.A., Osterbur, D. and Fristrom, J. (1982) Ecdysterone induces the transcription of four heat-shock genes in *Drosophila* S3 cells and imaginal discs. *Developmental Biology* 93, 498–507.

Jecock, R.M. and Devaney, E. (1992) Expression of small heat shock proteins by the third-stage larva of *Brugia pahangi*. *Molecular and Biochemical Parasitology* 56, 219–226.

Jedlicka, P., Mortin, M.A. and Wu, C. (1997) Multiple functions of *Drosophila* heat shock transcription factor *in vivo*. *EMBO Journal* 16, 2452–2462.

Jones, S.J., Riddle, D.L., Pouzyrev, A.T. *et al.* (2001) Changes in gene expression associated with developmental arrest and longevity in *Caenorhabditis elegans*. *Genome Research* 11, 1346–1352.

Kimura, K.D., Miyawaki, A., Matsumoto, K. and Mori, I. (2004) The *C. elegans* thermosensory neuron AFD responds to warming. *Current Biology* 14, 1291–1295.

Kiontke, K. and Sudhaus, W. (2006) Ecology of *Caenorhabditis* species. In: The *C. elegans* Research Community (ed.) *WormBook*, doi/10.1895/wormbook.1.37.1. http://www.wormbook.org.

Klass, M.R. (1977) Aging in the nematode *Caenorhabditis elegans*: major biological and environmental factors influencing life span. *Mechanics of Ageing and Development* 6, 413–429.

Knox, D.P., Geldhof, P., Visser, A. and Britton, C. (2007) RNA interference in parasitic nematodes of animals: a reality check? *Trends in Parasitology* 23, 105–107.

Kuhara, A., Okumura, M., Kimata, T., Tanizawa, Y., Takano, R., Kimura, K.D., Inada, H., Matsumoto, K. and Mori, I. (2008) Temperature sensing by an olfactory neuron in a circuit controlling behavior of *C. elegans*. *Science* 320, 803–807.

Kumari, S., Lillibridge, C.D., Bakeer, M., Lowrie, R.C., Jayaraman, K. and Philipp, M.T. (1994) *Brugia malayi*: the diagnostic potential of recombinant excretory/secretory antigens. *Experimental Parasitology* 79, 489–505.

La Volpe, A., Ciaramella, M. and Bazzicalupo, P. (1988) Structure, evolution and properties of a novel repetitive DNA family in *Caenorhabditis elegans*. *Nucleic Acid Research* 16, 8213–8231.

Lee, R.Y., Hench, J. and Ruvkun, G. (2001) Regulation of *C. elegans* DAF-16 and its human ortholog FKHRL1 by the *daf-2* insulin-like signaling pathway. *Current Biology* 11, 1950–1957.

Lee, S.-J. and Kenyon, C. (2009) Regulation of the longevity response to temperature by thermosensory neurons in *Caenorhabditis elegans*. *Current Biology* 19, 715–722.

Leroux, M.R., Ma, B.J., Batelier, G., Melki, R. and Candido, E.P. (1997) Unique structural features of a novel class of small heat shock proteins. *Journal of Biological Chemistry* 272, 12847–12853.

Li, J., Zhu, X., Boston, R., Ashton, F.T., Gamble, H.R. and Schad, G.A. (2000) Thermotaxis and thermosensory neurons in infective larvae of *Haemonchus contortus*, a passively ingested nematode parasite. *Journal of Comparative Neurology* 424, 58–73.

Lillibridge, C.D., Rudin, W. and Philipp, M.T. (1996) *Dirofilaria immitis*: ultrastructural localization, molecular characterization, and analysis of the expression of p27, a small heat shock protein homolog of nematodes. *Experimental Parasitology* 83, 30–45.

Lin, K., Hsin, H., Libina, N. and Kenyon C. (2001) Regulation of the *Caenorhabditis elegans* longevity protein DAF-16 by insulin/IGF-1 and germline signaling. *Nature Genetics* 28, 139–145.

Link, C.D., Cypser, J.R., Johnson, C.J. and Johnson, T.E. (1999) Direct observation of stress response in *Caenorhabditis elegans* using a reporter transgene. *Cell Stress & Chaperones* 4, 235–242.

Lithgow, G.J., White, T.M., Melov, S. and Johnson, T.E. (1995) Thermotolerance and extended life-span conferred by single-gene mutations and induced by thermal stress. *Proceedings of the National*

Academy of Sciences of the USA 92, 7540–7544.

Lopez, P.M., Boston, R., Ashton, F.T. and Schad, G.A. (2000) The neurons of class ALD mediate thermotaxis in the parasitic nematode *Strongyloides stercoralis*. *International Journal for Parasitology* 30, 1115–1121.

Martinez, J., Perez-Serrano, J., Bernadina, W.E. and Rodriguez-Caabeiro, F. (1999) In vitro stress response to elevated temperature, hydrogen peroxide and mebendazole in *Trichinella spiralis* muscle larvae. *International Journal for Parasitology* 29, 1457–1464.

Massey, H.C., Bhopale, M.K., Li, X., Castelletto, M. and Lok, J.B. (2006) The fork-head transcription factor FKTF-1b from *Strongyloides stercoralis* restores DAF-16 developmental function to mutant *Caenorhabditis elegans*. *International Journal for Parasitology* 36, 347–352.

McElwee, J., Bubb, K. and Thomas, J.H. (2003) Transcriptional outputs of the *Caenorhabditis elegans* forkhead protein DAF-16. *Aging Cell* 2, 111–121.

Mohri-Shiomi, A. and Garsin, D.A. (2008) Insulin signalling and the heat shock response modulate protein homeostasis in the *Caenorhabditis elegans* intestine during infection. *Journal of Biological Chemistry* 283, 194–201.

Mori, I. and Ohshima, Y. (1995) Neural regulation of thermotaxis in *Caenorhabditis elegans*. *Nature* 376, 344–348.

Mori, I., Sasakura, H. and Kuhara, A. (2007) Worm thermotaxis: a model system for analyzing thermosensation and neural plasticity. *Current Opinion in Neurobiology* 17, 712–719.

Morimoto, R.I. (1998) Regulation of the heat shock transcriptional response: cross talk between a family of heat shock factors, molecular chaperones, and negative regulators. *Genes & Development* 12, 3788–3796.

Morley, J.F. and Morimoto, R.I. (2004) Regulation of longevity in *C. elegans* by heat shock factor and molecular chaperones. *Molecular Biology of the Cell* 15, 657–664.

Newton-Howes, J., Heath, D.D., Shoemaker, C.B. and Grant, W.N. (2006) Characterisation and expression of an Hsp-70 gene from *Parastrongyloides trichosuri*. *International Journal for Parasitology* 36, 467–474.

Nikolaidis, N. and Nei, M. (2004) Concerted and nonconcerted evolution of the Hsp70 gene superfamily in two sibling species of nematodes. *Molecular Biology and Evolution* 21, 498–505.

Nolan, T.J., Brenes, M., Ashton, F.T., Zhu, X., Forbes, W.M., Boston, R. and Schad, G.A. (2004) The amphidial neuron pair ALD controls the temperature-sensitive choice of alternative developmental pathways in the parasitic nematode *Strongyloides stercoralis*. *Parasitology* 129, 753–759.

Ogawa, A., Streit, A., Antebi, A. and Sommer, R.J. (2009) A conserved endocrine mechanism controls the formation of dauer and infective larvae in nematodes. *Current Biology* 19, 67–71.

Ogg, S., Paradis, S., Gottlieb, S., Patterson, G.I., Lee, L., Tissenbaum, H.A. and Ruvkun, G. (1997) The fork head transcription factor DAF-16 transduces insulin-like metabolic and longevity signals in *C. elegans*. *Nature* 389, 994–999.

Page, T.J., Sikder, D., Yang, L., Pluta, L., Wolfinger, R.D., Kodadek, T. and Thomas, R.S. (2006) Genome-wide analysis of human HSF1 signaling reveals a transcriptional program linked to cellular adaptation and survival. *Molecular Biosystems* 2, 627–639.

Pauli, D., Tonka, C.H., Tissires, A. and Arrigo, A.P. (1990) Tissue-specific expression of the heat shock protein HSP27 during *Drosophila melanogaster* development. *Journal of Cell Biology* 111, 817–828.

Perisic, O., Xiao, H. and Lis, J.T. (1989) Stable binding of *Drosophila* heat shock factor to head-to-head and tail-to-tail repeats of a conserved 5bp recognition unit. *Cell* 59, 797–806.

Pirkkala, L., Nykanen, P. and Sistonen, L. (2001) Roles of the heat shock transcription factors in regulation of the heat shock response and beyond. *FASEB Journal* 15, 1118–1131.

Prahlad, V. and Morimoto, R.I. (2009) Integrating the stress response: lessons for neurodegenerative disease from *C. elegans*. *Trends in Cell Biology* 19, 52–61.

Prahlad, V., Cornelius, T. and Morimoto, R.I. (2008) Regulation of the cellular heat shock response in *Caenorhabditis elegans* by thermosensory neurons. *Science* 320, 811–814.

Raghavan, N., Ghosh, I., Eisinger, W.S., Pastrana, D. and Scott, A.L. (1999) Developmentally regulated expression of a unique small heat shock protein in *Brugia malayi*. *Molecular and Biochemical Parasitology* 104, 233–246.

Ramot, D., MacInnis, B.L. and Goodman, M.B. (2008) Bidirectional temperature sensing by a single thermosensory neuron in *C. elegans*. *Nature Neuroscience* 11, 908–915.

Rasmussen, E.B. and Lis, J.T. (1993) In vivo transcriptional pausing and cap formation on three *Drosophila* heat shock genes. *Proceedings of the National Academy of Sciences of the USA* 90, 7923–7927.

Ravi, V., Kubofcik, J., Bandopathyaya, S., Geetha, M., Narayanan, R.B., Nutman, T.B. and Kaliraj, P. (2004) *Wuchereria bancrofti*: cloning and characterisation of heat shock protein 70 from the human lymphatic filarial parasite. *Experimental Parasitology* 106, 1–10.

Riddihough, G. and Pelham, H.R. (1986) Activation of the *Drosophila hsp27* promoter by heat shock and by ecdysone involves independent and remote regulatory sequences. *EMBO Journal* 5, 1653–1658.

Ritossa, F. (1962) A new puffing pattern induced by temperature shock and DNP in *Drosophila*. *Experientia* 18, 571–573

Ritossa, F. (1996) Discovery of the heat shock response. *Cell Stress & Chaperones* 1, 97– 98.

Rothstein, N.M., Higashi, G., Yates, J. and Rajan, T.V. (1989) *Onchocerca volvulus* heat shock protein 70 is a major immunogen in amicrofilaremic individuals from a filariasis-endemic area. *Molecular and Biochemical Parasitology* 33, 229–235.

Satyal, S.H., Chen D., Fox, S.G., Kramer, J.M. and Morimoto, R.I. (1998) Negative regulation of the heat shock transcriptional response by HSBP1. *Genes & Development* 12, 1962–1974.

Selkirk, M.E., Denham, D.A., Partono, F. and Maizels, R.M. (1989) Heat shock cognate 70 is a prominent immunogen in Brugian filariasis. *Journal of Immunology* 143, 299–308.

Shirasu, K. (2009) The Hsp90–SGT1 chaperone complex for NLR immune sensors. *Annual Review of Plant Biology* 60, 139–164.

Singh, V. and Aballay, A (2006) Heat-shock transcription factor (HSF)-1 pathway required for *Caenorhabditis elegans* immunity. *Proceedings of the National Academy of Sciences of the USA* 103, 13092–13097.

Skantar, A.M. and Carta, L.K. (2004) Molecular characterization and phylogenetic evaluation of the Hsp90 gene from selected nematodes. *Journal of Nematology* 36, 466–480.

Snutch, T.P. and Baillie, D.L. (1983) Alterations in the pattern of gene expression following heat shock in the nematode *Caenorhabditis elegans*. *Canadian Journal of Biochemistry and Cell Biology* 61, 480–487.

Stiernagle, T. (2006) Maintenance of *C. elegans*. In: The *C. elegans* Research Community (ed.) *WormBook*, doi/10.1895/wormbook.1.101.1, http://www.wormbook.org.

Stringham, E.G., Dixon, D.K., Jones, D. and Candido, E.P. (1992) Temporal and spatial expression patterns of the small heat shock (*hsp16*) genes in transgenic *Caenorhabditis elegans*. *Molecular Biology of the Cell* 3, 221–233.

Tazir, Y., Steisslinger, V., Soblik, H., Younis, A.E., Beckmann, S., Grevelding, C.G., Steen, H., Brattig, N.W. and Erttmann, K.D. (2009) Molecular and functional characterisation of the heat shock protein 10 of *Strongyloides ratti*. *Molecular and Biochemical Parasitology* 168, 149–157.

Thompson, F.J., Martin, S.A.M. and Devaney, E. (1996) *Brugia pahangi*:

characterisation of a small heat shock protein cDNA clone. *Experimental Parasitology* 83, 259–266.

Thompson, F.J., Cockroft, A.C., Wheatley, I., Britton, C. and Devaney, E. (2001) Heat shock and developmental expression of *hsp83* in the filarial nematode *Brugia pahangi*. *European Journal of Biochemistry* 268, 5808–5815.

Tsan, M.F. and Gao, B. (2009) Heat shock proteins and immune system. *Journal of Leukocyte Biology* 85, 905–910.

Tweedie, S., Grigg, M.E., Ingram, L. and Selkirk, M.E. (1993) The expression of a small heat shock protein homologue is developmentally regulated in *Nippostrongylus brasiliensis*. *Molecular and Biochemical Parasitology* 61, 149–153.

Vercauteren, I., De Maere, V., Vercruysse, J., Stevens, M., Gevaert, K. and Claerebout, E. (2006) A small heat shock protein of *Ostertagia ostertagi*: stage-specific expression, heat inducibility, and protection trial. *Journal of Parasitology* 92, 1244–1250.

Viney, M.E. (1996) Developmental switching in the parasitic nematode *Strongyloides ratti*. *Proceedings of the Royal Society of London B* 263, 201–208.

Walker, G.A. and Lithgow, G.J. (2003) Lifespan extension in *C. elegans* by a molecular chaperone dependent upon insulin-like signals. *Aging Cell* 2, 131–139.

Walker, G.A., Thompson, F.J., Brawley, A., Scanlon, T. and Devaney, E. (2003) Heat shock factor functions at the convergence of the stress response and developmental pathways in *Caenorhabditis elegans*. *FASEB Journal* 17, 1960–1962.

Wang, Z., Zhou, X.E., Motola, D.L. *et al.* (2009) Identification of the nuclear receptor DAF-12 as a therapeutic target in parasitic nematodes. *Proceedings of the National Academy of Sciences of the USA* 106, 9138–9143.

Westwood, J.T., Clos, J. and Wu, C. (1991) Stress-induced oligomerization and chromosomal relocalization of heat-shock factor. *Nature* 353, 822–827.

Wu, Z., Nagano, I., Boonmars, T. and Takahashi, Y. (2007) Thermally induced and developmentally regulated expression of a small heat shock protein in *Trichinella spiralis*. *Parasitology Research* 101, 201–212.

Xiao, H., Perisic, O. and Lis, J.T. (1991) Cooperative binding of *Drosophila* heat shock factor to arrays of a conserved 5 bp unit. *Cell* 64, 585–593.

Xiao, X., Zuo, X., Davis, A.A., McMillan, D.R., Curry, B.B., Richardson, J.A. and Benjamin, I.J. (1999) HSF1 is required for extra-embryonic development, postnatal growth and protection during inflammatory responses in mice. *EMBO Journal* 18, 5943–5952.

Yao, J., Munson, K.M., Webb, W.W. and Lis, J.T. (2006) Dynamics of heat shock factor association with native gene loci in living cells. *Nature* 442, 1050–1053.

Yokoyama, K., Fukumoto, K., Murakami, T., Harada, S., Hosono, R., Wadhwa, R., Mitsui, Y. and Ohkuma, S. (2002) Extended longevity of *Caenorhabditis elegans* by knocking in extra copies of hsp70F, a homolog of mot-2 (mortalin)/mthsp70/GRP75. *FEBS Letters* 516, 53–57.

Yoneda, T., Benedetti, C., Urano, F., Clark, S.G., Harding, H.P. and Ron, D. (2004) Compartment-specific perturbation of protein handling activates genes encoding mitochondrial chaperones. *Journal of Cell Science* 117, 4055–4066.

Zhong, M.G., Orosz, A. and Wu, C. (1998) Direct sensing of heat and oxidation by *Drosophila* heat shock transcription factor. *Molecular Cell* 2, 101–108.

Zhong, M., Kim, S.-J. and Wu, C. (1999) Sensitivity of *Drosophila* heat shock transcription factor to low pH. *Journal of Biological Chemistry* 274, 3135–3140.

11 Osmotic and Ionic Regulation

DAVID A. WHARTON[1] AND ROLAND N. PERRY[2]

[1]*Department of Zoology, University of Otago, Dunedin, New Zealand;*
[2]*Rothamsted Research, Harpenden, UK and Biology Department, Ghent University, Ghent, Belgium*

11.1 Introduction

Nematodes inhabit a wide variety of habitats, both free-living and parasitic, but they are essentially aquatic in nature, requiring at least a film of water for activity and reproduction. Water in nature has salts and other solutes dissolved in it and the nematode will have to regulate its internal ionic and osmotic composition if it maintains its internal solutes at different concentrations to those in the surrounding medium. The degree to which this is necessary depends upon the habitat of the nematode. Marine nematodes are thought to be isosmotic to the surrounding seawater (Wright, 2004) and do not have to face much variation in the osmotic and ionic composition of their surroundings. Nematodes living in fresh waters (lakes, rivers, etc.) are exposed to a medium that is hyposmotic to their internal fluids but often relatively stable in composition (Wright, 2004). Estuarine species are exposed to a habitat that varies in osmotic and ionic composition as the relative contribution from marine and fresh waters varies during the day (Forster, 1998).

For nematodes living in soil, or in marine or freshwater sediments that are sometimes exposed to the air, the osmotic and ionic composition of their medium will vary as the soil becomes saturated with water by the incoming tide or following rainfall. When exposed, the solutes become concentrated upon evaporation of the water. Animal- and plant-parasitic nematodes will face markedly different physiochemical conditions during their life cycle as they move from a parasitic to a free-living phase and back again.

The techniques available to study osmotic and ionic regulation in nematodes have been limited by the small size of most species. In this chapter we describe a technique that may help overcome at least some of these problems, summarize what is currently known of the osmotic and ionic regulatory abilities of nematodes, and describe osmoregulatory mechanisms, in particular recent work on the molecular mechanisms of osmoregulation and osmotic avoidance in *Caenorhabditis elegans*.

11.2 Osmotic and Ionic Regulation in Nematodes

11.2.1 Measuring internal osmotic concentration, water flux and volume changes

Most nematodes are small (<1 mm in length) and hence extracting fluids for analysis is difficult. In addition, the body cavity of the nematode (the pseudocoel) may be only a small proportion of the total volume (perhaps 2%; Wright and Newall, 1976). The ability to extract pseudocoelomic fluid has thus been restricted to large parasitic nematodes, particularly ascarids and anisakids (Davey, 1995). Studies on osmoregulation in most nematodes have thus relied on indirect methods, such as measuring changes in water content and length and/or volume changes in response to osmotic stress (Wright and Newall, 1976).

It is difficult to make accurate measurements of length and diameter in nematodes that may be moving rapidly in some test solutions. The measurement of length alone is not sufficient, since the relationship between length and volume varies between species (Wright and Newall, 1980) or even in the same species under different conditions, such as exposure to chemicals (Wright and Newall, 1976). Any errors in measuring diameter (d), however, will have a serious effect on volume errors (since volume $\propto d^2$). The nematode may have to be restricted in a small volume of water to ensure that it is in focus under a microscope (Wright and Newall, 1980). There is also the possibility that water flux may result in changes in hydrostatic pressure in the nematode rather than being reflected in volume changes, due to the restriction on expansion imposed by the cuticle (Harris and Crofton, 1957). Where volume changes do occur, it may result in uneven collapse and expansion in different parts of the body, making estimating volume changes from length and diameter inaccurate (Wharton, 1986).

Water flux may be measured by comparing wet and dry weights or by using tritiated water. These techniques, however, need large numbers of nematodes for accurate results and do not allow the variation between individuals to be measured. Interference microscopy is perhaps the best current technique for determining the osmotic responses of small nematodes, since it allows the water content of individuals to be measured (Ellenby, 1968). Whilst it is to be expected that there will be a relationship between water content and osmolality, it would be preferable if osmotic concentration could be measured directly.

Samples of pseudocoelomic fluid have been obtained from large adults and larvae of some parasitic nematodes. The pseudocoelomic fluid of adult *Ascaris* (which are up to 35 cm long) can be collected by making an incision in the body wall and gently squeezing. Its osmotic pressure has been measured by determining its freezing point depression (Hobson *et al.*, 1952a). A solution with an osmolality of 1 mmol/kg produces a freezing point depression of 0.00186°C. The freezing point depression of quite small volumes of liquid (~5 nl) can be measured by determining its melting point using a nanolitre osmometer. This has been performed on the pseudocoelomic fluid of third-stage infective larvae (6 cm long) of the anisakid nematode *Pseudoterranova decipiens* (Fusé *et al.*, 1993a) by nicking the body wall with a scalpel and collecting the fluid with a fine pipette. Even smaller amounts of pseudocoelomic fluid were collected from the swimbladder nematode *Anguillicola crassus* (3.75 mm long; Anderson, 2000) recovered from its host the European eel. Nematodes were immersed in oil and the cuticle cut behind the pharynx. The droplet of fluid which emerged from the wound was taken up in a microcapillary tube and 10–20 nl samples transferred to the sample holder of a nanolitre osmometer for melting point determination (Kirk *et al.*, 2002).

The osmotic concentration of pseudocoelomic fluid from *Angusticaecum* sp., a parasite from the colon of a tortoise, has been measured using a thermoelectric method that compares the vapour pressure of the fluid with that of known concentrations of NaCl (Pannikar and Sproston, 1941). Nematodes stored in tap water had a vapour pressure equivalent to 1.16–1.28% NaCl (370–408 mmol/kg). However, the nematodes were in dubious physiological condition, having been recovered from the host several days after its death and then stored in tap water for several more days.

Most free-living nematodes are too small for fluid to be collected. However, nematodes are transparent and if they are frozen on a microscope cold stage and their melting point determined, that could provide a measure of their osmolality. To that end, Wharton (2010) has modified the sample holder of a nanolitre osmometer to enable it to be used as a microscope cold stage. Melting points can be determined with this instrument to an accuracy of ±0.02°C (Wharton, 2010). Figure 11.1 shows how the internal osmotic concentration of *Panagrolaimus davidi* (up to 1 mm long) varies after it has been exposed to various external osmotic concentrations for 24 h (on the surface of balanced salt solution agars with their osmolality adjusted with NaCl and measured using a Wescor vapour pressure osmometer). This nematode maintains its internal osmotic concentration hyperosmotic to the external medium across a range of concentrations. The difference in osmolality (internal – external) is fairly constant (238 mmol/kg at an external concentration of 0 mmol/kg and 325 mmol/kg at 1000 mmol/kg). Since the internal concentration was measured by determining the melting point of the last ice crystal in the nematode, it represents its minimum osmolality, which occurs in the pseudocoelomic fluid since

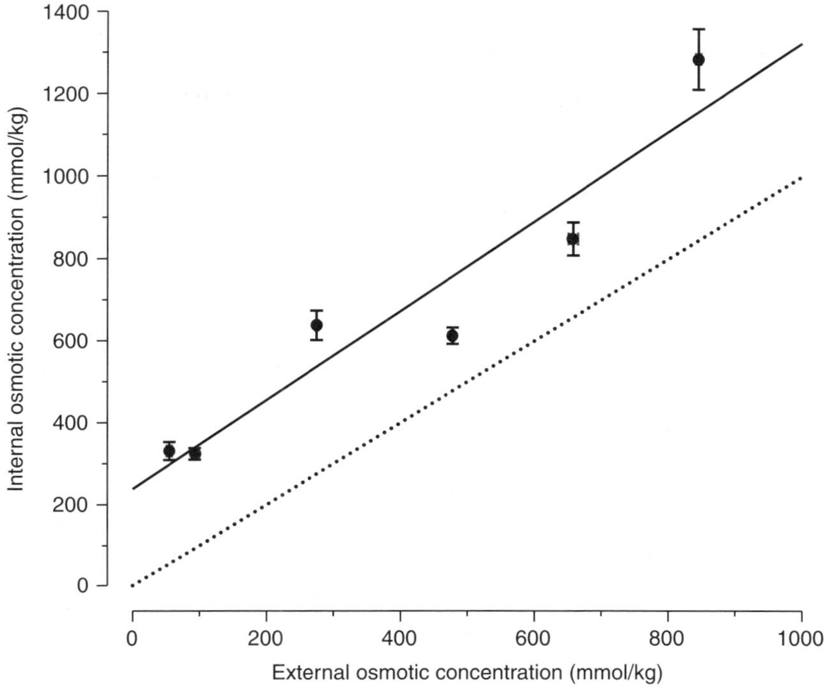

Fig. 11.1. The Antarctic nematode *Panagrolaimus davidi* maintains its internal osmotic concentration hyperosmotic to the external medium (Wharton, 2010). The dotted line is that obtained if the internal osmotic concentration equalled that of the external medium. Measurements of external osmotic concentration were made using a Wescor vapour pressure osmometer of liquid on the surface of agar plates, made up with a balanced salt solution (Piggott *et al.*, 2000) or distilled water with the osmolality adjusted using various concentrations of sodium chloride. Measurements of internal osmotic concentration were made after transfer of nematodes to immersion oil (Cargille's A) on a nanolitre osmometer modified as a cold microscope stage. Each nematode sample included calibration samples of Milli-Q water and a 1000 mmol/kg standard (Wharton, 2010). (Reproduced with the permission of the Company of Biologists.)

it melts last (Wharton, 2010). It may be possible to determine differences between different compartments in the nematode if they have different melting points.

In Table 11.1, measurements of the internal osmotic concentration of nematodes are listed and compared with that of their external medium, which was, in most cases, chosen to mirror that of their natural environment. Internal osmotic concentration has only been measured in six species of nematode to date and previous studies have been limited to large parasitic species. Measurements of internal osmotic concentration are in a fairly narrow range in this data set (296–471 mmol/kg; despite including parasitic and free-living nematodes, six species from four different orders and external

Table 11.1. Measurements of internal osmotic concentration in nematodes.

Species	External medium	External osmotic concentration (mmol/kg)	Internal[a] osmotic concentration (mmol/kg)	Reference
Ascaris suum[b]	Pig intestinal fluid	468	355	Hobson *et al.* (1952a)
Aspiculuris tetraptera	Mouse intestinal fluid	Not determined	371	Anya (1966)
Pseudoterranova decipiens	40% ASW 1% IO[c]	400	471	Fusé *et al.* (1993a)
		296	237	Stormo *et al.* (2009)
Anguillicola crassus	Seawater[d]	1000	311	Kirk *et al.* (2002)
Panagrolaimus davidi	BSSA[e]	93	324	Wharton (2010)
Angusticaecum sp.	Tap water	1	389	Pannikar and Sproston (1941)

[a]For *A. suum*, *P. decipiens*, and *A. crassus* these are measurements of pseudocoelomic fluid. For *P. davidi* it is the minimum internal concentration.
[b]The pig species of *Ascaris* is now known to be *A. suum* and not *A. lumbricoides* (Anderson, 2000).
[c]ASW, artificial seawater, nematodes exposed for various periods; IO, instant ocean, 3 days.
[d]Two weeks in natural seawater; plasma from uninfected eels maintained in similar fashion had an osmolality of 330 mmol/kg.
[e]Balanced salt solution agar: 1% agar made up with a balanced salt solution (Piggott *et al.*, 2000).

osmotic concentrations varying from 1 to 1000 mmol/kg). It would be interesting to extend these observations to a greater range of species, orders and environments.

11.2.2 The importance of balanced salt solutions

Many studies of nematode osmoregulation, including some recent studies, have used non-ionic or single salt solutions as incubation media (such as sucrose or different concentrations of NaCl). This can give a misleading picture of osmoregulatory abilities. *Enoplis brevis* and *Enoplis communis*, for example, fail to regulate their volume in 0.55 M NaCl but do show volume regulation when Ca^{2+} and K^{+} are present (Wright and Newall, 1976, 1980). *Heterorhabditis* sp. showed length regulation in balanced salt solutions but only partial regulation, and lower survival, in single salt solutions (Piggott *et al.*, 2002). Non-ionic solutions (such as urea, sucrose, glycerol and distilled water) are also unlikely to give results that reflect the nematode's natural osmoregulatory behaviour.

Balanced salt solutions that reflect the osmotic concentrations and ionic conditions experienced in the nematode's natural environment (e.g. the sea, soil or fluids of their host) should be used. For example, Piggott *et al.* (2000) developed a balanced salt solution based on the analysis of the ionic composition of soil water. This should prove a suitable medium for the physiological analysis of free-living nematodes, and the free-living stages of parasitic nematodes, that live in similar soil environments. Other solutions or media that have been used, depending on the natural environment of the nematode, include artificial seawater (Kester *et al.*, 1967), artificial tap water (Greenaway, 1970), Fenwick's salt solution (Myers, 1966) and nematode growth medium (Lamitina *et al.*, 2004).

11.2.3 Osmoconformers or osmoregulators?

The internal osmotic concentration of osmoconformers follows changes in the osmotic concentration of their environment, whilst osmoregulators keep their internal osmotic concentration constant in the face of changing external concentrations (Willmer *et al.*, 2005). To determine the strategy in nematodes it is, of course, necessary to measure internal osmotic concentrations. This has only been achieved in a few species (Table 11.1). In *P. davidi* there is some evidence for osmoregulation under hyposmotic stress but the nematode appears to be a strict osmoconformer under hyperosmotic stress (Fig. 11.1).

Third-stage larvae of *P. decipiens* can maintain a constant internal osmotic concentration under both hyposmotic and hyperosmotic conditions, and hence it is an osmoregulator (Fusé *et al.*, 1993a). However, this ability is lost if the nematode is stored in the test medium for 10 days (Fusé *et al.*, 1993b); presumably this is related to a progressive loss of physiological function after removal from their fish intermediate host.

Adults and parasitic larvae of *A. crassus* can rely, to a large extent, on the osmoregulatory abilities of their host (the European eel, *Anguilla anguilla*), which is capable of regulating the osmotic and ionic concentration of its body fluids across a wide range of salinities. The nematodes themselves are osmoconformers with respect to the osmotic concentration of the host's body fluids (Kirk *et al.*, 2002). The osmotic concentration of the pseudocoelomic fluid of *Ascaris suum* increases as the external (seawater) concentration increases, suggesting that they osmoconform (Hobson *et al.*, 1952b). The physiological condition of these parasitic nematodes in the test medium (and after removal from their host) is, however, likely to have been poor (Wright and Newall, 1976).

11.2.4 Hyperosmotic or hyposmotic regulation?

It might be expected that most nematodes would maintain their body fluids hyperosmotic to the external medium. Nematodes rely on a high turgor pressure for locomotion (Harris and Crofton, 1957), and hyperosmotic regulation may be necessary to maintain a positive turgor pressure. If the nematode was hyposmotic to its surroundings, water would be lost, turgor pressure reduced and the nematode would become inactive. There are reports of nematodes becoming inactive in media of high osmotic concentration (e.g. Wharton *et al.*, 1983), although this could be due to the effect of high ionic concentrations on nerve and muscle function, rather than a loss of turgor. In *C. elegans*, body stiffness is affected more by the mechanical properties of the cuticle than by internal hydrostatic pressure (Park *et al.*, 2007).

Panagrolaimus davidi maintains its internal osmotic concentration hyperosmotic to its surroundings across a range of external concentrations (Fig. 11.1). This also appears to be the case in *P. decipiens* (Fusé, *et al.*, 1993b), at least for the osmotic concentrations likely to be experienced within the body fluids of its marine teleost intermediate host. *Anguillicola crassus* is hyposmotic to seawater but isosmotic to the body fluids of its eel host, maintaining the osmolality of their pseudocoelomic fluid within ±20 mmol/kg of the osmolality of the host fluid (Kirk *et al.*, 2002). Measurements of the osmotic concentration of the pseudocoelomic fluid of *A. suum* generally indicate that it is hyposmotic to the intestinal fluid of its host (Wright and Newall, 1976). However, the physiological condition of the nematode may deteriorate upon removal from its host and the osmolalities that have been measured may not be an accurate reflection of those *in situ*. This may be less of a problem in *P. decipiens* since larvae from fish are in a state of developmental arrest and can be kept in artificial seawater for many weeks without dying (Fusé *et al.*, 1993b). *Angusticaecum* sp. appears to maintain its internal osmotic concentration hyperosmotic to the external medium (Pannikar and Sproston, 1941), although the nematodes were likely to be in poor physiological condition in these experiments (Wright and Newall, 1976).

In the absence of direct measurements of osmotic concentrations, osmoregulation has been inferred from changes in water flux and/or volume or length (Wright and Newall, 1976, 1980). If a decrease in water content or volume upon transfer to a medium that is hyperosmotic to the control medium (which mirrors that of the natural environment) is then followed by a recovery to normal levels, the nematode is considered to be displaying hyposmotic regulation (maintaining the body fluids hyposmotic with respect to the external medium). Similarly, an increase upon transfer to hyposmotic conditions followed by recovery indicates hyperosmotic regulation. Where there is no change in volume or water flux, the medium is assumed to be isosmotic to the body fluids.

Using interference microscopy to determine the water content of individual nematodes, Forster (1998) compared the ability of species of marine nematodes from different ecological zones to regulate their water content when exposed to different concentrations of artificial seawater. The four species were *Axonolaimus paraspinosus* and *Cervonema tenuicauda* (both from upper intertidal regions), *Daptonema oxycerca* (from the lower intertidal region) and *Sabatieria punctata* (a subtidal or sublittoral species). The nematodes from the upper intertidal zone, where salinities can fluctuate widely, showed the greatest capacity to regulate in hyposmotic conditions as well as being able to regulate in hyperosmotic conditions. The ability of *A. paraspinosus* to regulate was especially notable, with almost no change in water content. This contrasted with extreme expansion and eventual rupture of the cuticle of *D. oxycerca* in hypotonic solutions. Where the ability to regulate in hypotonic solutions was limited, a cuticle resistant to expansion stresses is important in osmotic stress tolerance (Forster, 1998). Forster (1998) discussed his data in the context of previous work and concluded that the osmoregulatory ability determines the horizontal distribution of nematode species in littoral habitats. Most marine nematodes show only limited ability to regulate in hyperosmotic conditions, nematode diversity being reduced in areas where salinity was greater than 100 parts per thousand (Olafsson, 1995; see Section 11.4). Studies on nematode osmoregulation, and the techniques used, are summarized in Table 11.2.

11.2.5 Ionic regulation

All animals regulate the ionic composition of their cells and, to a lesser extent, extracellular compartments via a variety of mechanisms (Willmer *et al.*, 2005). Studies of ionic regulation have also been limited by the small size of most nematodes and the difficulty of extracting extracellular fluid. Measurements of the ionic composition of the pseudocoelomic fluid have thus been restricted to large parasitic nematodes, such as ascarids (Wright and Newall, 1980). For small nematodes it is possible to measure element concentrations in single nematodes, or small quantities of nematodes, using flame emission spectroscopy, atomic absorption spectroscopy and radioactive assay techniques. However, these methods do not distinguish between intracellular and extracellular concentrations and between free and bound forms of the element concerned. The activity of specific ions in solution may be measured using ion-selective microelectrodes (Parri *et al.*, 1993) but these may be difficult to use with small nematodes. Scanning electron microscopy and energy-dispersive X-ray spectroscopy could be used to measure element concentrations in different nematode compartments (Wright and Newall, 1980; Wharton, 1986; Cook *et al.*, 1992) but has yet to be applied to the study of ionic regulation in nematodes.

The ionic composition of the pseudocoelomic fluid of *A. suum* is relatively constant in the face of changing external ionic concentrations (Hobson *et al.*, 1952b), implying that ionic regulation is occurring (Wright and Newall, 1980). The cuticle/epidermal complex is a possible site for such regulation. The cuticle possesses large aqueous pores, which are negatively charged and which present a barrier to the transport of large organic anions but not to small inorganic anions (Ho *et al.*, 1990; Thompson and Geary, 2002). The cuticle/epidermal complex is selectively permeable to both organic and inorganic ions, suggesting the active transport of ions (Pax *et al.*, 1995). Large conductance anion channels have been characterized in both the muscle- and cuticle-facing epidermal membranes. These may play a role in the excretion of organic acids produced as the end point of carbohydrate catabolism in the muscle cells (Blair *et al.*, 2003). *Pseudoterranova decipiens* fail to osmoregulate upon exposure to metabolic inhibitors (Fusé *et al.*, 1993b), providing indirect evidence for energy-requiring active ion transport (Wright, 2004). There is evidence that inorganic ions are transported across the intestine of *A. suum* (Thompson and Geary, 2002) but it is unclear what role a Na^+/glucose symporter or a Na^+/K^+ ATPase system plays in this process (Thompson and Geary, 2002).

Table 11.2. Osmotic relations in nematodes.[a]

Species	Environment	Relation to environment	Hyperosmotic regulation[b]	Hyposmotic regulation[b]	Methods[c]	Reference
Deontostoma spp.	Marine	Isosmotic?[d]	–	+++	VL/SS	Croll and Viglierchio (1969); Viglierchio (1974)
Monhystera disjuncta	Marine	Isosmotic?	–	–	VL/SS	Viglierchio (1974)
Enoplis communis	Marine	Isosmotic?	–	–	VL/BS	Wright and Newall (1976)
Sabatieria punctata	Marine	Isosmotic?	+	+++	IM/BS	Forster (1998)
Enoplis brevis	Estuarine	Isosmotic/hyperosmotic?	+++	++	VL/BS	Wright and Newall (1976)
Daptonema oxycerca	Intertidal	Hyposmotic/hyperosmotic?	++	++	IM/BS	Forster (1998)
Axonolaimus paraspinosus	Intertidal	Hyposmotic/hyperosmotic?	+++	+++	IM/BS	Forster (1998)
Cervonema tenuicaudata	Intertidal	Hyposmotic/hyperosmotic?	++	+++	IM/BS	Forster (1998)
Panagrolaimus davidi	Soil	Hyperosmotic	+++	–	MP/BS	Wharton (2010)
Rhabditis terrestris	Soil	Hyperosmotic?	+++	++	VL/BS	Stephenson (1942)
Rhabditis strongyloides	Soil	Hyperosmotic?	ND	+++	G/NS	Charwat *et al.* (2002)
Aphelenchus avenae	Soil	Hyperosmotic?	ND	+	G/NS	Charwat *et al.* (2002)
Panagrellus redivivus	Soil	Hyperosmotic?	+	++	G/SS/NS	Myers (1966)
Caenorhabditis elegans	Soil	Hyperosmotic?	+++	–	VL/BS	Lamitina *et al.* (2004)
Globodera rostochiensis	Soil[e]	Hyperosmotic?	+++	++	IM/NS	Clarke *et al.* (1978)
Heterodera schachtii	Soil[e]	Hyperosmotic?	+++	++	IM/NS	Perry *et al.* (1980)
Haemonchus contortus	Soil[e]	Hyperosmotic?	+++	ND	IM/BS	Atkinson and Onwuliri (1981)
Nippostrongylus brasiliensis	Soil[e]	Hyperosmotic?	+++	ND	IM/BS	Atkinson and Onwuliri (1981)
Trichostrongylus colubriformis	Soil[e]	Hyperosmotic?	+++	–	IM/BS	Wharton *et al.* (1983)
Steinernema spp.	Soil[e]	Hyperosmotic?	ND	++/+++	IM/BS	Piggott *et al.* (2000, 2002)
Heterorhabditis spp.	Soil[e]	Hyperosmotic?	ND	+/++/+++	IM/BS	Piggott *et al.* (2000, 2002)
Heterorhabditis bacteriophora	Soil[e]	Hyperosmotic?	ND	+	G/NS	Charwat *et al.* (2002)
Anguina funesta	Plant-parasitic	Hyperosmotic?	ND	+	G/NS	Charwat *et al.* (2002)

Pseudoterranova decipiens	Animal-parasitic	Hyperosmotic	+++	+++	MP/BS	Fusé *et al.* (1993a)
Anguillicola crassus	Animal-parasitic	Isosmotic	–	–	MP/BS	Kirk *et al.* (2002)
Ascaris suum	Animal-parasitic	Hyposmotic/hyperosmotic	–	–	MP/BS	Hobson *et al.* (1952a)
Hammerschmidtiella diesingi	Animal-parasitic	Isosmotic?	–	+	VL/SS	Lee (1960)
Aspiculuris tetraptera	Animal-parasitic	Isosmotic?	++	++	MP/VL/BS	Anya (1966)

[a]Studies that look at effects of osmotic stress on survival or motility only are not included.
[b]Maintaining the body fluids hyposmotic or hyperosmotic to their surroundings (if the nematode is a strict osmoregulator it would switch from hyperosmotic to hyposmotic regulation as the osmotic concentration of the external medium increases above that of its body fluids): –, no regulation; +, weak regulation; ++, some regulation, +++, strong regulation; ND, not determined.
[c]VL, volume/length measurement; IM, interference microscopy; G, gravimetric; MP, melting point of body fluids; BS, balanced salt solutions; SS, single salt solutions; NS, non-ionic solutions.
[d]?, indicates that the relation of internal osmotic concentration to the external medium and whether hyposmotic/hyperosmotic regulation occurs has been inferred from changes in volume/length or water flux.
[e]The free-living stage of a parasitic nematode.

In small nematodes, only total concentrations have been measured. In the marine nematode *E. communis*, its total sodium concentration follows the external concentration, whilst the estuarine *E. brevis* shows some ability to regulate its sodium content under hyposmotic conditions (Wright and Newall, 1976). *Panagrellus redivivus* and *Aphelenchus avenae* can regulate their sodium, but not their potassium, content under hyposmotic conditions (Myers, 1966). *Globodera rostochiensis* second-stage larvae can regulate their sodium content across a range of NaCl concentrations (Wright and Newall, 1976).

All animal cell membranes possess ion channels, which are responsible for the selective permeability of the cell to various ions and which allow the flow of ions down their electrochemical gradient. In addition, many cells have ion pumps that transport ions against their electrochemical gradient in an energy-consuming process of active transport. The best-known ion pump is the Na^+/K^+-ATPase, or sodium pump, which is responsible for maintaining the high K^+ and the low Na^+ concentrations within cells and for, among other things, regulating cell volume (Willmer *et al.*, 2005). Ion channels and pumps are presumably involved in the ionic regulation mechanisms of nematodes but the evidence, at present, is limited.

Several ion channels and pumps have been demonstrated in various tissues of *C. elegans* (Thompson and Geary, 2002; Strange, 2003). Those specifically associated with tissues that are likely to be involved in ionic regulation include CLH-4b, a ClC chloride channel homologue, associated with the excretory system (Nehrke *et al.*, 2000); CLH-1 in the hypodermis (Petalcorin *et al.*, 1999); and vacuolar-type ATPases in the hypodermis, excretory cell and rectal area (Oka *et al.*, 1997, 1998; Liegeois *et al.*, 2007).

11.3 Avoidance of Osmotic Stress

One way of dealing with osmotic stress is to avoid it by moving to a place where conditions are less stressful. This involves sensing and responding to changes in environmental conditions. *Caenorhabditis elegans* is attracted to low concentrations of salts, sugars and other substances, but high concentrations produce an avoidance reaction (Choe and Strange, 2007). Infective larvae of the plant-parasitic nematode *Meloidogyne javanica* are repelled by high salinity (Prot, 1978), as are those of *Meloidogyne incognita* (Le Saux and Queneherve, 2002). *Enoploides longispiculosus* and *Daptonema normandicum*, estuarine marine nematodes, migrate up and down the tidal flat according to the tide (Steyaert *et al.*, 2001). Although this migration pattern could be associated with changes in osmolality, other environmental factors or feeding behaviour may be involved (Wright, 2004).

In a salt concentration gradient nematodes migrate to an area of preferred concentration. Infective larvae of *Strongyloides stercoralis* migrate up and down a NaCl gradient, depending on where they are placed, eventually preferring a concentration of 0.03–0.07 M (Forbes *et al.*, 2003). Other nematodes that are attracted to low concentrations of NaCl, but repelled by high concentrations, include *Strongyloides ratti* (Tobata-Kudo *et al.*, 2000), *Steinernema carpocapsae* (Pye and Burman, 1981), *Rotylenchulus reniformis* (Riddle and Bird, 1985) and *C. elegans* (Ward, 1973). In skin-penetrating infective larvae of animal-parasitic nematodes, this response may be related to attraction to the sweat of their host (Forbes *et al.*, 2003).

The main chemosensory organs of nematodes are the amphids and phasmids (Jones, 2002). In *C. elegans*, mutants that are defective in their ability to sense high osmolarity (OSM) have been isolated (Culotti and Russell, 1978) and are linked to defects in their amphidial sensory neurons (Hart *et al.*, 1999). One OSM mutant (*osm-9*) is defective in both its ability to detect hyperosmotic stress and its response to touch. The protein this gene encodes is a member of the transient receptor potential family of cation channels and is expressed in the superficial sensory cilia of the amphidial neurons. Several other genes (including *ocr-2*, *odr-3* and *osm-10*) are also essential for osmotic avoidance behaviour and are expressed in the amphidial sensory neurons (Choe and Strange, 2007). In *S. stercoralis*, laser ablation of the ASE amphidial neuron pair disrupts the attraction of the nematode to its preferred NaCl concentration, whilst ablation of the ASH amphidial neuron pair abolishes hyperosmotic avoidance behaviour (Forbes *et al.*, 2004). In *C. elegans*, the ASE neurons mediate the response to sodium and chloride ions, whilst the ASH neurons mediate hyperosmotic avoidance behaviour (Bargmann and Mori, 1997); these responses may thus be widespread in nematodes.

11.4 Survival of Extreme Osmotic/Ionic Stress

What is an extreme osmotic stress for nematodes? Marine nematodes have no problem living in the sea, which is about 3.5% salinity (corresponding to about 0.6 M NaCl or 1000 mmol/kg). We might thus consider environments containing more than the equivalent of 0.6 M NaCl to be extreme. This occurs mainly in terrestrial environments where salts have become concentrated by evaporation. However, extreme for a species is defined by the salinity it normally experiences. A marine nematode will find exposure to fresh water an extreme stress, whilst for a freshwater nematode it would be exposure to seawater that was extreme.

Nematodes have been found in Australian saline lakes at salinities up to the equivalent of 9.3% NaCl (Bayly and Williams, 1966; Deckker and Geddes, 1980). The nematode fauna of these lakes includes *Monhystera*, *Prodesmodora*, *Mesodorylaimus* and a plectid (Hodda *et al.*, 2006). An unidentified nematode has been found in salt streams, springs and water holes in Namibia. Some nematodes are found in terrestrial mineral springs, which contain high concentrations of particular ions (Hodda *et al.*, 2006). Antarctic terrestrial soils in ice-free locations (such as the dry valleys of southern Victoria Land) can be very saline, but nematodes are apparently limited to areas of lower salinity (Nkem *et al.*, 2006). Antarctic saline lakes are of marine origin and contain nematodes at salinities up to 3.5% (Andrassy and Gibson, 2007).

Coastal marine sediments can contain interstitial water that reaches high salinity when exposed to the air for long periods. Three genera of nematodes (*Microlaimus*, *Theristus* and *Bathylaimus*) have been found in an intertidal mangrove plateau on the west coast of Zanzibar where hypersaline sediments reach over 10% salinity (Olafsson, 1995). Chinese salterns (where seawater is concentrated by evaporation to harvest salt) have been reported to contain the nematode *Prochromadora orleji* (Tarjan *et al.*, 1991).

Some organisms can tolerate such high osmotic pressures that they enter an ametabolic state of osmobiosis (Keilin, 1959). This may be considered to be similar to anhydrobiosis (Cooper and Van Gundy, 1971; see Perry and Moens, Chapter 1, this volume), where dehydration occurs evaporatively rather than as a result of osmotic stress. Some nematodes can tolerate very high osmotic stresses; for example, *Ditylenchus dipsaci* can survive exposure to 1.6 M NaCl for 24 h (Viglierchio *et al.*, 1969). This may be related to a low cuticular permeability that restricts water flux (Wharton *et al.*, 1988). Similarly, the restricted permeability of the eggshell of many nematodes protects the enclosed embryo or larva against osmotic stress (Wharton, 1980). The mechanisms by which nematodes survive extreme osmotic stress are otherwise unknown. Brine shrimps (*Artemia* and related genera) inhabit extreme hypersaline environments, and studies on this organism may indicate some of the adaptations that are necessary (Clegg and Gajardo, 2009).

11.5 Mechanisms of Osmotic Regulation

11.5.1 Excretory structures and osmoregulation

When a nematode is immersed in distilled water, or other hyposmotic media, water will enter and if not removed the nematode will swell and die. The behaviour and structure of the excretory system in some nematodes suggests that it is involved in the removal of excess water. A variety of nematode structures have been suggested to have an excretory function (Wright and Newall, 1976; Wright, 2004) but water removal has been most closely associated with the tubular type of excretory system found in secernentean nematodes. This excretory system often has associated glands and is referred to as a secretory–excretory system (Wright, 2004).

In *C. elegans* the secretory–excretory system consists of four cells: the excretory, duct, pore and gland cells. The large H-shaped excretory cell extends processes that enclose the lateral excretory ducts, lie adjacent to the lateral epidermal chords and travel anteriorly and posteriorly for almost the entire length of the worm. These are connected via a cell body and excretory sinus to the excretory duct (Nelson *et al.*, 1983). A similar arrangement is seen in third-stage larvae of *Haemonchus contortus* (Wharton and Sommerville, 1984), and with modifications in other species that may have only one lateral excretory duct or where the excretory duct may only extend posteriorly in relation to the gland cell (Wright, 1998). The lateral excretory duct has been shown, in some species, to be lined with fine tubules or canaliculi (Wright and Newall, 1976; Nelson *et al.*, 1983; Wharton and Sommerville, 1984), suggesting that it has a transport function (Wright, 1998). The similar appearance of this structure and the spongiome and contractile vacuole system of some protists (Lee, 1970; Wharton and Sommerville, 1984) suggests that it may play a role in water removal.

The excretory sinus or ampulla, which connects the lateral excretory ducts to the excretory pore, appears to pulsate as it fills with and empties of water. The rate of this pulsation can be quite irregular but varies according to the degree of osmotic stress. In general, the rate of pulsation is high in solutions that are likely to be hyposmotic to the nematode's contents and low in hyperosmotic solutions (Weinstein, 1952; Waddell, 1968; Croll *et al.*, 1972; Atkinson and Onwuliri, 1981; Nelson and Riddle, 1984). The excretory ampulla is not, however, pumping the water out but undergoing a passive cycle of filling and emptying, controlled by the opening of the excretory pore (Wharton and Sommerville, 1984). Laser ablation of the pore, duct or excretory cell of *C. elegans* (but not the gland cell) causes the animal to fill with water and die, suggesting that these cells are important in removing excess water (Nelson and Riddle, 1984).

The excretory duct or ampulla, which connects the lateral excretory canals to the excretory pore, is surrounded by a duct cell, the membrane of which is extensively folded (Nelson *et al.*, 1983; Wharton and Sommerville, 1984). This could be involved in selective solute reabsorption from fluid within the ampulla (Nelson *et al.*, 1983). This suggests that the excretory system could function in a similar fashion to the acini and ducts of mammalian salivary and sweat glands (Strange, 2003) and the excretory systems of some invertebrates that produce a hyposmotic urine via active secretion followed by selective reabsorption of solutes (Willmer *et al.*, 2005). Some types of chloride channels are expressed specifically in the excretory cell of *C. elegans* (Strange, 2003), as are some types of aquaporin (Mah *et al.*, 2007). This suggests some specialization of water and ionic flux in this cell that may be related to osmoregulatory function.

In *C. elegans*, the NHR-31 nuclear receptor is responsible for the correct growth, morphology and function of the excretory cell, via its promotion of vacuolar ATPase gene expression. When the expression of this gene is suppressed, the excretory cell has an abnormal appearance, particularly in the number and extent of the canaliculi that surround the lumen, and the ability of the nematode to recover after hyperosmotic stress is impaired (Hahn-Windgassen and Van Gilst, 2009). Other factors thought to be involved in the formation and function of the excretory system in *C. elegans* include: WNK kinases, CLIC-like proteins, patched-related proteins, mucins, cysteine-rich intestinal protein and the CEH-6 homeobox protein (Buechner, 2002; Berry *et al.*, 2003; Perens and Shaham, 2005; Liegeois *et al.*, 2007; Mah *et al.*, 2007; Hisamoto *et al.*, 2008; Tong and Buechner, 2008).

In some species it may be the intestine, rather than the excretory system, that is the site of water removal. Increased rates of filling and emptying of the intestine in distilled water have been observed in *Rhabditis terrestris* (Stephenson, 1942), and ligature of the anal region reduces the ability of *E. brevis* to regulate its volume in 50% seawater (Wright and Newall, 1976).

11.5.2 Cuticular permeability

The cuticle is likely to be a major route of water flux under osmotic stress. The permeability of the cuticle to water, and to osmotically active solutes, is thus an important factor in the dynamics of water stress. Marine nematodes are usually isosmotic to seawater and have high cuticular water permeabilities. Soil nematodes, some plant-parasitic nematodes and the free-living stages of animal-parasitic nematodes are more likely to be exposed to osmotic stress and have much lower cuticular permeabilities (Wright and Newall, 1976; Wharton *et al.*, 1988). Slowing down the rate of water flux allows time for osmoregulatory mechanisms to operate. Low cuticular permeability is associated with the ability to survive other environmental stress, such as desiccation (Wharton *et al.*, 1988; see Perry and Moens, Chapter 1, this volume). The dauer larva of *C. elegans* has a cuticle that is different in structure to that of a normal third-stage larva, with a reduced permeability, which facilitates survival of environmental stress (Cassada and Russell, 1975; Popham and Webster, 1979).

The epicuticle is one of the outermost layers of the cuticle and consists of lipid and protein. It is often overlain by a glycoprotein surface coat (Lee, 2002). The lipid content of the epicuticle suggests that it is a major permeability barrier. The decreased ability of *Caenorhabditis briggsae* to withstand osmotic stress with age has been associated with changes in the epicuticle and an increase in permeability (Searcy *et al.*, 1976). Extracuticular lipid material covers the cuticle of some nematodes, forming an accessory layer that may be important in controlling water flux (Wharton *et al.*, 2008). In *A. suum*, water-filled pores allow water to pass through the cuticle, and the epidermis may act as the major permeability barrier (Ho *et al.*, 1990). *Pseudoterranova decipiens* also transports water directly across the cuticle (Fusé *et al.*, 1993a,b).

11.5.3 The operation and control of osmoregulatory mechanisms

Most recent work on the osmoregulatory mechanisms of nematodes has been conducted on the model organism *C. elegans*. The natural ecology of *C. elegans* is poorly understood but it is a free-living bacterial feeder that appears to be associated with organically rich soil, compost and similar habitats (Fitch, 2005). In response to crowding and nutrient depletion it forms a dauer larva, which is developmentally arrested and resistant to a variety of stresses, including osmotic stress (Cassada and Russell, 1975). Dauer larvae have been found associated with a variety of invertebrates, with which they have a phoretic relationship, utilizing their host for transport to a fresh habitat. The nematode may spend most of its time in the dauer larval stage (Fitch, 2005). The nematode is exposed to hyposmotic stress if its habitat becomes flooded and to hyperosmotic stress when it dries out. Given its genetic and experimental tractability, *C. elegans* is an important model system in which to study osmoregulation and other problems in integrative physiology (Strange, 2007).

Caenorhabditis elegans survives hyperosmotic stress and can retain activity in agar media containing up to 0.5 M NaCl. The ability to survive at higher osmotic concentrations (0.5 M NaCl) is enhanced by prior exposure to a lesser hyperosmotic stress (0.2 M NaCl; Lamitina *et al.*, 2004). If first acclimated to 0.2–0.4 M NaCl, the nematodes swell and then shrink again upon exposure to hyposmotic stress (0.021 M NaCl). Under hyperosmotic stress the nematodes shrink, suggesting a loss of water from their bodies, but then return to a more normal volume (Lamitina *et al.*, 2004; Strange, 2007). The nematodes accumulate glycerol in response to hyperosmotic stress (Lamitina *et al.*, 2004).

These observations are consistent with a general model of the response of animal cells to hyperosmotic stress (Strange, 2007; see Barrett, Chapter 12, this volume). Water is rapidly lost upon exposure to hyperosmotic stress and the cell shrinks. The cell responds by the activation of regulatory cell volume increase by salt uptake, involving $Na^+/K^+/2Cl^-$ cotransporters or Na^+/H^+ and Cl^-/HCO_3^- exchange mechanisms. Water follows the osmotic gradient thus established and cell volume returns to normal levels (Lang *et al.*, 1998). The high ionic concentrations that result can damage cells, so the inorganic ions are replaced by compatible organic osmolytes, such as glycerol. The final process is the repair of molecular damage, including DNA breaks and protein denaturation (Strange, 2007).

The accumulation of glycerol in *C. elegans* is mediated by glycerol-3-phosphate dehydrogenase (GPD), which is involved in glycerol synthesis. This enzyme is encoded by two genes (*gpdh-1* and *gpdh-2*). Expression of *gpdh-1* is upregulated by hyperosmotic stress, whilst *gpdh-2* shows only a weak and transient increase in expression levels (Lamitina *et al.*, 2006). Upon transfer from hyperosmotic to normal media, glycerol levels return to those of controls by the glycerol being excreted to the external medium, perhaps via the excretory system and/or intestine (Lamitina *et al.*, 2004).

Double-knockout mutants (to *gpdh-1* and *gpdh-2*, in which the product of the genes are not produced) have greatly reduced glycerol levels when exposed to hyperosmotic stress (Lamitina *et al.*, 2006). Transgenic nematodes where *gpdh-1* and *gpdh-2* are tagged with a green fluorescent protein (GFP) reporter gene show that *gpdh-2* is constitutively (continually) expressed under normal conditions in the intestine, hypodermis and excretory cell, whilst *gpdh-1* is expressed in the intestine and hypodermis only during hyperosmotic stress (Lamitina *et al.*, 2006). If glycerol levels are being regulated by the expression of *gpdh-1*, what controls the expression of this gene?

Using the *gpdh-1*–GFP reporter system, Lamitina *et al.* (2006) performed a genome-wide RNA interference (RNAi) feeding screen to look for genes that controlled *gpdh-1* expression. RNAi involves feeding an *Escherichia coli* clone expressing double-stranded RNA to the nematode, resulting in degradation of the corresponding mRNA and suppression of the gene in question (Ahringer, 2006). One hundred and six genes were identified whose knock-down induced *gpdh-1* expression in the absence of hyperosmotic stress. Since this technique can produce false-negative results, Lamitina *et al.* (2006) also used the *C. elegans* interactome (a genome-wide map of protein–protein interactions; Li *et al.*, 2004) to identify genes that interacted with those found in the RNAi screen. This increased the number of regulators of *gpdh-1* expression identified to 122. Most of these genes (44%) are involved in the regulation of protein homeostasis, encoding proteins involved in RNA processing and protein synthesis, folding and degradation.

Increased levels of damaged or denatured proteins, as a result of the high ionic concentrations produced by hyperosmotic stress, may thus trigger osmoprotective gene expression (*gpdh-1*) and the accumulation of glycerol (Fig. 11.2). Further evidence for this has come from a genome-wide RNAi screen of *C. elegans* that has identified 40 genes that are essential for hyperosmotic survival (Choe and Strange, 2008). Half of these genes encode proteins that are involved in the detection, transportation and degradation of damaged proteins.

Other genes that may be involved in, or regulate the response to, hyperosmotic stress in *C. elegans* include collagen genes, genes encoding a germinal centre kinase-3 and a with-no-lysine (K) protein kinase (Choe and Strange, 2007), genes involved in insulin signalling pathways (Lamitina and Strange, 2005), *che-14* and other genes whose mutations produce phenotypes with defects in osmoregulation and sense organs (Liegeois *et al.*, 2007), *Ce-lea-1* encoding a late embryogenesis abundant protein with roles in the response to water stress (Gal *et al.*, 2004), *cyk-3* encoding a ubiquitin C-terminal hydrolase in embryos (Kaitna *et al.*, 2002) and novel genes (*osm-7, osm-11, osr-1*) associated with hyperosmotic stress resistance, high glycerol levels and long defecation cycles (Solomon *et al.*, 2004; Wheeler and Thomas, 2006).

There have been few studies of the molecular mechanisms underlying osmoregulatory responses in nematodes other than *C. elegans*. Glycerol is widely used as a compatible osmolyte in invertebrates (Yancey, 2005). The infective larvae of the entomopathogenic nematode *S. carpocapsae* synthesize glycerol and trehalose during osmotic dehydration (Qiu and Bedding, 2002),

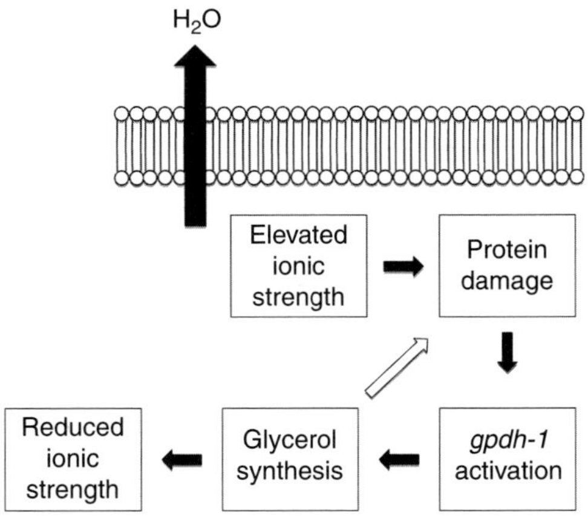

Fig. 11.2. A model proposed by Strange (2007) for the regulation of osmosensitive gene expression by the disruption of protein homeostasis in *Caenorhabditis elegans*. Hyperosmotic stress produces water loss, which raises intracellular ionic concentrations. This disrupts protein synthesis and folding (Lamitina *et al.*, 2006). Damaged proteins act as a signal that triggers *grph-1* expression and glycerol synthesis. Glycerol is a compatible osmolyte that replaces inorganic ions in the cytoplasm and acts as a chemical chaperone that aids the refolding of proteins. Stress-induced loss of function in genes involved in protein homeostasis also results in the accumulation of damaged proteins and the activation of *gpdh-1*. (Redrawn from Strange, 2007.)

whilst in several other species of nematode glycerol synthesis is associated with anaerobic conditions (see Barrett, Chapter 12, this volume).

The patterns of protein synthesis of osmotically dehydrated *Steinernema feltiae* infective larvae (exposed to 24% glycerol) and unstressed controls have been compared using two-dimensional electrophoresis (Chen *et al.*, 2005). Ten novel proteins were produced and a further ten upregulated in response to osmotic dehydration. The proteins that could be identified by mass spectrometry included actin (filament formation and muscle fibres), proteasome regulatory particle (involved in protein degradation), and GroEL and GroES (members of the Hsp60 family of chaperones involved in protein folding). A further study compared the response to osmotic dehydration (24% glycerol) and evaporative dehydration (97% then 85% relative humidity) in *S. feltiae* (Chen *et al.*, 2006). Many of the proteins identified were expressed in response to both osmotic and evaporative dehydration, including proteins known to be involved in stress responses, such as Hsp60, coenzyme Q (ubiquinone), biosynthesis protein 4, inositol monophosphatase and fumarate lyase, and also proteins involved in the regulation of the cell cycle, gene transcription and the organization of macromolecular structures. The stress response is thus complex. Two proteins, an unidentified protein and a putative nuclear protein of unknown function, were upregulated only by evaporative dehydration.

Another approach is to look for changes in gene expression in response to an environmental stress, involving RNA extraction of the stressed nematodes, conversion into complementary DNA (cDNA), subtractive hybridization against cDNA from control nematodes, cloning, sequencing and the identification of genes that are up- or downregulated in response to the stress (or a cDNA microarray can be used; Clark *et al.*, 2009). This approach has been used in *S. feltiae*, and other nematodes, to look at the response to evaporative dehydration (Gal *et al.*, 2003, see also Burnell and Tunnacliffe, Chapter 6, and Adams and Adhikari, Chapter 9, this volume) and in *Heterorhabditis bacteriophora* in response to heat and cold stress (see Grewal *et al.*, Chapter 7, this volume), but there have not yet been any studies of this kind that investigate the response to osmotic stress.

There is evidence for a K-Cl cotransporter protein in *M. incognita*, which may be involved in osmotic and ionic regulation; the gene was expressed in mobile and sedentary stages of the life cycle, and Neveu *et al.* (2002) hypothesized that the gene is involved in the regulation of osmotic pressure of cells in order to keep body fluids hyperosmotic to the environment.

11.5.4 Aquaporins

Aquaporins (AQPs) are proteins that are embedded in cell membranes, forming channels that regulate water flow (Campbell *et al.*, 2008). Eight AQP genes have been identified in *C. elegans* (Kuwahara *et al.*, 2000), with a further three present in the *C. elegans* genome (Mah *et al.*, 2007). The genes *aqp1–8* have been expressed in *Xenopus* oocytes to study the transport properties of the AQPs. Three of these AQPs transport water only, one AQP transports glycerol only, two AQPs transport both water and glycerol, and two AQPs did not transport any of the substances tested (Huang *et al.*, 2007). AQPs labelled with GFP are expressed in various tissues, including those associated with water transport and excretion: the intestine, hypodermis and excretory cell. Mutants where AQP expression was suppressed, however, showed little effect on resistance to osmotic stress, suggesting that they are not essential for osmoregulation (Huang *et al.*, 2007). The only AQP that is exclusively expressed in the excretory system is *aqp-8*. The expression of *aqp-8* is regulated by CEH-6, which is a POU homeobox transcription factor (Mah *et al.*, 2007).

Nematode aquaporins have also been reported from *Plectus murrayi* (Adhikari *et al.*, 2009) and from various tissues of both larvae and adults of *Toxocara canis* (Loukas *et al.*, 1999).

11.6 Conclusions and Future Directions

Progress on the study of the physiology of both osmotic and ionic regulation continues to be hampered by the difficulty of extracting pseudocoelomic fluid from small nematodes. Hopefully the nanolitre osmometer technique outlined here and by Wharton (2010) will allow us to move forwards in this area of osmoregulation.

Molecular studies on *C. elegans* are providing much information of relevance to osmotic and ionic regulation and the general homeostatic mechanisms of the animal. So far these studies have focused on the response to hyperosmotic conditions. The response to hyposmotic conditions may involve different mechanisms. There is also the question of how widely applicable studies on *C. elegans* are to nematodes in general. *Caenorhabditis elegans* is adapted to a particular ecological situation – the rapid colonization of organically rich soil, compost and similar habitats (Fitch, 2005). We might expect, but not necessarily find, that free-living nematodes, and the free-living stages of parasitic nematodes, that live in similar soil habitats have similar mechanisms to *C. elegans*. However, nematodes live in a much wider range of habitats than this. The problems of osmotic and ionic regulation faced by a marine nematode, for example, are obviously different from those faced by a soil nematode. The parasitic stages of animal- and plant-parasitic nematodes experience very different environmental conditions from those faced by a nematode in the soil (Geary and Thompson, 2001), and the infective stages of these parasitic species have to cope with marked changes in osmotic and ionic environments from outside to inside the host.

The study of osmotic and ionic regulation in nematodes may seem to be of rather narrow academic interest. However, being relatively simple animals, nematodes can help us understand the processes that occur in more complex animals, including humans. Studies on the excretory system of *C. elegans* can be applied to the development of the human kidney and the study of renal diseases (Barr, 2005) and the development of tubular systems in general

(Buechner, 2002). Nematodes are the most numerous and diverse members of the meiofauna in marine, freshwater and soil environments (Yeates *et al.*, 2009). Understanding their environmental physiology will thus contribute to our knowledge of how these ecosystems function.

11.7 Acknowledgements

We would like to thank Denis Wright, Imperial College London, for his comments on this chapter.

11.8 References

Adhikari, B.N., Wall, D.H. and Adams, B.J. (2009) Desiccation survival in an Antarctic nematode: molecular analysis using expressed sequenced tags. *BMC Genomics* 10, 69.

Ahringer, J. (ed.) (2006) Reverse genetics. In: The *C. elegans* Research Community (ed.) *WormBook*, available at: http://www.wormbook.org./chapters/www_introreversegenetics/introreversegenetics.html (accessed 30th October 2009).

Anderson, R.C. (2000) *Nematode Parasites of Vertebrates: their Development and Transmission*. CAB International, Wallingford, UK.

Andrassy, I. and Gibson, J.A.E. (2007) Nematodes from saline and freshwater lakes of the Vestfold Hills, East Antarctica, including the description of *Hypodontolaimus antarcticus* sp. n. *Polar Biology* 30, 669–678.

Anya, A.O. (1966) Investigations on osmotic regulation in the parasitic nematode, *Aspiculuris tetraptera*. *Parasitology* 56, 583–588.

Atkinson, H.J. and Onwuliri, C.O.E. (1981) *Nippostrongylus brasiliensis* and *Haemonchus contortus*: function of the excretory ampulla of the third stage larva. *Experimental Parasitology* 52, 191–198.

Bargmann, C.I. and Mori, I. (1997) Chemotaxis and thermotaxis. In: Riddle, D.L., Blumenthal, T., Meyer, B.J. and Priess, J.R. (eds) C. elegans *II*. Cold Spring Harbor Press, Cold Spring Harbor, New York, pp. 717–738.

Barr, M.M. (2005) *Caenorhabditis elegans* as a model to study renal development and disease: sexy cilia. *Journal of the American Society of Nephrology* 16, 305–312.

Bayly, I.A.E. and Williams, W.D. (1966) Chemical and biological studies on some saline lakes of south-east Australia. *Australian Journal of Marine and Freshwater Research* 17, 177–228.

Berry, K.L., Bulow, H.E., Hall, D.H. and Hobert, O. (2003) A *C. elegans* CLIC-like protein required for intracellular tube formation and maintenance. *Science* 302, 2134–2137.

Blair, K.L., Geary, T.G., Mensch, S.K., Vidmar, T.J., Li, S.K., Ho, N.F.H. and Thompson, D.P. (2003) Biophysical characterization of a large conductance anion channel in hypodermal membranes of the gastrointestinal nematode, *Ascaris suum*. *Comparative Biochemistry and Physiology A – Molecular & Integrative Physiology* 134, 805–818.

Buechner, M. (2002) Tubes and the single *C. elegans* excretory cell. *Trends in Cell Biology* 12, 479–484.

Campbell, E.M., Ball, A., Hoppler, S. and Bowman, A.S. (2008) Invertebrate aquaporins: a review. *Journal of Comparative Physiology B – Biochemical Systemic and Environmental Physiology* 178, 935–955.

Cassada, R.C. and Russell, R.L. (1975) The dauer larvae, a postembryonic developmental variant of the nematode *Caenorhabditis elegans*. *Developmental Biology* 46, 326–342.

Charwat, S.M., Fisher, J.M. and Wyss, U. (2002) The effect of osmotic stress on desiccation survival and water content of four nematode species. *Nematology* 4, 89–97.

Chen, S., Gollop, N. and Glazer, I. (2005) Cross-stress tolerance and expression of stress-related proteins in osmotically desiccated entomopathogenic *Steinernema feltiae* IS-6. *Parasitology* 131, 695–703.

Chen, S.B., Glazer, I., Gollop, N., Cash, P., Argo, E., Innes, A., Stewart, E., Davidson, I. and Wilson, M.J. (2006) Proteomic analysis of the entomopathogenic nematode *Steinernema feltiae* IS-6 IJs under evaporative and osmotic stresses. *Molecular and Biochemical Parasitology* 145, 195–204.

Choe, K.P. and Strange, K. (2007) Molecular and genetic characterization of osmosensing and signal transduction in the nematode *Caenorhabditis elegans*. *FEBS Journal* 274, 5782–5789.

Choe, K.P. and Strange, K. (2008) Genome-wide RNAi screen and *in vivo* protein aggregation reporters identify degradation of damaged proteins as an essential hypertonic stress response. *American Journal of Physiology – Cell Physiology* 295, C1488–C1498.

Clark, M.S., Thorne, M.A.S., Purac, J., Burns, G., Hillyard, G., Popovic, Z.D., Grubor-Lajsic, G. and Worland, M.R. (2009) Surviving the cold: molecular analyses of insect cryoprotective dehydration in the Arctic springtail *Megaphorura arctica* (Tullberg). *BMC Genomics* 10, 328.

Clarke, A.J., Perry, R.N. and Hennessy, J. (1978) Osmotic stress and the hatching of *Globodera rostochiensis*. *Nematologica* 24, 384–392.

Clegg, J.S. and Gajardo, G. (2009) Two highly diverged New World *Artemia* species, *A. franciscana* and *A. persimilis*, from contrasting hypersaline habitats express a conserved stress protein complement. *Comparative Biochemistry and Physiology A – Molecular & Integrative Physiology* 153, 451–456.

Cook, R., Thomas, B.J. and Mizen, K.A. (1992) X-ray microanalysis of feeding syncytia induced in plants by cyst nematodes. *Nematologica* 38, 36–49.

Cooper, A.F.J. and Van Gundy, S.D. (1971) Senescence, quiescence, and cryptobiosis. In: Zuckerman, B.M., Mai, W.F. and Rhode, R.A. (eds) *Plant Parasitic Nematodes*, Vol. 2. Academic Press, New York and London, pp. 297–317.

Croll, N.A. and Viglierchio, D.R. (1969) Osmoregulation and the uptake of ions in a marine nematode. *Proceedings of the Helminthological Society of Washington* 36, 1–9.

Croll, N.A., Slater, A. and Smith, J.M. (1972) *Ancylostoma tubaeforme*: osmoregulatory ampulla of larvae. *Experimental Parasitology* 31, 356–360.

Culotti, J.G. and Russell, R.L. (1978) Osmotic avoidance defective mutants of the nematode *Caenorhabditis elegans*. *Genetics* 90, 243–256.

Davey, K.G. (1995) Water, water compartments and water regulation in some nematodes parasitic in vertebrates. *Journal of Nematology* 27, 433–440.

Deckker, P. and Geddes, M.C. (1980) Seasonal fauna of ephemeral saline lakes near the Coorong Lagoon, South Australia. *Australian Journal of Marine and Freshwater Research* 31, 677–699.

Ellenby, C. (1968) Determination of the water content of nematode worms by interference microscopy. *Experimentia* 24, 84–85.

Fitch, D.H.A. (2005) Evolution: an ecological context for *C. elegans*. *Current Biology* 15, R655–R658.

Forbes, W.M., Ashton, F.T., Boston, R. and Schad, G.A. (2003) Chemotactic behaviour of *Strongyloides stercoralis* infective larvae on a sodium chloride gradient. *Parasitology* 127, 189–197.

Forbes, W.M., Ashton, F.T., Boston, R., Zhu, X. and Schad, G.A. (2004) Chemoattraction and chemorepulsion of *Strongyloides stercoralis* infective larvae on a sodium chloride gradient is mediated by amphidial neuron pairs ASE and ASH, respectively. *Veterinary Parasitology* 120, 189–198.

Forster, S.J. (1998) Osmotic stress tolerance and osmoregulation of intertidal and subtidal nematodes. *Journal of Experimental Marine Biology and Ecology* 224, 109–125.

Fusé, M., Davey, K.G. and Sommerville, R.I. (1993a) Osmoregulation in the parasitic nematode *Pseudoterranova decipiens*. *Journal of Experimental Biology* 175, 127–142.

Fusé, M., Davey, K.G. and Somerville, R.I. (1993b) Water compartments and osmoregulation in the parasitic nematode *Pseudoterranova decipiens*. *Journal of Experimental Biology* 175, 143–152.

Gal, T.Z., Glazer, I. and Koltai, H. (2003) Differential gene expression during desiccation stress in the insect-killing nematode *Steinernema feltiae* IS-6. *Journal of Parasitology* 89, 761–766.

Gal, T.Z., Glazer, I. and Koltai, H. (2004) An LEA group 3 family member is involved in survival of *C. elegans* during exposure to stress. *FEBS Letters* 577, 21–26.

Geary, T.G. and Thompson, D.P. (2001) *Caenorhabditis elegans*: how good a model for veterinary parasites? *Veterinary Parasitology* 101, 371–386.

Greenaway, P. (1970) Sodium regulation in the freshwater mollusc *Limnaea stagnalis* (L) (Gastropoda, Pulmonata). *Journal of Experimental Biology* 53, 147–163.

Hahn-Windgassen, A. and Van Gilst, M.R. (2009) The *Caenorhabditis elegans* HNF4 alpha homolog, NHR-31, mediates excretory tube growth and function through coordinate regulation of the vacuolar ATPase. *Plos Genetics* 5, e1000553.

Harris, J.E. and Crofton, H.D. (1957) Internal pressure and cuticular structure in *Ascaris*. *Journal of Experimental Biology* 34, 116–130.

Hart, A.C., Kass, J., Shapiro, J.E. and Kaplan, J.M. (1999) Distinct signaling pathways mediate touch and osmosensory responses in a polymodal sensory neuron. *Journal of Neuroscience* 19, 1952–1958.

Hisamoto, N., Moriguchi, T., Urushiyama, S., Mitani, S., Shibuya, H. and Matsumoto, K. (2008) *Caenorhabditis elegans* WNK-STE20 pathway regulates tube formation by modulating ClC channel activity. *Embo Reports* 9, 70–75.

Ho, N.F.H., Geary, T.G., Raub, T.J., Barsuhn, C.L. and Thompson, D.P. (1990) Biophysical transport properties of the cuticle of *Ascaris suum*. *Molecular & Biochemical Parasitology* 41, 153–166.

Hobson, A.D., Stephenson, W. and Beadle, L.C. (1952a) Studies on the physiology of *Ascaris lumbricoides*. I. The relation of the total osmotic pressure, conductivity and chloride content of the body fluid to that of the external environment. *Journal of Experimental Biology* 29, 1–21.

Hobson, A.D., Stephenson, W. and Eden, A. (1952b) Studies on the physiology of *Ascaris lumbricoides*. II. The inorganic composition of the body fluids in relation to that of the external environment. *Journal of Experimental Biology* 29, 22–29.

Hodda, M., Ocaña, A. and Traunspurger, W. (2006) Nematodes from extreme freshwater habitats. In: Euyalem-Abebe, Andrassy, I. and Traunspurger, W. (eds) *Freshwater Nematodes: Ecology and Taxonomy*. CAB International, Wallingford, UK, pp. 179–210.

Huang, C.G., Lamitina, T., Agre, P. and Strange, K. (2007) Functional analysis of the aquaporin gene family in *Caenorhabditis elegans*. *American Journal of Physiology – Cell Physiology* 292, C1867–C1873.

Jones, J. (2002) Nematode sense organs. In: Lee, D.L. (ed.) *The Biology of Nematodes*. Taylor & Francis, London, pp. 353–368.

Kaitna, S., Schnabel, H., Schnabel, R., Hyman, A.A. and Glotzer, M.A. (2002) A ubiquitin C-terminal hydrolase is required to maintain osmotic balance and execute actin-dependent processes in the early *C. elegans* embryo. *Journal of Cell Science* 115, 2293–2302.

Keilin, D. (1959) The Leeuwenhoek Lecture. The problem of anabiosis or latent life: history and current concept. *Proceedings of the Royal Society of London B* 150, 149–191.

Kester, D.R., Duedall, I.W., Connors, D.N. and Pytkowicz, R.M. (1967) Preparation of artificial seawater. *Limnology & Oceanography* 12, 176–179.

Kirk, R.S., Morritt, D., Lewis, J.W. and Kennedy, C.R. (2002) The osmotic relationship of the swimbladder nematode *Anguillicola crassus* with seawater eels. *Parasitology* 124, 339–347.

Kuwahara, M., Asai, T., Sato, K., Shinbo, I., Terada, Y., Marumo, F. and Sasaki, S. (2000) Functional characterization of a water channel of the nematode *Caenorhabditis elegans. Biochimica et Biophysica Acta – Gene Structure and Expression* 1517, 107–112.

Lamitina, S.T. and Strange, K. (2005) Transcriptional targets of DAF-16 insulin signaling pathway protect *C. elegans* from extreme hypertonic stress. *American Journal of Physiology – Cell Physiology* 288, C467–C474.

Lamitina, S.T., Morrison, R., Moeckel, G.W. and Strange, K. (2004) Adaptation of the nematode *Caenorhabditis elegans* to extreme osmotic stress. *American Journal of Physiology – Cell Physiology* 286, C785–C791.

Lamitina, T., Huang, C.G. and Strange, K. (2006) Genome-wide RNAi screening identifies protein damage as a regulator of osmoprotective gene expression. *Proceedings of the National Academy of Sciences of the USA* 103, 12173–12178.

Lang, F., Busch, G.L., Ritter, M., Volkl, H., Waldegger, S., Gulbins, E. and Haussinger, D. (1998) Functional significance of cell volume regulatory mechanisms. *Physiological Reviews* 78, 247–306.

Lee, D.L. (1960) The effect of changes in the osmotic pressure upon *Hammerschmidtiella diesingi* (Hammerschmidt, 1838) with reference to the survival of the nematodes during moulting of the cockroach. *Parasitology* 50, 241–246.

Lee, D.L. (1970) The fine structure of the excretory system in adult *Nippostrongylus brasiliensis* (Nematoda) and a suggested function for the 'excretory glands'. *Tissue & Cell* 2, 225–231.

Lee, D.L. (2002) Cuticle, moulting and exsheathment. In: Lee, D.L. (ed.) *The Biology of Nematodes*. Taylor & Francis, London, pp. 171–209.

Le Saux, R. and Quénéhervé, P. (2002) Differential responses of two plant-parasitic nematodes, *Meloidogyne incognita* and *Rotylenchus reniformis*, to some inorganic ions. *Nematology* 4, 99–105.

Li, S.M., Armstrong, C.M., Bertin, N. *et al.* (2004) A map of the interactome network of the metazoan *C. elegans. Science* 303, 540–543.

Liegeois, S., Benedetto, A., Michaux, G., Belliard, G. and Labouesse, M. (2007) Genes required for osmoregulation and apical secretion in *Caenorhabditis elegans. Genetics* 175, 709–724.

Loukas, A., Hunt, P. and Maizels, R.M. (1999) Cloning and expression of an aquaporin-like gene from a parasitic nematode. *Molecular and Biochemical Parasitology* 99, 287–293.

Mah, A.K., Armstrong, K.R., Chew, D.S., Chu, J.S., Tu, D.K., Johnsen, R.C., Chen, N., Chamberlin, H.M. and Baillie, D.L. (2007) Transcriptional regulation of AQP-8, a *Caenorhabditis elegans* aquaporin exclusively expressed in the excretory system, by the POU homeobox transcription factor CEH-6. *Journal of Biological Chemistry* 282, 28074–28086.

Myers, R.F. (1966) Osmoregulation in *Panagrellus redivivus* and *Aphelenchus avenae. Nematologica* 12, 579–586.

Nehrke, K., Begenisich, T., Pilato, J. and Melvin, J.E. (2000) *C. elegans* ClC-type chloride channels: novel variants and functional expression. *American Journal of Physiology – Cell Physiology* 279, C2052–C2066.

Nelson, F.K. and Riddle, D.L. (1984) Functional study of the *Caenorhabditis elegans* secretory–excretory system using laser microsurgery. *Journal of Experimental Zoology* 231, 45–56.

Nelson, F.K., Albert, P.S. and Riddle, D.L. (1983) Fine structure of the *Caenorhabditis elegans* secretory–excretory system. *Journal of Ultrastructural Research* 82, 156–171.

Neveu, C., Semblat, J.P., Abad, P. and Castagnone-Sereno, P. (2002) Characterization of a cDNA related to K-Cl cotransporters in the root-knot nematode *Meloidogyne incognita. DNA Sequence* 13, 117–121.

Nkem, J.N., Virginia, R.A., Barrett, J.E., Wall, D.H. and Li, G. (2006) Salt tolerance and survival thresholds for two species of Antarctic soil nematodes. *Polar Biology* 29, 643–651.

Oka, T., Yamamato, R. and Futai, M. (1997) Three *vha* genes encode proteolipids of *Caenorhabditis elegans* vacuolar-type ATPase. Gene structures and preferential expression in an H-shaped excretory cell and rectal cells. *Journal of Biological Chemistry* 272, 24387–24392.

Oka, T., Yamamato, R. and Futai, M. (1998) Multiple genes for vacuolar-type ATPase proteolipids in *Caenorhabditis elegans*. A new gene, *vha-3*, has a distinct gel-specific distribution. *Journal of Biological Chemistry* 273, 22570–22576.

Olafsson, E. (1995) Meiobenthos in mangrove areas in Eastern Africa with emphasis on assemblage structure of free-living marine nematodes. *Hydrobiologia* 312, 47–57.

Pannikar, N.K. and Sproston, N.G. (1941) Osmotic relations of some metazoan parasites. *Parasitology* 33, 214–223.

Park, S.J., Goodman, M.B. and Pruitt, B.L. (2007) Analysis of nematode mechanics by piezoresistive displacement clamp. *Proceedings of the National Academy of Sciences of the USA* 104, 17376–17381.

Parri, H.R., Djamgoz, M.B.A., Holdendye, L. and Walker, R.J. (1993) An ion-sensitive microelectrode study on the effect of a high-concentration of ivermectin on chloride balance in the somatic muscle bag cells of *Ascaris suum. Parasitology* 106, 421–427.

Pax, R.A., Geary, T.G., Bennett, J.L. and Thompson, D.P. (1995) *Ascaris suum* – characterization of transmural and hypodermal potentials. *Experimental Parasitology* 80, 85–97.

Perens, E.A. and Shaham, S. (2005) *C. elegans daf-6* encodes a patched-related protein required for lumen formation. *Developmental Cell* 8, 893–906.

Perry, R.N., Clarke, A.J. and Hennessy, J. (1980) The influence of osmotic pressure on the hatching of *Heterodera schachtii. Revue de Nématologie* 3, 3–9.

Petalcorin, M.I., Oka, T., Koga, M., Ogura, K., Wada, Y., Ohshima, Y. and Futai, M. (1999) Disruption of clh-1, a chloride channel gene, results in a wider body of *Caenorhabditis elegans. Journal of Molecular Biology* 294, 347–355.

Piggott, S.J., Perry, R.N. and Wright, D.J. (2000) Hypo-osmotic regulation in entomopathogenic nematodes: *Steinernema* spp. and *Heterorhabditis* spp. *Nematology* 2, 561–566.

Piggott, S.J., Liu, Q.Z., Glazer, I. and Wright, D.J. (2002) Does osmoregulatory behaviour in entomopathogenic nematodes predispose desiccation tolerance? *Nematology* 4, 483–487.

Popham, J.D. and Webster, J.M. (1979) Aspects of the fine structure of the dauer larvae of *Caenorhabditis elegans. Canadian Journal of Zoology* 57, 794–800.

Prot, J.C. (1978) Influence of concentrations of salts on the movement of second stage juveniles of *Meloidogyne javanica. Revue de Nématologie* 1, 12–26.

Pye, A.E. and Burman, M. (1981) *Neoplectana carpocapsae*: nematode accumulation on chemical and bacterial gradients. *Experimental Parasitology* 51, 13–20.

Qiu, L.H. and Bedding, R.A. (2002) Characteristics of protectant synthesis of infective juveniles of *Steinernema carpocapsae* and importance of glycerol as a protectant for survival of the nematodes during osmotic dehydration. *Comparative Biochemistry & Physiology B – Biochemistry & Molecular Biology* 131, 757–765.

Riddle, D.L. and Bird, A.F. (1985) Responses of the plant parasitic nematodes *Rotylenchulus reniformis*, *Anguina agrostis* and *Meloidogyne javanica* to chemical attractants. *Parasitology* 91, 165–195.

Searcy, D.G., Kisiel, M.J. and Zuckerman, B.M. (1976) Age-related increase of cuticle permeability in the nematode

Caenorhabditis briggsae. Experimental Aging Research 2, 293–301.

Solomon, A., Bandhakavi, S., Jabbar, S., Shah, R., Beitel, G.J. and Morimoto, R.I. (2004) *Caenorhabditis elegans* OSR-1 regulates behavioral and physiological responses to hyperosmotic environments. *Genetics* 167, 161–170.

Stephenson, W. (1942) The effects of variation in osmotic pressure upon a free-living soil nematode. *Parasitology* 34, 253–265.

Steyaert, M., Herman, P.M.J., Moens, T., Widdows, J. and Vincx, M. (2001) Tidal migrations of nematodes on an estuarine tidal flat (the Molenplaat, Schelde Estuary, SW Netherlands). *Marine Ecology Progress Series* 224, 299–304.

Stormo, S.K., Praebel, K. and Elvevoll, E.O. (2009) Cold tolerance in sealworm (*Pseudoterranova decipiens*) due to heat-shock adaptations. *Parasitology* 136, 1317–1324.

Strange, K. (2003) From genes to integrative physiology: ion channel and transporter biology in *Caenorhabditis elegans. Physiological Reviews* 83, 377–415.

Strange, K. (2007) Revisiting the Krogh Principle in the post-genome era: *Caenorhabditis elegans* as a model system for integrative physiology research. *Journal of Experimental Biology* 210, 1622–1631.

Tarjan, A.C., Davis, J.S. and Nguyen, K.B. (1991) The genus *Prochromadora* with a redescription of *P. orleji* from a marine saltern in the People's Republic of China. *Journal of Nematology* 23, 491–501.

Thompson, D.P. and Geary, T.G. (2002) Excretion/secretion, ionic and osmotic regulation. In: Lee, D.L. (ed.) *The Biology of Nematodes*. Taylor & Francis, London, pp. 291–320.

Tobata-Kudo, H., Higo, H., Koga, M. and Tada, I. (2000) Chemokinetic behavior of the infective third-stage larvae of *Strongyloides ratti* on a sodium chloride gradient. *Parasitology International* 49, 183–188.

Tong, X. and Buechner, M. (2008) CRIP homologues maintain apical cytoskeleton to regulate tubule size in *C. elegans. Developmental Biology* 317, 225–233.

Viglierchio, D.R. (1974) Osmoregulation and electrolyte uptake in Antarctic nematodes. *Transactions of the American Microscopical Society* 93, 325–338.

Viglierchio, D.R., Croll, N.A. and Gortz, J.H. (1969) The physiological response of nematodes to osmotic stress and an osmotic treatment for separating nematodes. *Nematologica* 15, 15–21.

Waddell, H. (1968) The excretory system of the kidney worm *Stephanurus dentatus* (Nematoda). *Parasitology* 58, 907–919.

Ward, S. (1973) Chemotaxis by the nematode *Caenorhabditis elegans*: identification of attractants and analysis of the response by the use of mutants. *Proceedings of the National Academy of Sciences of the USA* 70, 817–821.

Weinstein, P. (1952) Regulation of water balance as a function of excretory system of the filariform larvae of *Nippostrongylus muris* and *Ancylostoma caninum. Experimental Parasitology* 1, 363–376.

Wharton, D. (1980) Nematode egg-shells. *Parasitology* 81, 447–463.

Wharton, D.A. (1986) *A Functional Biology of Nematodes*. Croom Helm, London and Sydney.

Wharton, D.A. (2010) Osmoregulation in the Antarctic nematode *Panagrolaimus davidi. Journal of Experimental Biology* 213, 2025–2030.

Wharton, D.A. and Sommerville, R.I. (1984) The structure of the excretory system of the infective larva of *Haemonchus contortus. International Journal for Parasitology* 14, 591–600.

Wharton, D.A., Perry, R.N. and Beane, J. (1983) The effect of osmotic stress on behaviour and water content of the infective larvae of *Trichostrongylus colubriformis. International Journal for Parasitology* 13, 185–190.

Wharton, D.A., Preston, C.M., Barrett, J. and Perry, R.N. (1988) Changes in cuticular permeability associated with recovery from anhydrobiosis in the plant parasitic nematode, *Ditylenchus dipsaci. Parasitology* 97, 317–330.

Wharton, D.A., Petrone, L., Duncan, A. and McQuillan, A.J. (2008) A surface lipid may control the permeability slump associated with entry into anhydrobiosis in the plant parasitic nematode *Ditylenchus dipsaci*. *Journal of Experimental Biology* 211, 2901–2908.

Wheeler, J.M. and Thomas, J.H. (2006) Identification of a novel gene family involved in osmotic stress response in *Caenorhabditis elegans*. *Genetics* 174, 1327–1336.

Willmer, P., Stone, G. and Johnston, I. (2005) *Environmental Physiology of Animals*. Blackwell Science, Malden, Massachusetts.

Wright, D.J. (1998) Respiratory physiology, nitrogen excretion and osmotic and ionic regulation. In: Perry, R.N. and Wright, D.J. (eds) *The Physiology and Biochemistry of Free-living and Plant-parasitic Nematodes*. CAB International, Wallingford, UK, pp. 103–131.

Wright, D.J. (2004) Osmoregulatory and excretory behaviour. In: Gaugler, R. and Bilgrami, A.R. (eds) *Nematode Behaviour*. CAB International, Wallingford, UK, pp. 177–196.

Wright, D.J. and Newall, D.R. (1976) Nitrogen excretion, osmotic and ionic regulation in nematodes. In: Croll, N.A. (ed.) *The Organisation of Nematodes*. Academic Press, New York and London, pp. 163–210.

Wright, D.J. and Newall, D.R. (1980) Osmotic and ionic regulation in nematodes. In: Zuckerman, B.M. (ed.) *Nematodes as Biological Models*. Academic Press, New York and London, pp. 143–164.

Yancey, P.H. (2005) Organic osmolytes as compatible, metabolic and counteracting cytoprotectants in high osmolarity and other stresses. *Journal of Experimental Biology* 208, 2819–2830.

Yeates, G.W., Ferris, H., Moens, T. and Van Der Putten, W.H. (2009) The role of nematodes in ecosystems. In: Wilson, M.J. and Kakouli-Duarte, T. (eds) *Nematodes as Environmental Indicators*. CAB International, Wallingford, UK, pp. 1–44.

12 Biochemistry of Survival

John Barrett

Institute of Biological, Environmental and Rural Sciences, Aberystwyth University, Aberystwyth, UK

12.1 Introduction

Nematodes inhabit a vast range of environments, ranging from the dry valleys of Antarctica to hot springs, to German beer mats. In addition, the free-living stages of some parasitic nematodes and some of the life cycle stages of free-living nematodes also show extraordinary abilities to survive extreme environmental conditions (Womersley *et al.*, 1998; see Perry and Moens, Chapter 1, this volume). Nematodes even survived when the Columbia spacecraft broke up on re-entry in 2003 (Szewczyk and Lamb, 2005).

Organisms from different environments face different problems, whilst organisms in the same environment may respond to the same stress in different ways. The tolerance limits of an organism can be visualized as a star plot, each axis representing a different environmental parameter (Fig. 12.1). The plot gives an inner 'life zone', within which the organism can complete its life cycle. The 'life zone' is surrounded by a larger 'survival zone', where the organism can survive for varying lengths of time but not complete its life cycle. Most organisms have relatively similar 'life zones', but a few,

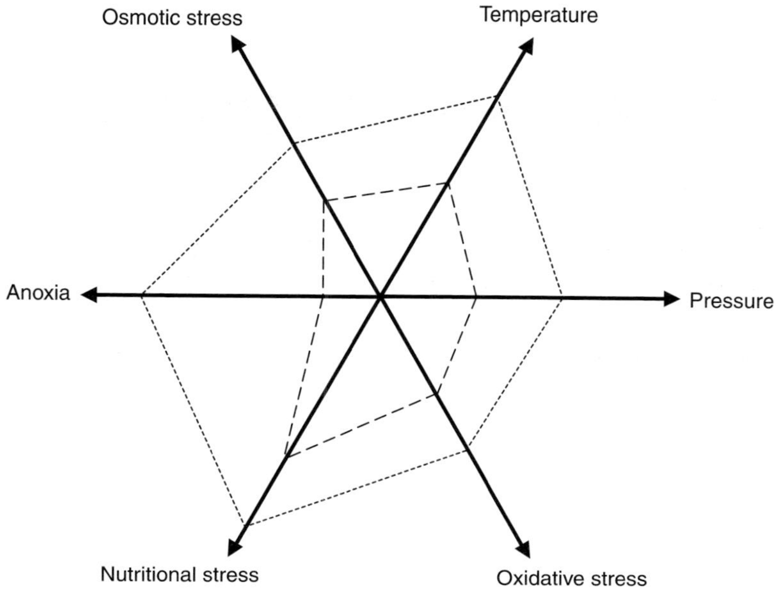

Fig. 12.1. Hypothetical plot of tolerance limits. The inner zone, bounded by the dashed line, is the 'life zone', in which the organism can complete its life cycle; surrounding it is the larger 'survival zone', where the organism can survive for varying periods of time but not complete its life cycle.

the extremophiles, have 'life zones' that are firmly in or well beyond the 'survival zone' of most other organisms. However, in reality these plots are too simplistic since they ignore any interaction between factors.

Core biochemical constituents – proteins, lipids and nucleic acids – are susceptible to direct perturbation by environmental factors. However, organisms usually die well before their proteins become denatured or their lipids undergo a phase change. The key factor in survival under adverse conditions is the need to maintain the structural and functional integrity of the cell membranes. If membrane function is not maintained, ionic gradients (and hence electrical potentials) disappear, metabolic compartmentation breaks down and key intermediates are lost to the environment. So how have nematodes managed to adapt their common set of biochemical processes to cope with the wide spectrum of environmental challenges?

12.2 Proteins and Enzymes

Both temperature and pressure have direct effects on protein structure.

12.2.1 Temperature and protein stability

As might be expected, there is a strong correlation between protein stability and adaptation temperature (McFall-Ngai and Horowitz, 1990). Most proteins are only marginally stable *in vivo* and a fine balance is struck between stabilizing and destabilizing forces. If the proteins become too rigid or too flexible they can no longer undergo the conformational changes required to maintain their biological function. Hydrophobic forces resulting from the tendency of non-polar side chains to bury themselves in the interior of the molecule are balanced by the ionic and hydrogen bond forces between exposed amino acid side chains and the aqueous environment. Further stabilization comes from covalent links, particularly sulfhydryl links. The effect of temperature on protein conformation is complex. Increasing the temperature raises the kinetic energy, and the increased molecular motion causes the protein to unfold. At the same time, increasing temperature strengthens hydrophobic interactions and weakens ionic interactions and hydrogen bonds. As the temperature is lowered the reverse happens: hydrophobic interactions are weakened and ionic and hydrogen bonds are strengthened. So, depending on their amino acid composition, proteins can exhibit either cold or heat denaturation.

Stability studies are usually carried out on isolated proteins in solution. However, in the cell, interactions with other cellular constituents can strongly influence protein stability. Hydrophilic interactions are modulated by solutes. At low ionic strengths ions associate with the surface of the protein. This effectively increases water 'activity' and results in an increase in protein solubility. At higher ionic strengths more water becomes associated with the ions, protein–protein interactions increase, and protein aggregation and precipitation follows. This is the basis of 'salting in' and 'salting out'.

Polyols increase protein conformational stability by strengthening hydrophilic interactions (Fields *et al.*, 2001). High protein concentrations per se also have a stabilizing effect and may be an important factor in anhydrobiotic species. Some extreme thermophiles have high internal concentrations of potassium (up to 2.3 M). The 'salting out' effect at high ionic strength helps to balance the thermal-induced unfolding. It is not known if nematodes from hot springs have a similar adaptation.

12.2.2 Enzymes in hot- and cold-adapted animals

A survey of animals adapted to warm and cold environments shows that, at any given temperature, the k_{cat} (turnover number) for enzymes from animals adapted to low temperature is higher than the k_{cat} for the equivalent enzyme from warm-adapted animals. There is, for example, a four- to fivefold difference in k_{cat} between equivalent enzymes in Antarctic fish and mammals measured at the same temperature (Fields and Somero, 1998). Increasing temperature leads to an increase in K_m (i.e. a decrease in affinity for substrate). But again, at any given temperature, the K_m for enzymes from animals adapted to low temperatures is lower (i.e. the substrate affinity is higher) than the K_m for the equivalent enzymes from warm-adapted animals. There is, however, no change in the basic catalytic method. Animals seem to adapt to lower temperatures not by increasing the absolute amount of enzyme, but rather by increasing the 'efficiency' of the enzyme.

To function at low temperatures cold-adapted enzymes have to be more flexible than their warm-adapted orthologues. In general, cold-adapted enzymes have fewer hydrophobic amino acids and more polar ones than warm-adapted enzymes, although the number of amino acid substitutions is usually small and confined to surface loops or hinge regions. The number of sulfhydryl links tends to be the same in warm- and cold-adapted enzymes, partly because the number of sulfydryl links in proteins is small anyhow and partly because sulfydryl links are often associated with the active site. The active sites of enzymes are, of necessity, highly conserved and this limits the sites where amino acid substitutions can take place.

When the free-living stages of mammalian or avian parasites invade their final hosts there is a sudden temperature jump. Isoenzyme and microarray analyses show that different genes are expressed in the free-living and adult stages of parasites (Thompson *et al.*, 2006). In the case of *Ascaris lumbricoides* it was found that the isoenzymes of malate dehydrogenase from eggs and adults had different temperature optima (Barrett and Fairbairn, 1971). In the infective stages of nematodes, there is evidence that the isoenzymes which will function in the final host are probably synthesized prior to infection.

12.2.3 Proteins and hydrostatic pressure

Hydrostatic effects can become significant at depths of only a few hundred metres and the effect on protein structure and function can be complex. In general, biochemical processes are accompanied by changes in volume, in part because of changes in protein conformation and in part due to changes in water organization at the protein surface. These volume changes can be positive or negative, substantial or insignificant. Processes with positive volume changes will be inhibited by high pressures; processes with negative volume changes will be enhanced at high pressures (Le Chatelier's principle). Protein denaturation and protein subunit dissociation are accompanied by volume changes of the order of -10 to $-200\,cm^3/mol$ and so are favoured at high pressures. Some enzyme reactions have negative volume changes, for example trypsin ($-10\,cm^3/mol$) and enolase ($-65\,cm^3/mol$), and are activated by high pressure; others, such as fumarase ($+28\,cm^3/mol$) and mitochondrial ATPase ($+30\,cm^3/mol$), have positive volume changes and are inhibited by pressure. Thus, high pressure can cause protein dissociation and denaturation, inhibit some enzymes and speed up others. Enzymes from deep-sea organisms have reduced sensitivity to pressure (Somero, 1990). It is not entirely clear how this is achieved but it seems to involve relatively few amino acid substitutions. Nematodes are well represented in the deep-sea benthos, with up to 200 per $10\,cm^2$ at depths of over 3000 m, but nothing is known of their physiology.

12.2.4 Stress proteins

In most organisms the initial response to stress is to suspend translation of most RNAs and start transcribing a limited set of genes which encode proteins that increase cellular survival. The best known of these are the heat shock proteins (see also Devaney, Chapter 10, this volume).

12.2.4.1 Heat shock proteins (molecular chaperones)

Despite their name, heat shock proteins (Hsps) are general stress proteins and are synthesized in response to a wide range of stressors. In nematodes, synthesis of Hsps has been shown to be triggered by osmotic stress, desiccation, anoxia, oxidative stress, ionizing radiation and pathogens, as well as temperature stress. The synthesis of Hsps is under the control of a heat shock transcription factor (HSTF). In the resting state, RNA polymerase II initiates transcription of Hsps, then pauses after approximately 25 nucleotides. Following heat shock, HSTF binds to the upstream promoter region and this restarts transcription, resulting in a rapid response. A homologue of HSTF has been shown to occur in *Caenorhabditis elegans* and *Brugia pahangi* (Hajdu-Cronin *et al.*, 2004; Devaney, 2006), suggesting that this signalling system is highly conserved throughout the animal kingdom.

Hsps show tissue specificity and are divided into a number of families, depending on their molecular weight. The principal families are: Hsp100, Hsp90, Hsp70, Hsp60 and the small Hsps (sHsps; also called α crystallins). Hsps are highly conserved across the animal kingdom. With the possible exception of Hsp100, all of the classes of Hsps have been reported from nematodes.

Although their roles are not completely understood, Hsps bind to hydrophobic amino acids in partially unfolded proteins, preventing further denaturation. Binding activates the intrinsic Hsp's ATPase, and this may power refolding of the partially denatured protein. Some classes of Hsps are expressed constitutively at low to moderate concentrations. These have a variety of cellular functions, including assisting protein folding, preventing unwanted protein aggregation, assisting protein transport across membranes and directing damaged proteins towards recycling.

12.2.4.2 Late embryogenesis abundant proteins and anhydrins

Late embryogenesis abundant (LEA) proteins and other hydrophilins have long been associated with seeds, pollen and desiccation-resistant plants. Desiccation stress in *Aphelenchus avenae* causes the upregulation of three genes encoding a group 3 LEA protein, a novel protein called anhydrin and a glutaredoxin (Browne *et al.*, 2004). Anhydrin is a small basic polypeptide with no counterpart in the sequence databases. Both the LEA protein and the novel anhydrin are highly hydrophilic and lack secondary structure in solution. On drying, the LEA protein has been shown to undergo reversible coiling, consistent with a role in desiccation resistance, and the protein also appears to be cleaved to form smaller peptides (Goyal *et al.*, 2003, 2005; Li and He, 2009). LEA proteins have been variously proposed to protect cellular structures from the effects of desiccation by acting as a hydration buffer, sequestering ions, directly protecting proteins and membranes or by aiding the renaturation of unfolded proteins (see Burnell and Tunnacliffe, Chapter 6, this volume). However, there is little, if any, supporting evidence for any of these suggestions.

LEA-like proteins have been reported in *Steinernema feltiae* and *Steinernema carpocapsae* (Solomon *et al.*, 2000; Tyson *et al.*, 2007) and there are several LEA-related sequences in the genomes of *Ditylenchus africanus*, *C. elegans* and *Caenorhabditis briggsae* (Haegeman *et al.*, 2009). Despite their obvious association with water stress, LEA proteins and anhydrins, like Hsps, may form part of a more general stress response (Gal *et al.*, 2004).

12.2.4.3 Ice-active and antifreeze proteins

Many 'cold-tolerant animals' synthesize proteins that contribute to their survival by interacting with ice. Ice-nucleating proteins initiate freezing and so prevent super-cooling. Antifreeze proteins (thermal hysteresis proteins) inhibit the growth of ice crystals and, in the presence of ice, lower the freezing point of water without altering the melting point. Finally, recrystallization proteins do not produce any significant lowering of the freezing point but inhibit ice recrystallization and alter crystal shape. Recrystallization proteins probably help to reduce membrane damage caused by growing ice crystals (see Wharton, Chapter 8, this volume). To date, ice-active proteins have been reported in the freeze-tolerant Antarctic nematode *Panagrolaimus davidi* (Wharton *et al.*, 2005) and a homologue of type II fish antifreeze protein has been found in the expressed sequence tag (EST) database from the Antarctic nematode *Plectus murrayi* (Adhikari *et al.*, 2009).

12.3 Detoxification Mechanisms

As well as physical stressors, organisms must cope with chemical stressors. These include xenobiotics from a variety of sources (pollutants, plant secondary metabolites, anthelmintics, nematicides, etc.), as well as heavy metals and free radicals. After a xenobiotic compound has been ingested, either it can be excreted essentially unchanged or it can change spontaneously in a non-enzymatic reaction; this can occur, for example, when the molecule is exposed to particular physiological conditions, such as high or low pH in the gut. However, the majority of xenobiotic molecules are metabolized enzymatically into other, usually less toxic, derivatives.

12.3.1 Xenobiotic metabolism

Animals metabolize xenobiotics in three phases (Table 12.1). In phase 1, hydrolytic, oxidative or reductive enzymes introduce reactive groups such

Table 12.1. The three phases of detoxification.

Phase 1 (bioactivation)	Phase 2 (conjugation with)	Phase 3 (elimination)
Oxidation[a]	Glutathione[b]	Excretion[b]
Dehalogenation[c]	Glucuronides[d]	Sequestration[e]
Hydroxylation[a]	Methyl groups[d]	Further metabolism[f]
Reduction[b]	Acetyl groups[d]	
Hydrolysis[b]	Amino groups[c]	
	Sulfate[d]	
	Phosphate[d]	
	Thiosulfate (Rhodanese)[b]	

[a]Large number of cytochrome P450 sequences found in nematode genomes.
[b]Demonstrated in nematodes.
[c]Not demonstrated in nematodes.
[d]Limited evidence in nematodes.
[e]High levels of binding proteins found in nematodes including metallothioneins, but detoxification roles not established.
[f]Low activities of cysteine conjugate β-lyase have been found in nematodes.
(From Barrett, 1997.)

as hydroxyl, amino, carboxyl or sulfhydryl into the molecule. In phase 2, the activated molecule is conjugated with a low molecular weight organic compound such as an amino acid, a sugar, glutathione, acetate or propionate or an inorganic ion such as sulfate or phosphate. Compounds which already possess active groups can be conjugated directly without phase 1 metabolism. In phase 3, conjugates may be further metabolized before being excreted or sequestered. Conjugation normally increases water solubility, making excretion easier, and reduces chemical activity; however, some compounds, such as ethylene chloride, are chemically activated by conjugation.

In vertebrates and most invertebrates, the principal phase 1 reactions are oxidative, catalysed by cytochrome P450. Cytochrome P450 monooxygenases catalyse the oxidation of a wide range of organic compounds, the cytochrome acting as a terminal oxidase accepting electrons from NADPH via NADPH cytochrome P450 reductase or from NAD via cytochrome b_5. Some 75 full-length cytochrome P450 gene sequences, comprising 18 families, have been identified in the genome of *C. elegans*, as well as an NADP cytochrome P450 reductase. At least one of these cytochrome P450 families (CYP35) is strongly induced by xenobiotics, and two other families (CYP29, CYP33) are involved in eicosapentanoic acid metabolism (Kulas *et al.*, 2008). The genome of *Pristionchus pacificus* has 198 CYP gene sequences (see Herrmann and Sommer, Chapter 4, this volume), whilst *Meloidogyne incognita*, by contrast, has only 27 full or partial CYP genes (Abad *et al.*, 2008). An aryl hydrocarbon receptor (AHR-1) and its nuclear translocator (AHRNT) have been identified in *C. elegans* (Qin and Powell-Coffman, 2004). Aryl hydrocarbon receptors are ligand-activated transcription factors. In mammals they bind to xenobiotics in the cytoplasm and are then translocated into the nucleus, where they mediate a range of biochemical and physiological responses. A second 'xenobiotic sensing' receptor gene, *nhr-8*, has also been found in *C. elegans* and may regulate P450 expression (Lindblom and Dodd, 2006). Both of these receptors belong to a signalling system that seems to be highly conserved across the animal kingdom.

Low levels of cytochrome P450 activity have been reported in the free-living larvae of *Heligmosomoides polygyrus* (Kerboeuf *et al.*, 1995) and CYP homologues are present in the genome of *Haemonchus contortus*. However, attempts to demonstrate cytochrome P450 activity enzymatically in a number of adult animal-parasitic helminths (*A. lumbricoides*, *H. contortus*, *H. polygyrus*, *Onchocerca gutterosa*) have proved unsuccessful. It is possible that in adult animal-parasitic nematodes, peroxidases or some other oxidative enzymes can substitute for the P450 monooxygenases. Alternatively, the absence of P450 activity might be related to the need to reduce lipid peroxidation in the parasites' membranes, since P450 monooxygenases are an important source of superoxide radicals.

A wide range of hydrolases have been found in nematodes, including *O*- and *N*-deacetylases and amidases. They also have enzyme systems capable of reducing aldehydes and ketones, as well as some unusual azo- and nitro-reductases, including a reductase which can reduce nitric oxide (Barrett, 1997).

A gene for cyanate lyase has been demonstrated in *Meloidogyne hapla* (Opperman *et al.*, 2008). Cyanate lyase is a bacterial enzyme that breaks down cyanate to ammonia and carbon dioxide and it is thought to have been acquired by *M. hapla* through horizontal gene transfer.

Phase 2 in nematodes seems to be relatively limited when compared to mammals. The principle phase 2 reaction is conjugation with glutathione catalysed by glutathione transferase, although there is evidence for some conjugation with glucose, phosphate and sulfate as well as *N*-acetylation (Hattori *et al.*, 2006). Genes coding for UDP-glucoronosyl transferase, glycosyl transferase and methyltransferase have been detected in *C. elegans*, but these genes are not necessarily involved in detoxification (Reichert and Menzel, 2005). Glutathione transferases are a group of multifunctional proteins found in virtually all plant and animal cells. Glutathione transferase activity seems to be universally distributed in nematodes and accounts for 2–5% of the total cytosolic protein (Barrett, 1995; Persaud *et al.*, 1997). Seven species-independent classes of glutathione transferases (alpha, mu, pi, theta, sigma, zeta and omega) are currently recognized, based on substrate and inhibitor specificity and amino acid sequence. However, although nematodes have alpha- and pi-type isoenzymes, the majority of nematode glutathione transferases belong to one of two novel, nematode-specific classes (Van Rossum *et al.*, 2004; Schuller *et al.*, 2005). The evolution of multi-gene forms seems to be characteristic of xenobiotic metabolizing enzymes and is a way of dealing with a wide range of different substrates.

Another glutathione-dependent detoxification pathway found in nematodes is the glyoxalase system. This consists of two enzymes, glyoxalase 1 and glyoxalase 2, which catalyse the breakdown of 2-oxoaldehydes. Rhodanese, which catalyses the detoxification of cyanide and sulfide by conjugation with thiosulfate, has also been found in nematodes (Barrett, 1997).

12.3.2 ATP binding cassette (ABC) transporters

Conjugation with glutathione normally reduces chemical reactivity. However, glutathione conjugates are potent inhibitors of glutathione-dependent enzymes, including glutathione reductase and glutathione transferase itself. For this reason, glutathione conjugates have to be removed by phase 3 metabolism. Mammals can metabolize glutathione conjugates by a variety of pathways, ultimately to mercapturic acids. There is little evidence that nematodes can metabolize glutathione conjugates and they are probably transported out of cells by an ABC transporter. The ABC transporters are a protein superfamily with representatives in all phyla from prokaryotes to mammals. P-glycoprotein belongs to the MDR/TAP subfamily, members of which are involved in multidrug resistance. Approximately 60 ABC genes have been identified in *C. elegans* (Lindblom and Dodd, 2006) and 129 in the genome of *P. pacificus* (see Herrmann and Sommer, Chapter 4, this volume). P-glycoprotein genes have also been shown in *M. incognita, Onchocerca volvulus* and *H. contortus* (Riou *et al.*, 2005; Ardelli *et al.*, 2006; Abad *et al.*, 2008).

12.3.3 Xenobiotic binding proteins

An alternative to excretion for xenobiotics is sequestration. Insects have what appear to be specific lipoprotein binding proteins, but no such specific binding proteins have been found in nematodes. However, two groups of proteins found in nematodes, the glutathione transferases and lipid binding proteins, are capable of binding xenobiotic compounds non-covalently (Barrett, 1997; McDermott *et al.*, 2002). Since both of these proteins occur in the cytoplasm in high concentrations (up to 50 μM), sequestration could be a significant route for detoxification in nematodes. Compounds bound to proteins can have a kinetic advantage as substrates for other enzymes, as compared to unbound compounds, and so may be preferentially metabolized.

12.3.4 Heavy metals

The toxicity of heavy metals is due to the formation of complexes with organic molecules and in particular with sulfydryl groups. Cells respond to heavy metals by increasing the steady-state levels of a range of proteins. The function of the upregulated proteins is to repair cellular damage and remove the toxicant. Heavy metals are sequestered by chelation or are removed by specific metal ion pumps. A secondary effect of metal toxicity is often the production of reactive oxygen species, and the metal stress response can also involve antioxidant enzymes. Exposure of *C. elegans* to cadmium results in the upregulation of some 48 genes encoding, among other things, metallothioneins, glutathione transferase, superoxide dismutase, Hsp70, ubiquitin, catalase and ABC transporters (Liao *et al.*, 2002). The metallothioneins, of which there are two isoforms in *C. elegans*, are low molecular weight, cysteine-rich proteins whose primary function is to bind heavy metals. As well as their role in xenobiotic metabolism, glutathione transferases are also able to sequester heavy metals. *Caenorhabditis elegans* may also be able to synthesize class III metallothioneins (phytochelatins), which are usually more associated with plants. These are polymeric peptides which bind heavy metals and consist of just three amino acids – glycine, glutamine and cysteine (Vatamaniuk *et al.*, 2005).

The heavy metal stress response in *C. elegans* is mediated via the mitogen-activated protein (MAP) kinase pathway (Kim *et al.*, 2004). However, unlike other metallothionein genes, the promoter regions of the *C. elegans* genes do not seem to have a functional response element (MRE) nor does the genome of *C. elegans* contain a copy of the transcriptional factor MTF-1. In *C. elegans*, an arsenite-inducible gene, *aip1*, appears to be a homologue of mammalian AIRAP (arsenite-inducible RNA-associated protein). This is a histidine- and cysteine-rich RNA-associated protein which binds to the proteasome 19S regulatory cap and may facilitate the removal of misfolded proteins formed as the result of arsenite poisoning (Yun *et al.*, 2008).

12.3.5 Antioxidant systems

All aerobic organisms require antioxidant enzymes to cope with reactive oxygen species (ROS) such as the superoxide anion ($O_2^{\bullet-}$), the hydroxyl radical ($^{\bullet}OH$) and hydrogen peroxide generated during normal metabolism. The metabolism of some xenobiotics such as redox cycling drugs, heavy metal poisoning and exposure to radiation can also generate ROS. In addition, animal-parasitic nematodes have to deal with ROS generated by the effector arm of the host's immune system. ROS can cause damage to all major classes of biological material, giving rise to protein oxidation, lipid peroxidation and DNA strand breaks. Nematodes infecting the swim bladder of fishes could, at least in theory, be exposed to hyperbaric oxygen.

Superoxide dismutases (SODs), which deal with the superoxide radical, have been found in every nematode species examined so far. The SODs are a family of metalloenzymes that catalyse the conversion of superoxide to hydrogen peroxide. Since hydrogen peroxide also causes oxidative damage, SODs have to act in concert with systems to remove hydrogen peroxide. In nematodes there are two systems that can remove hydrogen peroxide: catalase and peroxyredoxin. The latter, which can also convert alkylhydroperoxides to the corresponding alcohols, occurs in high concentrations in cells (up to 20 μM) and is probably the more important of the two (Henkle-Dührsen and Kampkötter, 2001). In addition to their antioxidant function, the peroxyredoxins may have an important redox sensing role. With one exception so far (*Setaria cervi*), selenium-containing glutathione peroxidases have not been found in nematodes, although selenium-independent glutathione peroxidases are present (Singh and Rathur, 2005). The selenium-independent glutathione peroxidases remove lipid hydroperoxides but, unlike their selenium-dependent counterparts, have little or no activity with hydrogen peroxide. Glutathione transferases rapidly conjugate the products of lipid peroxidation (lipid epoxides, α-β unsaturated aldehydes) with glutathione and are an important component of the oxidative stress response in nematodes (Liebau *et al.*, 2000). Other components of the oxidative stress response include enzymes involved in glutathione synthesis (γ-glutamine cysteine synthase) and various disulfide reductases (glutathione reductase, thioredoxin, glutaredoxin). Non-enzymatic protection against oxidative damage is provided by antioxidants such as ascorbate, α-tocophoral, glutathione and cysteine.

In several animal-parasitic nematodes, antioxidant enzymes have been found in the secretory/excretory products and so may provide protection against host-derived ROS. Antioxidant enzymes in plant-parasitic nematodes may similarly provide protection from plant defences (Jones *et al.*, 2004).

In *C. elegans*, the oxidative stress response is coordinated by the transcription factor SKN-1 (An *et al.*, 2005). In response to oxidative stress, SKN-1 accumulates in the cell nucleus and induces genes related to phase 2 metabolism and glutathione synthesis. Although *C. elegans* SKN-1 binds to DNA via a unique monomeric mechanism, it is closely related to its functional counterpart in mammals.

Upregulation of antioxidant enzymes occurs not only in response to oxidative stress but also as part of a general stress response. Upregulation of antioxidant enzymes also seems to be a central factor in extended lifespan in *C. elegans* (Vanfleteren, 1993).

12.4 Energy Metabolism

A key strategy for dormant or resting stages is conservation of energy reserves, but they still need to maintain an adequate level of energy production to maintain ionic and metabolic gradients, and a limited amount of protein turnover is inevitable. Dormant and resting stages of animals are able to reduce their metabolic rate to between one-fifth and one-twentieth of their basic metabolic rate. Following recovery, there is often a temporary burst of metabolism correlated with the recovery of membrane function (Barrett, 1991a; Wharton *et al.*, 2000). In anhydrobiotic nematodes, where metabolism has almost certainly halted, the dry tissue still contains appreciable amounts of ATP (Barrett, 1982). This ATP is necessary to 'prime' the catabolic pathways when the worms are rehydrated and metabolism starts once more.

12.4.1 Aerobic metabolism

Free-living, entomopathogenic and plant-parasitic nematodes, as well as the free-living stages of animal-parasitic nematodes, break down carbohydrate by the usual steps of glycolysis to give pyruvate, NADH and ATP. This takes place in the cytoplasm:

$$\text{Glucose} + 2\text{ADP} + 2\text{Pi} + 2\text{NAD}^+ \rightarrow 2\text{Pyruvate} + 2\text{ATP} + 2\text{NADH} + 2\text{H}^+ + 2\text{H}_2\text{O}$$

Pyruvate is transported into the mitochondrion, where it is completely broken down by the TCA cycle to CO_2 and H_2O, with the formation of reducing equivalents in the form of NADH and $FADH_2$ (and a substrate-level GTP). The mitochondrial reducing equivalents (plus the glycolytic NADH, which is also transported into the mitochondria) are oxidized by the electron transport chain to yield ATP. The net reaction for the oxidation of glucose via glycolysis, the TCA cycle and the electron transport chain is:

$$\text{Glucose} + 36\text{ADP} + 36\text{Pi} + 6O_2 \rightarrow 36\text{ATP} + 6CO_2 + 42H_2O$$

The ATP calculations ignore any energy cost involved in transporting intermediates into the mitochondrion. When glycogen is the substrate rather than glucose, the net production of ATP is increased by 1, to 37 ATP/glucose unit. The main energy reserve in nematodes is usually glycogen (3–20% of the dry weight). In addition, many nematodes contain appreciable amounts of trehalose (0.1–5% of the dry weight). Before it can enter the glycolytic sequence, trehalose must be hydrolysed into two glucose molecules by the enzyme trehalase (Pellerone *et al.*, 2003). Trehalose has a particular advantage in dormant and resting stages, in that it provides a way of storing glucose as a non-reducing, soluble, low molecular weight molecule. Being non-reducing means that trehalose is relatively unreactive and so does not take part in adventitious reactions; it also may have a role as part of a general stress response, stabilizing macromolecular structures (see Section 12.8).

Lipids (triacylglycerides) are first hydrolysed to glycerol (which can enter the glycolytic pathway) and long chain fatty acids. The fatty acids are then broken down by the β-oxidation spiral. This pathway, which is also located in the mitochondrion, generates acetyl-CoA and reducing equivalents, which are then metabolized by the TCA cycle and the electron transport chain. The overall equation for fatty acid oxidation is:

$$\text{Palmitate} + 23O_2 + 129\text{ADP} + 129\text{Pi} \rightarrow 129\text{ATP} + 16CO_2 + 145H_2O$$

Compared with carbohydrate, lipids are a more reduced energy source and, weight for weight, have a higher calorific value (saturated lipids having a higher calorific value than unsaturated ones). Lipids are therefore the ideal energy store for resting or dormant stages. The infective stages of entomopathogenic and animal-parasitic nematodes in particular have up to 35–40% dry weight lipid (Barrett and Wright, 1998). Relatively high levels of free fatty acids have been found in some plant-parasitic nematodes and infective larvae (Barrett, 1968; Womersley *et al.*, 1982; Holz *et al.*, 1997). A possible advantage of storing free fatty acids is that they are membrane permeable and so do not need a specific membrane-located carrier. High levels of free fatty acid could also be related to membrane remodelling (see Section 12.6). An increase in lipid stores frequently precedes the formation of dormant stages (Narbonne and Roy, 2009). Anhydrobiotic nematodes also typically have very high lipid contents, and in this case the lipid may provide a supportive matrix for the intracellular organelles when the tissues are dehydrated.

Amino acids are not thought to be significant substrates for oxidative metabolism in nematodes. Amino acids can be catabolized by pathways which ultimately feed into the TCA cycle, where the intermediates can be fully metabolized. Being at the same oxidation state as carbohydrates, the ATP yields from amino acid catabolism are also similar (e.g. 15 ATP/mol for alanine, 27 ATP/mol for glutamate).

12.4.2 Anaerobic metabolism

Most of the energy released by carbohydrate or lipid catabolism is generated in the electron transport chain. In the absence of oxygen, the electron transport chain cannot function and the TCA cycle and β-oxidation spiral come to a halt. There is no evidence in nematodes for the co-fermentation of amino acids and carbohydrate (in molluscs aspartate is broken down anaerobically to succinate or propionate and this is stoichiometrically coupled with glycolysis). Glycolysis can function anaerobically by linking the re-oxidation of NADH with the reduction of a suitable metabolic intermediate, thus maintaining the intracellular redox potential. In mammals this involves the reduction of pyruvate to lactate, with a net yield of 2 ATP/mol of glucose (3 ATP when glycogen is the substrate):

$$\text{Glucose} + 2\text{ADP} + 2\text{Pi} \rightarrow 2\text{Lactate} + 2\text{ATP} + 2\text{H}_2\text{O}$$

How entry into anaerobiosis is controlled at the cellular level in nematodes is not clear. A phosphorylation/dephosphorylation mechanism is one possibility, and in *C. elegans* embryos a switch to anaerobic metabolism is accompanied by the dephosphorylation of cell cycle proteins (Padilla *et al.*, 2002). Several oxygen-sensing neurones have been described in *C. elegans* and these involve a hexa-coordinated neuroglobin, GLB-5, in conjunction with an atypical guanylate kinase (Persson *et al.*, 2009). However, it is not clear if GLB-5 acts as an oxygen receptor or as an oxygen sink. In mammals neuroglobins and the related cytoglobins are thought to have a role in protecting nervous tissue during hypoxia, either by facilitating the diffusion of oxygen through tissues or by scavenging reactive oxygen intermediates.

As organisms enter anaerobiosis, their metabolism may go through a transition phase before entering a steady state, and there is evidence, again in *C. elegans*, that the change from aerobic to anaerobic metabolism is a two-stage process (Paul *et al.*, 2000). In mammals, a hypoxia-inducible factor (HIF) regulates cellular and systemic responses to low oxygen levels. When oxygen levels are high, the HIF-1α subunit is hydroxylated and targeted for degradation by the von Hipper-Lindau tumour suppressor protein (VHL). Genes encoding homologues of HIF (*hif*-1) and VHL (*vhl*-1) have been found in *C. elegans*. Some 110 hypoxia-regulated genes have also been identified, of which 63 appear to depend on HIF signalling (Jiang *et al.*, 2001; Shen *et al.*, 2005). This suggests that the mechanisms for hypoxia signalling are probably highly conserved among the metazoa. The *DAF-2* insulin receptor homologue also seems to be involved (Jiang *et al.*, 2001; Scott *et al.*, 2002).

Most free-living and plant-parasitic nematodes can survive at least brief periods of anaerobiosis. In contrast to animal-parasitic nematodes, which tend to excrete a wide range of end products (Barrett, 1984), free-living, entomopathogenic and plant-parasitic nematodes produce a relatively limited range of end products under anaerobic conditions. Ethanol is the main end product in *A. avenae* and *Bursaphelenchus xylophilus* (Cooper and Van Gundy, 1971; Mendis and Evans, 1984; Bolla *et al.*, 1987). In addition, *A. avenae* excretes small amounts of glycerol, lactate and succinate. *Meloidogyne incognita* excretes mainly lactate (Anil *et al.*, 1998). Under anaerobic conditions *Panagrellus redivivus* excretes lactate and ethanol, with small amounts of alanine, acetate, propionate, glycerol and acetoin (Barrett and Butterworth, 1984); glycerol is also produced by *C. briggsae* and *Turbatrix aceti* (Liu and Rothstein, 1976). Finally, *C. elegans* excretes lactate, acetate, succinate and propionate (Föll *et al.*, 1999). In many free-living invertebrates, lactate dehydrogenase is replaced by a functionally analogous imino acid dehydrogenase, resulting in the formation of an imino acid (octopine, alanopine, strombine, etc.) rather than lactate. Imino dehydrogenases have not, however, been demonstrated in nematodes. Alanine is also often a major anaerobic end product in free-living molluscs and annelids but, again, not in nematodes. Nematodes are complex organisms; different tissues may well produce different end products and under anaerobic conditions nematodes usually excrete a mixture of major and minor end products.

When organisms are only periodically exposed to anaerobic conditions, the resynthesis of carbohydrate from the anaerobic end products is important for substrate conservation. Consequently, the accumulated end products must be tissue compatible. Products such as lactate, glycerol and acetate can be readily resynthesized into carbohydrate; resynthesis from ethanol requires a specialized anapleuric cycle called the glyoxalate cycle. This cycle consists of two enzymes, isocitrate lyase and malate synthase, which effectively short circuit the TCA cycle. First demonstrated in nematodes in developing *A. lumbricoides* eggs, where it catalyses the net conversion of lipid to carbohydrate (Barrett *et al.*, 1970), the cycle has subsequently been demonstrated in a range of free-living, entomopathogenic and plant-parasitic nematodes (Colonna and McFadden, 1975; Madin *et al.*, 1985; Gordon, 1987; O'Riordan and Burnell, 1990), where it is involved in gluconeogenesis.

The entomopathogenic nematode *S. carpocapsae* more closely resembles the animal-parasitic nematodes, and under anaerobic conditions excretes primarily succinate (Thompson *et al.*, 1991).

12.4.3 Animal-parasitic nematodes

In their final hosts, adult animal-parasitic nematodes often inhabit sites which have a low or fluctuating oxygen tension, although few, if any, of these sites are strictly anaerobic. Adult animal-parasitic nematodes have adopted an essentially anaerobic energy metabolism. There is no β-oxidation of fatty acids and again no catabolism of amino acids, so they have an absolute requirement for carbohydrate for their energy needs. On the basis of their end products, animal-parasitic nematodes can be broadly divided into two groups. First, there are those, mostly blood and tissue parasites, that rely solely on glycolysis and excrete primarily lactate. There are few reports of ethanol as an anaerobic end product in animal-parasitic nematodes, although alcohol dehydrogenase has been found in several species. Second, there are those nematodes that fix carbon dioxide and have what may be described as a 'partial, reverse TCA cycle' (Barrett, 1984), the best-known example of which is *A. lumbricoides* (Fig. 12.2). In the '*Ascaris* type' there is a normal glycolytic sequence as far as phosphoenolpyruvate; pyruvate kinase activity is low in these nematodes, so instead of forming pyruvate, carbon dioxide fixation takes place and phosphoenolpyruvate is converted to oxaloacetate by a cytoplasmic phosphoenolpyruvate carboxykinase. The oxaloacetate is reduced to malate by a cytoplasmic malate dehydrogenase and this serves to re-oxidize the glycolytic NADH. Malate is then transported into the mitochondrion, where it undergoes a dismutation reaction. Part is oxidatively decarboxylated to pyruvate via an NAD-linked malic enzyme; malate is also in equilibrium with fumarate via fumarase and the fumarate is reduced to succinate. So the immediate end products of this pathway are succinate and pyruvate, although these are usually further metabolized to propionate and acetate and eventually to short chain fatty acids such as valerate, 2-methylbutyrate and 2-methylvalerate. The route from oxaloacetate to succinate is identical to the second span of the TCA cycle but operating in reverse.

The advantage of the '*Ascaris*-type' scheme is that the reduction of fumarate to succinate is linked to the electron transport chain and results in the formation of additional ATP. This increases the yield of ATP per mol of glucose metabolized from two in anaerobic glycolysis to three. Still low, but a 50% increase! The decarboxylation of succinate to propionate and of pyruvate to acetate are both potential energy-yielding steps, as is one of the steps in the synthesis of the short chain fatty acids (Komuniecki *et al.*, 1981), so in *Ascaris* the ATP yield per mol of glucose could be as high as four or five.

There are a limited number of metabolic intermediates required for synthesis in animals and of the ten key intermediates needed (triose phosphate, tetrose phosphate, pentose phosphate, hexose phosphate, phosphoenolpyruvate, pyruvate, acetyl-CoA, 2-oxoglutarate, succinyl-CoA and oxaloacetate), anaerobic glycolysis can provide only four. The '*Ascaris*-type' system, by contrast, provides eight.

Most, if not all, nematodes probably require at least some oxygen for normal sustained activity but can survive under anaerobic conditions for varying lengths of time (von Brand, 1966). They all use oxygen when it is available; that is, at least in air, they have a measurable oxygen consumption and are capable of oxidative phosphorylation. In addition, they contain several enzymes, proline-4-monoxygenase, cysteine dioxygenase, tyrosine hydroxylase, phenylalanine hydroxylase and tryptophan hydroxylase, which have an absolute requirement for molecular oxygen (Barrett, 1991b). So perhaps

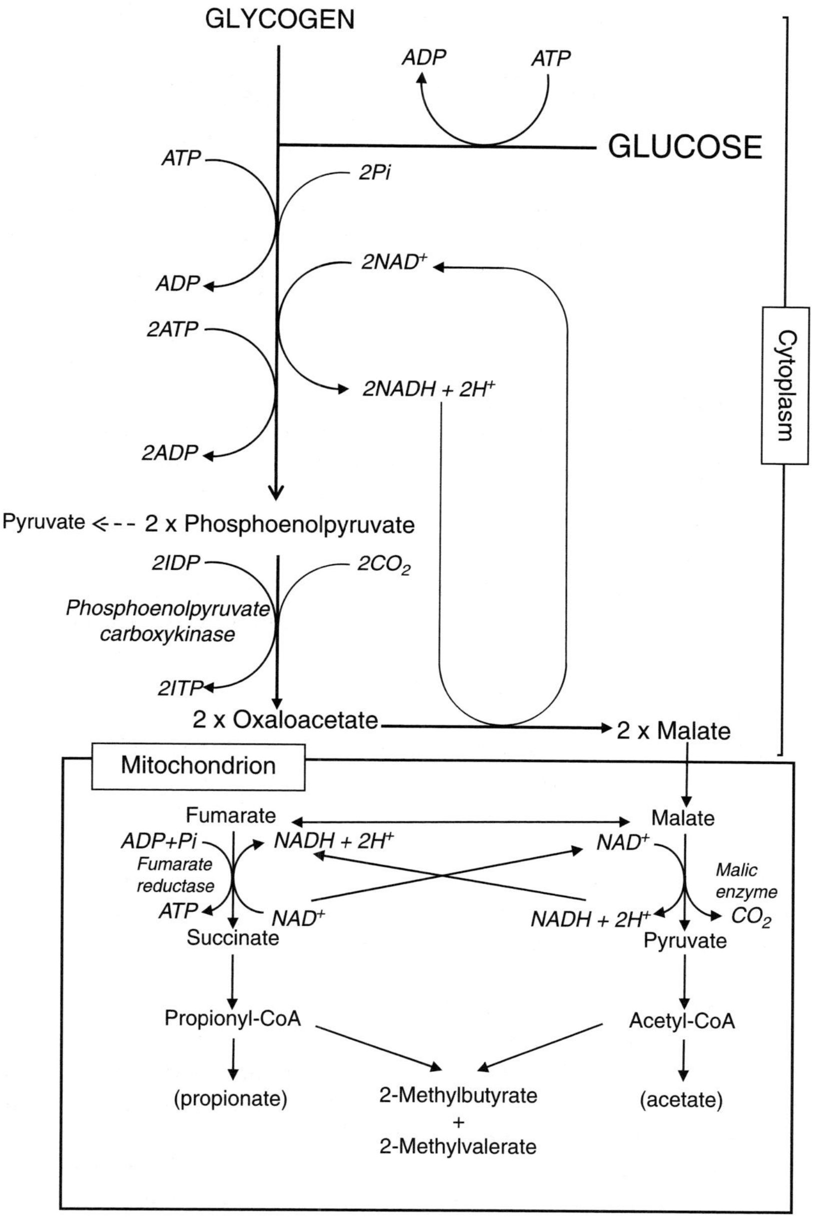

Sum: Glucose + 2IDP + ADP + 3Pi + CO_2 = Succinate + Pyruvate + 2ITP + ATP + H_2O

Fig. 12.2. Carbohydrate catabolism in adult *Ascaris lumbricoides*. Glycogen is broken down by the normal glycolytic sequence as far as phosphoenolpyruvate. Pyruvate kinase activity is low and instead carbon dioxide fixation takes place and phosphoenolpyruvate is converted to oxaloacetate by an IDP-dependent cytoplasmic phosphoenolpyruvate carboxykinase. The oxaloacetate is then reduced to malate by a cytoplasmic malate dehydrogenase. Malate is transported into the mitochondrion, where a dismutation takes place; part is oxidatively decarboxylated to pyruvate and part is reduced to succinate. Pyruvate and succinate are then the starting point for branched chain fatty acid synthesis. (After Barrett, 1984.)

animal-parasitic nematodes should be thought of as micro-aerobic rather than anaerobic.

12.4.4 Anaerobic metabolism in an aerobic environment

In most aerobic organisms, the switch from aerobic to anaerobic metabolism is accompanied by an increase in the rate of carbohydrate utilization. This 'Pasteur effect' reflects the relatively low yield of ATP/mol of glucose catabolized in aerobic pathways compared with aerobic ones, so to maintain the same rate of ATP production more substrate must be used. The single largest ATP sink during anoxia (up to 75% of the ATP usage) is ion pumps. Hypoxia-tolerant cells tend to have a low inherent permeability due to a low ion channel density and/or low ion channel activity. During anoxia, ion channel activity may be further decreased by 'channel arrest' (Boutilier and St Pierre, 2000).

In anaerobes, suppression of the 'Pasteur effect' is a key adaptation for substrate conservation. In animal-parasitic nematodes, anaerobiosis is a steady-state condition and persists even in the presence of air, an extremely effective way of suppressing the 'Pasteur effect'.

The '*Ascaris*-type' pathway is characteristic of nematodes which live in the vertebrate intestine. High levels of carbon dioxide occur in the vertebrate intestine, especially the rumen (up to 10 mM). So, possibly, the '*Ascaris*-type' pathway may also be an adaptation to high ambient carbon dioxide levels.

12.4.5 The thiobios

Underneath the surface of most marine and swamp sediments there exists an anaerobic sulfide layer, where the levels of hydrogen sulfide can reach low mM concentrations. This region is colonized by several groups of organism, including molluscs, annelids and nematodes. Hydrogen sulfide is extremely poisonous (more so than hydrogen cyanide), inhibiting cytochrome oxidase and binding to haemoglobin, although the fact that the thiobios is anaerobic serves to reduce the impact of hydrogen sulfide on metabolism. Nevertheless, the organisms that live there show a number of specializations which help them survive (Vetter *et al.*, 1991). There is no evidence that the enzyme systems of these organisms are especially resistant to inhibition by sulfide, nor are the organisms impermeable to it. Many of the thiobios organisms have symbiotic sulfide-oxidizing bacteria. Nematodes of the subfamily Stilbonematinae have surface-bound sulfide-oxidizing bacteria and there is evidence that the nematodes actively 'graze' on the bacteria (Powell *et al.*, 1979; Ott *et al.*, 2004). Endosymbiotic sulfide-oxidizing bacteria occur in nematodes of the genera *Astomonema*, *Parastomonema* and *Rhabdothyreus*. Nematodes of these three genera all lack a gut and it is assumed that they are dependent on the chemoautotrophic bacteria for their nutrition (Musat *et al.*, 2007). Other adaptations found in annelids and molluscs from the thiobios are sulfide-binding proteins and an endogenous sulfide-oxidizing system, neither of which have yet been reported from nematodes.

Several species of nematodes occur around hydrothermal vents, where they have to cope with up to 30 mM sulfide and temperatures possibly as high as 60°C (Flint *et al.*, 2006). Their physiology is unknown, but it has been suggested that the high hydrostatic pressures at these depths (2500 m) may offset some of the effects of high temperatures (Hazel and Williams, 1991).

12.5 Membranes and Lipids

Membranes serve a wide range of essential functions, as physical barriers, in bioenergetics, in cell signalling, in neuronal activity and as an organizing matrix for multi-enzyme systems. The forces that stabilize the lipid bilayer in membranes are the same forces that stabilize proteins in solution. Hydrophobic interactions are responsible for the arrangement of the non-polar acyl chains in the internal core, whilst the polar head groups face the aqueous phase on either side of the membrane. Cholesterol, which has both polar and non-polar groups, intercalates in the non-polar part of the bilayer but retains interactions with the polar head groups. Temperature, pressure and desiccation all affect membrane structure. In order to maintain their biological function, membranes have to keep their physical properties within narrow limits. Membranes can adapt to environmental change in two ways: intrinsic adaptations and extrinsic adaptations. Intrinsic changes involve alterations in lipid composition and in the relative proportions of lipid and protein. Extrinsic adaptations involve changes in the surrounding medium, such as changes in pH or interactions with polyols or stress proteins.

12.6 Membranes and Temperature

Membrane processes are highly sensitive to temperature change and this is a key factor in determining an organism's thermal tolerance limits. At physiological temperatures membranes are in a fluid or liquid crystal phase. As temperatures fall the motion of the phospholipid acyl side chains is reduced and they begin to pack together more closely. At sufficiently low temperatures (below the phase transition temperature, T_m) the lipids enter a gel phase (Fig. 12.3). Because of heterogeneity in the membrane lipids there may be a transition phase where liquid crystal and gel phases coexist in different regions of the membrane. In the gel phase, membrane function is severely impaired, integral proteins become excluded from the membrane, the lateral movement of proteins (an essential component of cell signalling) is restricted and permeability barriers are breached.

When temperatures are raised the motion of the acyl side chains is increased, and the membrane becomes more fluid and moves towards an inverted hexagonal phase structure (Fig 12.3). In the inverted hexagonal phase (H_{II}) the phospholipids form tubular structures with the polar head groups towards the centre and the acyl chains to the outside. Not surprisingly, in the inverted hexagonal phase the membrane barrier properties are completely destroyed.

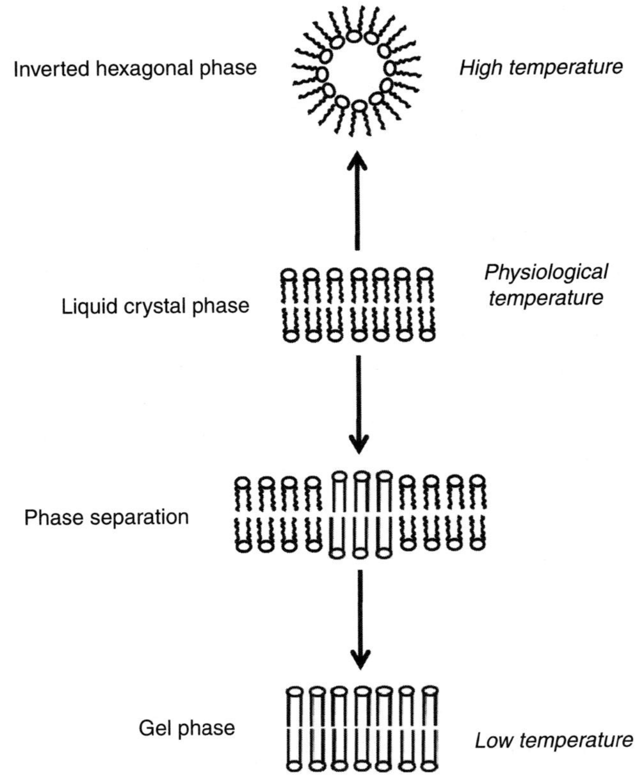

Fig. 12.3. Effect of temperature on membrane structure. As the temperature falls, gel phase regions develop, which may separate from the liquid crystal phase (phase separation). At sufficiently low temperatures the bulk of the phospholipids enter the gel phase. As the temperature rises above 'physiological', the tendency for lipids to enter the inverted hexagonal phase increases, in part because the acyl chains become more spread out.

Under normal conditions membranes possess some regions where hexagonal structures are formed, since these structures are necessary for normal membrane function such as endocytosis and membrane trafficking.

12.6.1 Intrinsic adaptations to temperature

Lipids are enzyme products, not gene products, and as such offer an almost unlimited range of permutations for remodelling. Changes in the phospholipid fatty acid chains in response to temperature change have probably received the most attention. There are at least five possibilities: (i) change in chain length; (ii) change in number of double bonds; (iii) change in the position of the double bonds in the chain; (iv) *cis* versus *trans* configuration of the double bonds; and (v) swapping position of the acyl chains on the glycerol backbone. The most common adaptation reported in relation to temperature change is a decrease in saturation (i.e. more double bonds) at lower temperature. This is brought about by fatty acid desaturases, which are activated by low temperatures. Nematodes are no exception, and a decrease in the proportion of unsaturated fatty acids as temperature increases has been shown in a number of species (Lower *et al.*, 1970; Lyons *et al.*, 1975; Tanaka *et al.*, 1996; Hatab and Gaugler, 1997; Jagdale and Gordon, 1997). Similarly, the fatty acids of parasites from cold-blooded hosts have been found to have a higher degree of unsaturation than those from mammals (Sidorov and Smirnov, 1980). Changes in chain length in response to temperature change seem more characteristic of bacteria and plants than animals.

The high levels of free fatty acids found in some free-living, infective nematode larvae may be related to the sudden temperature change experienced when they invade their final mammalian or avian host (Barrett, 1968). Odd-numbered and branched chain fatty acids have been found in animal-parasitic nematodes and again may be related to membrane remodelling (Joachim *et al.*, 2000).

Another frequent response to temperature change is a change in the relative proportions of phosphatidylethanolamine and phosphatidylcholine (PE/PC ratio) and this has also been found in nematodes (Hatab and Gaugler, 1997). Phosphatidylcholine, with its bulky head group, favours gel phase structures, whilst phosphatidylethanolamine favours hexagonal phase formation. So the PE/PC ratio tends to fall as temperatures increase and increase as temperatures fall. An increase in plasmalogen levels with increasing environmental temperature has also been reported (Roots and Johnston, 1968).

Cholesterol stabilizes the liquid crystal phase and inhibits gel phase formation. The cholesterol level in membranes tends to increase as the temperature rises and the cholesterol level in ectotherms is lower than that in endotherms. Interestingly, nematodes cannot synthesize cholesterol *de novo* and must obtain it from their diet. The overall lipid/protein ratio has also been noted to increase at lower temperatures.

Because of membrane heterogeneity, not all the membranes in an organism will necessarily show the same changes in lipid composition. In general, mitochondrial membranes show the most 'complete' temperature compensation, plasma membranes less so. The different changes in lipid composition also have different time constants; some of the changes can take place within hours of a temperature change, others may take place over several generations.

12.6.2 Extrinsic adaptations to temperature

Extrinsic factors also have a role in membrane stability. A rise in cellular pH (as occurs when temperatures fall) increases the ionization of the phospholipid head groups; this increases their hydration state and leads to higher membrane fluidity. Conversely, a decrease in pH results in a decrease in ionization; the phospholipids pack more closely, leading to a decrease in fluidity.

Inorganic cations bind to the charged phospholipid head groups, thus affecting their hydration state, whilst organic osmolytes such as trehalose may have an important role in membrane stabilization (see Burnell and Tunnacliffe, Chapter 6, this volume).

The synthesis of Hsps is a key component of the stress response and a subpopulation of sHsps is found either on the surface of, or inside, cell membranes. These Hsps influence membrane fluidity and stabilize the liquid crystal phase through specific lipid–protein interactions (Tsvetkova *et al.*, 2002).

12.6.3 Storage lipids

Lipids are primarily stored as triacylglycerides. Since storage lipids are not involved in membrane function it might have been assumed that storage lipids are less likely than membrane lipids to show temperature-adaptive changes. Indeed, the fatty acid composition of storage lipids usually closely resembles that of the organism's diet (see, for example, Rouse *et al.*, 1992; Hatab and Gaugler, 2001). However, storage lipids can only be mobilized if they are in a fluid state and there is usually a relationship between the melting point of the storage lipids and the organism's environmental temperature. How this control is achieved is not clear, but it may involve selective uptake of dietary fatty acids and differential incorporation into triacylglycerides, as well as fatty acid elongation (Horikawa *et al.*, 2008).

12.7 Membranes and Hydrostatic Pressure

Hydrostatic pressure has been shown to affect adversely a number of membrane-associated functions, including nerve conduction and excitability, synaptic transmission and active and passive transport (Macdonald, 1984). These pressure effects can be interpreted in terms of activation volumes, this time those involved with conformational changes in ion channels. High hydrostatic pressures cause compression of the bilayer. This results in changes in hydration and a transition from liquid crystal to gel phase in a manner similar to a drop in temperature (hence the trade-off between temperature and pressure; see Section 12.4.5).

Organisms from ocean depths have membranes with a higher percentage of unsaturated fatty acids and less cholesterol when compared with shallow-water relatives (Williams *et al.*, 2001). The situation in benthic nematodes is unknown.

12.8 Membranes and Desiccation

The interaction between membrane lipids and water is a fundamental feature of membrane organization. The head groups of phospholipids are normally strongly hydrated (10–12 mol of water per mol of phospholipid). As water is withdrawn, adjacent membranes tend to fuse and the head groups pack more tightly. Acyl chain interactions are strengthened and the membranes move towards the more rigid gel phase. When dried membranes are rehydrated, phase transitions take place and membrane integrity is lost. Again two types of adaptation can occur in response to desiccation: first, lipid remodelling (intrinsic adaptation) and, second, the synthesis of solutes, which stabilize membranes and proteins in the absence of water (extrinsic adaptation).

There are no reports of altered lipid composition in membranes of animals which have anhydrobiotic stages in their life cycles. However, lipid remodelling does occur in plants subjected to water stress, where the PE/PC ratios increase in response to desiccation, leading to the formation of the inverted hexagonal phase upon extreme dehydration.

The synthesis of trehalose in response to water stress has been widely reported (see Burnell and Tunnacliffe, Chapter 6, this volume). The hydroxyl groups of trehalose are sterically arranged such that they are able to form precise hydrogen bonds with polar lipid head groups. This stabilizes phospholipid spacing, preventing membrane fusion and inhibiting phase transitions. Trehalose probably stabilizes proteins in essentially the same way, replacing water at the surface and maintaining protein 3D structure. As a non-reducing sugar, trehalose is very stable and can be accumulated in high concentrations.

12.8.1 Osmotic stress

When exposed to hypertonic stress, cells accumulate compatible (non-perturbating) organic osmolytes. As the organic osmolytes are accumulated, inorganic ions are lost from the cytoplasm and the ionic strength is lowered. In *C. elegans*, glycerol, and to a lesser extent trehalose, have been shown to accumulate in response to hyperosmotic shock and there is an increase in glycerol-3-phosphate dehydrogenase activity (Lamitina *et al.*, 2004; Lamitina and Strange, 2005). In mammals and in yeast, the osmotic stress response is controlled by the MAP kinase pathway. In *C. elegans* the behavioural and physiological responses to osmotic stress have been shown to be under the control of the gene *osr-1*, which interacts with p38MAP kinase (Solomon *et al.*, 2004); however, the DAF-16 insulin signalling pathway may also be involved (Lamitina and Strange, 2005). Hyperosmotic stress induces protein damage, and of the 40 genes in *C. elegans* designated as essential to survive osmotic stress, half are involved in the detection, transport and degradation of damaged proteins (Choe and Strange, 2008).

Although there are similarities between the effects of osmotic stress, freezing and dehydration on membrane structure and function, they are different (Hazel and Williams, 1991). Osmotic stress, freezing and dehydration all remove freezable water from the system. Dehydration also removes the non-freezable water. This difference is reflected in the different solutes used to stabilize membranes and proteins during osmotic stress or freezing as opposed to desiccation. A wide range of different solutes can function as osmotic agents and antifreezes in invertebrates, including glycerol, proline, alanine, glycine, ethanol and glucose. Protection against desiccation requires much more specific interactions, the most efficient of which is trehalose, whilst myo-inositol and ribitol may also contribute but to a much lesser extent.

12.9 Conclusions and Future Directions

Adaptations that enable an organism temporarily to survive outside of its 'living zone' are referred to as resistance adaptations, whilst adaptations that enable an organism to live permanently in an extreme environment are called capacity adaptations. An interesting question with respect to any particular environmental stressor is 'Are the capacity adaptations and resistance adaptations the same?' In the case of nematodes, sometimes they are, as in lipid remodelling in response to an increase in temperature; in other cases they are different, as in the different response to anoxia in free-living nematodes and animal-parasitic nematodes.

Work with *C. elegans* is beginning to unravel the control of stress responses in nematodes. The pathways involved in responding to anoxia, desiccation, heavy metal toxicity, and temperature and oxidative stress appear to be highly conserved across the animal kingdom. In all cases, the stress response in nematodes involves the upregulation of a whole series of potentially protective enzymes and proteins and it is possible that there is a single generic response involved.

A number of factors contribute to the remarkable survival abilities of nematodes. The inert cuticle, composed of a unique cross-linked collagen, protects them from the immediate environment. Nematodes have a well-developed behavioural repertoire which enables them to avoid unfavourable environments. Finally, as a group, their metabolic pathways are relatively unspecialized, which gives them a great deal of biochemical flexibility in responding to environmental challenges. To date, none of the biochemical adaptations to stress found in nematodes have been found to be unique; however, nothing is known about the biochemistry of nematodes which inhabit the more extreme environments on the planet such as hot springs, black smokers or the abyssal depths. There may be more surprises in store!

12.10 References

Abad, P., Gouzy, J., Aury, J.-M. *et al.* (2008) Genome sequence of the metazoan plant-parasitic nematode *Meloidogyne incognita*. *Nature Biotechnology* 26, 909–915.

Adhikari, B.N., Wall, D.A. and Adams, B.J. (2009) Desiccation survival in an Antarctic nematode: molecular analysis using expressed sequence tags. *BMC Genomics* 10, 69.

An, J.H., Vranas, K., Licke, M., Inoue, H., Hisamoyo, N., Matsumoto, K. and Blackwell, T.K. (2005) Regulation of the *Caenorhabditis elegans* oxidative stress defense protein SKN-1 by glycogen synthase kinase-3. *Proceedings of the National Academy of Sciences of the USA* 102, 16275–16280.

Anil, S., Das, D. and Dasgupta, D.R. (1998) Effect of anoxic conditions on the activity of lactate dehydrogenase and alcohol dehydrogenase of the root-knot nematode, *Meloidogyne incognita. Annals of Plant Protection Sciences* 6, 182–185.

Ardelli, B.F., Guerriero, S.B. and Prichard, R.K. (2006) Ivermectin imposes selection pressure on P-glycoprotein from *Onchocerca volvulus*: linkage disequilibrium and genotype diversity. *Parasitology* 132, 375–386.

Barrett, J. (1968) Lipids of the infective and parasitic stages of some nematodes. *Nature* 218, 1267–1268.

Barrett, J. (1982) Metabolic responses to anabiosis in the fourth-stage juveniles of *Ditylenchus dipsaci* (Nematoda). *Proceedings of the Royal Society of London B* 216, 159–177.

Barrett, J. (1984) The anaerobic end-products of helminths. *Parasitology* 88, 179–198.

Barrett, J. (1991a) Anhydrobiotic nematodes. *Agricultural Zoology Reviews* 4, 161–176.

Barrett, J. (1991b) Amino acid metabolism in helminths. *Advances in Parasitology* 30, 39–105.

Barrett, J. (1995) Helminth glutathione transferases. *Helminthologia* 32, 125–128.

Barrett, J. (1997) Helminth detoxification mechanisms. *Journal of Helminthology* 71, 85–89.

Barrett, J. and Butterworth, P.E. (1984) Acetaldehyde formation by mitochondria from the free-living nematode *Panagrellus redivivus. Biochemical Journal* 221, 535–540.

Barrett, J. and Fairbairn, D. (1971) Effects of temperature on the kinetics of malate dehydrogenases in the developing eggs and adult muscle of *Ascaris lumbricoides* (Nematoda). *Journal of Experimental Zoology* 176, 169–178.

Barrett, J. and Wright, D.J. (1998) Intermediary metabolism. In: Perry, R.N. and Wright, D.J. (eds) *The Physiology and Biochemistry of Free-living and Plant-parasitic Nematodes*. CAB International, Wallingford, UK, pp. 331–353.

Barrett, J., Ward, C.W. and Fairbairn, D. (1970) The glyoxylate cycle and the conversion of triglycerides to carbohydrates in the developing eggs of *Ascaris lumbricoides. Comparative Biochemistry and Physiology* 35, 577–586.

Bolla, R.I., Weaver, C., Koslowski, P., Fitzsimmons, K. and Winter, R.E.K. (1987) Characterization of a nonparasitic isolate of *Bursaphelenchus xylophilus. Journal of Nematology* 19, 304–310.

Boutilier, R.G. and St Pierre, J. (2000) Surviving hypoxia without really dying. *Comparative Biochemistry and Physiology* 126A, 481–490.

Browne, J.A., Dolan, K.M., Tyson, T., Goyal, K., Tunnacliffe, A. and Burnell, A.M. (2004) Dehydration-specific induction of hydrophilic protein genes in the anhydrobiotic nematode *Aphelenchus avenae. Eukaryotic Cell* 3, 966–975.

Choe, K.P. and Strange, K. (2008) Genome wide RNAi screen and *in vivo* protein aggregation reporters identify degradation of damaged proteins as an essential hypertonic stress response. *American Journal of Physiology – Cell Physiology* 295, C1488–C1498.

Colonna, W.J. and McFadden, B.A. (1975) Isocitrate lyase from parasitic and free-living nematodes. *Archives of Biochemistry and Biophysics* 170, 608–619.

Cooper, A.F. and Van Gundy, S.D. (1971) Ethanol production and utilization by *Aphelenchus avenae* and *Caenorhabditis* sp. *Journal of Nematology* 3, 205–214.

Devaney, E. (2006) Thermoregulation in the life cycle of nematodes. *International Journal for Parasitology* 36, 641–649.

Fields, P.A. and Somero, G.N. (1998) Hot spots in cold adaptation: localized increases in conformational flexibility in lactate dehydrogenase A_4 orthologs of Antarctic notothenioid fishes. *Proceedings of the National Academy of Sciences of the USA* 98, 11476–11481.

Fields, P.A., Wahlstrand, B.D. and Somero, G.N. (2001) Intrinsic versus extrinsic stabilization of enzymes: the interaction of solutes and temperature on A_4-lactate dehdrogenase orthologs from warm- and cold-adapted marine tissues. *European Journal of Biochemistry* 268, 4497–4505.

Flint, H.C., Copley, J.T. and Ferrero, T.J. (2006) Patterns of nematode diversity at hydrothermal vents on the East Pacific Rise. *Cahiers de Biologie Marine* 47, 365–370.

Föll, R.L., Pleyers, A., Lewandovski, G. J., Wermter, C., Hegemann, V. and Paul, R.J. (1999) Anaerobiosis in the nematode *Caenorhabditis elegans. Comparative Biochemistry and Physiology B* 124, 269–280.

Gal, T.Z., Glazer, I. and Koltai, H. (2004) An LEA group 3 family member is involved in survival of *C. elegans* during exposure to stress. *FEBS Letters* 577, 21–26.

Gordon, R. (1987) Glyoxylate pathway in the free-living stages of the entomophilic nematode *Romanomermis culicivorax. Journal of Nematology* 19, 277–281.

Goyal, K., Tisi, L., Basran, A., Browne, J., Burnell, A., Zurdo, J. and Tunnacliffe, A. (2003) Transition from natively unfolded to folded state induced by desiccation in an anhydrobiotic nematode protein. *Journal of Biological Chemistry* 278, 12977–12984.

Goyal, K., Pinelli, C., Maslen, S.L., Rastogi, R.K., Stephens, E. and Tunnacliffe, A. (2005) Dehydration regulated processing of late embryogenesis abundant protein in a desiccation-tolerant nematode. *FEBS Letters* 579, 4093–4098.

Haegeman, A., Jacob, J., Vanholme, B., Kyndt, T., Mitreva, M. and Gheysen, G. (2009) Expressed sequence tags of the peanut pod nematode *Ditylenchus africanus*: the first transcriptome analysis of an Anguinid nematode. *Molecular and Biochemical Parasitology* 167, 32–40.

Hajdu-Cronin, Y.M., Chen, W.J. and Sternberg, P.W. (2004) The L-type cyclin CYL-1 and the heat-shock-factor HSF-1 are required for heat-shock-induced protein expression in *Caenorhabditis elegans. Genetics* 168, 1937–1949.

Hatab, M.A. and Gaugler, R. (1997) Influence of growth temperature on fatty acids and phospholipids of *Steinernema riobravis* infective juveniles. *Journal of Thermal Biology* 22, 237–244.

Hatab, M.A. and Gaugler, R. (2001) Diet composition and lipids of *in vitro* produced *Heterorhabditis bacteriophora. Biological Control* 20, 1–7.

Hattori, K., Inoue, M., Inoue, T., Arai, H. and Tamura, H. (2006) A novel sulfotransferase abundantly expressed in the dauer larvae of *Caenorhabditis elegans. Journal of Biochemistry (Tokyo)* 139, 355–362.

Hazel, J.R. and Williams, E.E. (1991) The role of alteration in membrane lipid composition in enabling physiological adaptation of organisms to their physical environment. *Progress in Lipid Research* 29, 167–227.

Henkle-Dührsen, K. and Kampkötter, A. (2001) Antioxidant enzyme families in parasitic nematodes. *Molecular and Biochemical Parasitology* 114, 129–142.

Holz, R.A., Wright, D.J. and Perry, R.N. (1997) The lipid content and fatty acid composition of hatched second stage juveniles of *Globodera rostochiensis* and *G. pallida. Fundamental and Applied Nematology* 20, 291–298.

Horikawa, M., Nokura, T., Hashimoto, T. and Sakamoto, K. (2008) Elongation and desaturation of fatty acids are critical in growth, lipid metabolism and ontogeny of *Caenorhabditis elegans. Journal of Biochemistry (Tokyo)* 144, 149–158.

Jagdale, G.B. and Gordon, R. (1997) Effect of temperature on the composition of fatty acids in the total lipids and phospholipids of entomopathogenic nematodes. *Journal of Thermal Biology* 22, 245–251.

Jiang, H., Guo, R. and Powell-Coffman, J.A. (2001) The *Caenorhabditis elegans hif-1* gene encodes a bHLH-PAS protein that is required for adaptation to hypoxia. *Proceedings of the National Academy of Sciences of the USA* 98, 7916–7921.

Joachim, A., Ryll, M. and Duagschies, A. (2000) Fatty acid patterns of different stages of *Oesophagostomum dentatum* and *Oesophagostomum quadrispinulatum* as revealed by gas chromatography. *International Journal for Parasitology* 30, 819–827.

Jones, J.T., Reavy, B., Smart, G. and Prior, A.E. (2004) Glutathione peroxidases of the potato cyst nematode *Globodera rostochiensis. Gene* 324, 47–54.

Kerboeuf, D., Soubieux, D., Guilluy, R., Brazier, J.L. and Riviere, J.L. (1995) *In vivo* metabolism of aminopyridine by the larvae of the helminth *Heligmosomoides polygyrus. Parasitology Research* 81, 302–304.

Kim, D.H., Liberati, N.T., Mizuno, T., Inoue, H., Hasamoto, N., Matsumoto, K. and Ausubel, F.M. (2004) Integration of *Caenorhabditis elegans* MAPK pathways mediating immunity and stress resistance by MEK-1, MAPK kinase and VHP-1 MAPK phosphatase. *Proceedings of the National Academy of Sciences of the USA* 101, 10990–10994.

Komuniecki, R., Komuniecki, P.R. and Saz, H.J. (1981) Pathway of formation of branched-chain fatty acids in *Ascaris* mitochondria. *Journal of Parasitology* 67, 841–846.

Kulas, J., Schmidt, C., Rothe, M., Schunck, W.-H. and Menzel, R. (2008) Cytochrome P450-dependent metabolism of eicosapentanoic acid in the nematode *Caenorhabditis elegans. Archives of Biochemistry and Biophysics* 472, 65–75.

Lamitina, S.T. and Strange, K. (2005) Transcriptional targets of DAF-16 insulin signalling pathway protect *C. elegans* from extreme hypertonic stress. *American Journal of Physiology – Cell Physiology* 288, C467–C474.

Lamitina, S.T., Morrison, R., Moeckel, G.W. and Strange, K. (2004) Adaptation of the nematode *Caenorhabditis elegans* to extreme osmotic stress. *American Journal of Physiology – Cell Physiology* 286, C785–C791.

Liao, V.H.-C., Dong, J. and Freedman, J.H. (2002) Molecular characterization of a novel cadmium inducible gene from the nematode *Caenorhabditis elegans. Journal of Biological Chemistry* 277, 42049–42059.

Li, D.X. and He, X.M. (2009) Desiccation induced structural alterations in a 66-amino acid fragment of an anhydrobiotic nematode late embryogenesis abundant (LEA) protein. *Biomacromolecules* 10, 1469–1477.

Liebau, E., Eschbach, M.-L., Tawe, W., Sommer, A., Fischer, P., Walter, R.D. and Henkle-Dührsen, K. (2000) Identification of a stress-responsive *Onchocerca volvulus* glutathione S-transferase (*Ov*-GST-3) by RT-PCR differential display. *Molecular and Biochemical Parasitology* 109, 101–110.

Lindblom, T.H and Dodd, A.K. (2006) Xenobiotic detoxification in the nematode *Caenorhabditis elegans. Journal of Experimental Zoology* 305, 720–730.

Liu, A. and Rothstein, M. (1976) Nematode biochemistry – XV. Enzyme changes related to glycerol excretion in *Caenorhabditis briggsae. Comparative Biochemistry and Physiology* 54B, 233–238.

Lower, W.R., Willett, J.D. and Hansen, E.L. (1970) Selection for adaptation to increased temperatures in free-living nematodes – II. Some lipid differences in *Panagrellus redivivus. Comparative Biochemistry and Physiology* 34, 473–479.

Lyons, J.M., Keith, A.D. and Thomason, I.T. (1975) Temperature-induced phase transitions in nematode lipids and their influence on respiration. *Journal of Nematology* 7, 98–104.

Macdonald, A.G. (1984) The effects of pressure on the molecular structure and physiological function of cell membranes. *Philosophical Transactions of the Royal Society, Series B* 304, 47–68.

Madin, K.A.C., Loomis, S.H. and Crowe, J.H. (1985) Anhydrobiosis in nematodes – control of carbon flow through the glyoxylate cycle. *Journal of Experimental Zoology* 234, 341–350.

McDermott, L., Kennedy, M.W., McManus, D.P., Bradley, J.E., Cooper, A. and Storch, J. (2002) How helminth lipid-binding proteins offload their ligands to membranes: differential mechanisms of fatty acid transfer by the ABA-1 polyprotein allergen and Ov-FAR-1 proteins of nematodes and Sj-FABPc of schistosomes. *Biochemistry* 41, 6706–6713.

McFall-Ngai, M. and Horowitz, J. (1990) A comparative study of the thermal stability of the vertebrate eye lens: Antarctic fish to the desert iguana. *Experimental Eye Research* 50, 703–709.

Mendis, A.H.W. and Evans, A.A.F. (1984) Major volatile metabolites produced by two isolates of the nematode *Aphelenchus avenae* under aerobic and anaerobic conditions. *Comparative Biochemistry and Physiology* 78B, 737–739.

Musat, N., Giere, O., Gieseke, A., Thiermann, F., Amann, R. and Dubilier, N. (2007) Molecular and morphological characterization of the association between bacterial endosymbionts and the marine nematode *Astomonema* sp. from the Bahamas. *Environmental Microbiology* 9, 1345–1353.

Narbonne, P. and Roy, R. (2009) *Caenorhabditis elegans* dauers need LKB1/AMPK to ration lipid reserves and ensure long-term survival. *Nature* 457, 210–214.

Opperman, C.H., Bird, D.McK., Williamson, V.W. *et al.* (2008) Sequence and genetic map of *Meloidogyne hapla*: a compact nematode genome for plant parasitism. *Proceedings of the National Academy of Sciences of the USA* 105, 14802–14807.

O'Riordan, V.B. and Burnell, A.M. (1990) Intermediary metabolism in the dauer larvae of *Caenorhabditis elegans* II. The glyoxylate cycle and fatty acid oxidation. *Comparative Biochemistry and Physiology* 95B 125–130.

Ott, J., Bright, M. and Bulgheresi, S. (2004) Symbioses between marine nematodes and sulphur-oxidizing chemoautotrophic bacteria. *Symbiosis* 36, 103–126.

Padilla, P., Nystul, T.G., Zager, R.A., Ali, C.M., Johnson, A.C.M. and Roth, M.B. (2002) Dephosphorylation of cell cycle-regulated proteins correlates with anoxia-induced suspended animation in *Caenorhabditis elegans*. *Molecular Biology of the Cell* 13, 1473–1483.

Paul, R.J., Gohla, J., Föll, R. and Schneckenburger, H. (2000) Metabolic adaptations to environmental changes in *Caenorhabditis elegans*. *Comparative Biochemistry and Physiology B* 127, 469–479.

Pellerone, F.I., Archer, S.K., Behm, C.A., Grant, W.N., Lacey, M.J. and Somerville, A.C. (2003) Trehalose metabolism genes in *Caenorhabditis elegans* and filarial nematodes. *International Journal for Parasitology* 11, 1195–1206.

Persaud, A.D., Perry, R.N. and Barrett, J. (1997) The purification and properties of glutathione S-transferase from *Ditylenchus myceliophagus*. *Fundamental and Applied Nematology* 20, 601–609.

Persson, A., Gross, E., Laurent, P., Busch, K.E., Bretes, H. and de Bono, M. (2009) Natural variation in a neural globin tunes oxygen sensing in wild *Caenorhabditis elegans*. *Nature* 458, 1030–1033.

Powell, E.N., Crenshaw, M.A. and Reiger, R.M. (1979) Adaptations to sulfide in the meiofauna of the sulfide system I. ^{35}S-sulfide accumulation and the presence of a sulphide detoxification system. *Journal of Experimental Marine Biology and Ecology* 37, 57–76.

Qin, H.T. and Powell-Coffman, J.A. (2004) The *Caenorhabditis elegans* aryl hydrocarbon receptor AHR-1 regulates neuronal development. *Developmental Biology* 270, 64–75.

Reichert, K. and Menzel, R. (2005) Expression profiling of five different xenobiotics using *Caenorhabditis elegans* whole genome array. *Chemosphere* 61, 229–239.

Riou, M., Koch, C., Delaleu, B., Berthon, P. and Kerboeuf, D. (2005) Immunolocalisation of an ABC transporter, P-glycoprotein, in the eggshells and cuticles of free-living and parasitic stages of *Haemonchus contortus*. *Parasitology Research* 96, 142–148.

Roots, B.I. and Johnston, P.V. (1968) Plasmalogens of the nervous system and environmental temperature. *Comparative Biochemistry and Physiology* 26, 553–560.

Rouse, D.B., Webster, C.D. and Radwin, I.A. (1992) Enhancement of the fatty acid composition of the nematode *Panagrellus redivivus* using three different media.

Journal of the World Aquaculture Society 23, 89–95.

Schuller, D.J., Liu, Q., Kriksunov, I.A., Campbell, A.M., Barrett, J., Brophy, P.M. and Hao, Q. (2005) The crystal structure of a new class of glutathione transferase from the model human hookworm nematode *Heligmosomoides polygyrus*. *Proteins: Structure, Function and Bioinformatics* 61, 1024–1031.

Scott, B.A., Avidan, M.S. and Crowder, C.M. (2002) Regulation of hypoxic death in *C. elegans* by the insulin/IGF receptor homolog DAF-2. *Science* 296, 2388–2391.

Shen, C., Nettleton, D., Jiang, M., Stuart, K., Kim, S.K. and Powell-Coffman, J.A. (2005) Roles of the HIF-1 hypoxia-inducible factor during hypoxia response in *Caenorhabditis elegans*. *Journal of Biological Chemistry* 280, 20580–20588.

Sidorov, V.S. and Smirnov, L.P. (1980) Fatty acid composition of some helminths of cold blooded and warm blooded vertebrates. *Journal of Evolutionary Biology and Physiology* 16, 551–555 (in Russian).

Singh, A. and Rathur, S. (2005) Identification and characterization of a selenium-dependent glutathione peroxidase in *Setaria cervi*. *Biochemical and Biophysical Research Communications* 331, 1069–1074.

Solomon, A., Salomon, R., Paperna, I. and Glazier, I. (2000) Desiccation stress of entomopathogenic nematodes induces the accumulation of a novel heat-stable protein. *Parasitology* 121, 409–416.

Solomon, A., Bandhakavi, S., Jabbar, S., Shah, R., Beitel, G.J. and Morimoto, R.I. (2004) *Caenorhabditis elegans* OSR-1 regulates behavioural and physiological responses to hyperosmotic environments. *Genetics* 167, 161–170.

Somero, G.N. (1990) Life at low volume change: hydrostatic pressure as a selective factor in the aquatic environment. *Integrative and Comparative Biology* 30, 123–135.

Szewczyk, N.J. and Lamb, W. (2005) Surviving atmospheric spacecraft breakup. *Wilderness and Environmental Medicine* 16, 27–32.

Tanaka, T., Ikita, K., Ashida, T., Motoyama, Y., Yamaguchi, Y. and Satouchi, K. (1996) Effects of growth temperature on the fatty acid composition of the free-living nematode *Caenorhabditis elegans*. *Lipids* 31, 1173–1178.

Thompson, F.J., Barker, G.L.A., Hughes, L., Wilkes, C.P., Coghill, J. and Viney, M. (2006) A microarray analysis of gene expression in the free-living stages of the parasitic nematode *Strongyloides ratti*. *BMC Genomics* 7, 157.

Thompson, S.N., Platzer, E.G. and Lee, R.W.K. (1991) Bioenergetics in a parasitic nematode, *Steinernema carpocapsae*, monitored *in vivo* by flow NMR spectroscopy. *Parasitology Research* 77, 86–90.

Tsvetkova, N.M., Horvath, I., Torok, Z. *et al.* (2002) Small heat-shock proteins regulate membrane lipid polymorphism. *Proceedings of the National Academy of Sciences of the USA* 99, 13504–13509.

Tyson, T., Reardon, W., Browne, J.A. and Burnell, A.M. (2007) Gene induction by desiccation stress in the entomopathogenic nematode *Steinernema carpocapsae* reveals parallels with drought tolerance mechanisms in plants. *International Journal for Parasitology* 37, 763–766.

Vanfleteren, J.R. (1993) Oxidative stress and ageing in *Caenorhabditis elegans*. *Biochemical Journal* 292, 605–608.

Van Rossum, A.J., Jefferies, J.R., Rijsewijk, F.A.M., La Course, E.J., Teesdale-Spittle, P., Barrett, J., Tait, A. and Brophy, P.M. (2004) Binding of haematin by a new class of glutathione transferase from the blood feeding parasitic nematode *Haemonchus contortus*. *Infection and Immunity* 72, 2780–2790.

Vatamaniuk, O.K., Bucher, A., Sundaram, M.V. and Rea, P.A. (2005) CeHMT-1a putative phytochelatin transporter is required for cadmium tolerance in *Caenorhabditis elegans*. *Journal of Biological Chemistry* 280, 23684–23690.

Vetter, R.D., Powell, M.A. and Somero, G.N. (1991) Metazoan adaptations to hydrogen sulphide. In: Bryant, C. (ed.) *Metazoan Life without Oxygen*. Chapman & Hall, London, pp109–128.

von Brand, T. (1966) *Biochemistry of Parasites*. Academic Press, New York.

Wharton, D.A., Rolfe, R.N. and Perry, R.N. (2000) Electrophysiological activity during recovery from anhydrobiosis in fourth stage juveniles of *Ditylenchus dipsaci*. *Nematology* 2, 881–886.

Wharton, D.A., Barrett, J., Goodall, J.G., Marshall, C.J. and Ramløv, H. (2005) Ice-active proteins from the Antarctic nematode *Panagrolaimus davidi*. *Cryobiology* 51, 198–207.

Williams, E.E., Stewart, B.S., Beuchat, C.A., Somero, G.N. and Hazel, J.R. (2001) Hydrostatic pressure and temperature effects on the molecular order of erythrocyte membranes from deep-, shallow- and non-diving mammals. *Canadian Journal of Zoology* 79, 888–894.

Womersley, C., Thompson, S.N. and Smith, L. (1982) Anhydrobiosis in nematodes II: carbohydrate and lipid analysis in undesiccated and desiccated nematodes. *Journal of Nematology* 14, 145–153.

Womersley, C.Z., Wharton, D.A. and Higa, L.M. (1998) Survival biology. In: Perry, R.N. and Wright, D.J. (eds) *The Physiology and Biochemistry of Free-living and Plant-parasitic Nematodes*. CAB International, Wallingford, UK, pp 271–302.

Yun, C., Stanhill, A., Yang, Y. *et al.* (2008) Proteasomal adaptation to environmental stress links resistance to proteotoxicity with longevity in *Caenorhabditis elegans*. *Proceedings of the National Academy of Sciences* of the *USA* 105, 7094–7099.

Gene Index

Species Index

General Index